Mechanisms and Machine Science

Volume 40

Series editor

Marco Ceccarelli
Laboratory of Robotics and Mechatronics
University of Cassino and South Latium
Cassino, Italy

More information about this series at http://www.springer.com/series/8779

Antonio Simón Mata · Alex Bataller Torras
Juan Antonio Cabrera Carrillo
Francisco Ezquerro Juanco
Antonio Jesús Guerra Fernández
Fernando Nadal Martínez
Antonio Ortiz Fernández

Fundamentals of Machine Theory and Mechanisms

 Springer

Antonio Simón Mata
Department of Mechanical Engineering
   and Fluid Mechanics
University of Málaga
Málaga
Spain

Alex Bataller Torras
Department of Mechanical Engineering
   and Fluid Mechanics
University of Málaga
Málaga
Spain

Juan Antonio Cabrera Carrillo
University of Málaga
Málaga
Spain

Francisco Ezquerro Juanco
University of Málaga
Málaga
Spain

Antonio Jesús Guerra Fernández
University of Málaga
Málaga
Spain

Fernando Nadal Martínez
University of Málaga
Málaga
Spain

Antonio Ortiz Fernández
University of Málaga
Málaga
Spain

ISSN 2211-0984   ISSN 2211-0992   (electronic)
Mechanisms and Machine Science
ISBN 978-3-319-81166-6   ISBN 978-3-319-31970-4   (eBook)
DOI 10.1007/978-3-319-31970-4

Printed on acid-free paper

This Springer imprint is published by Springer Nature
The registered company is Springer International Publishing AG Switzerland

Preface

This book is the result of many years of experience teaching kinematics and dynamics of mechanisms at Malaga University. The compilation of different class notes resulted in the first book by the authors in 2000. Along the last years, five improved editions have been printed, all of them only available in Spanish. At present, many universities in Spain and Latin America use this book as a teaching support.

In this first edition in English, we have included those chapters that we think are essential in a Theory of Mechanisms and Machines course. Instead of following a rigid order per topic, the chapters have been organized the main goal being to present the contents in a fluid way, trying not to cut the rhythm of the development. So, the first part completely develops the kinematic and dynamic analysis of linkages. Then, it continues with dynamics, studying flywheels and vibrations in systems with one degree of freedom. Back to kinematics, it studies the transmission of motion with gears. Finally, it presents the main concepts of the synthesis of mechanisms as well as the latest techniques in this field, such as an optimization method based on evolutionary algorithms and a new method to measure the error between two curves based on turning functions. Several examples are included to compare the results obtained following different synthesis methods.

At the end of the book, there are three addendums that complete the concepts developed in some of the chapters and that help the student to assimilate them. The first addendum develops the trigonometric method for the position analysis of different linkages. The second one includes Freudenstein's method applied to the resolution of Raven's position equations in a four-bar linkage. Finally, the last addendum solves the kinematic and dynamic analysis of a six-link mechanism using different methods explained in this book.

Because of its educational focus, we have included some graphical methods in this book. Although, nowadays, the use of analytical methods is basic in Theory of Mechanisms and Machines, according to our experience, the didactic aspect of

graphical methods is unquestionable. Therefore, we use them at the beginning
of the kinematic and dynamic analysis of linkages. This helps to consolidate new
concepts and to better understand the behavior of mechanisms. Then, once this has
been achieved by the student, modern and powerful analytical methods are
developed.

Contents

Chapter 1
Kinematic Chains

Abstract Mechanics is a branch of Physics that deals with scientific analysis of motion, time and forces. It is divided in two parts: statics and dynamics. Statics studies the equilibrium conditions of stationary systems (not taking motion into account). On the other side, dynamics studies moving systems that change with time and the forces acting on them. At the same time, dynamics is divided up into kinematics and kinetics. Kinematics analyzes the movement of bodies without considering the forces causing such motion while kinetics takes those forces into account. In this chapter we will describe the basic definitions that will help us to predict the kinematic behavior of a given mechanism as degrees of freedom, links and kinematic pairs, classification of kinematic pairs, kinematic chain, mechanisms, kinematic skeleton and machine, classification of link movement, kinematic inversion, Grashof's criterion, mechanical advantage and kinematic curves.

Mechanics (Fig. 1.1) is a branch of Physics that deals with scientific analysis of motion, time and forces. It is divided in two parts: statics and dynamics. Statics studies the equilibrium conditions of stationary systems (not taking motion into account). On the other side, dynamics studies moving systems that change with time and the forces acting on them. At the same time, dynamics is divided up into kinematics and kinetics. Kinematics analyzes the movement of bodies without considering the forces causing such motion while kinetics takes those forces into account.

1.1 Basic Concepts

Mechanism and Machine Theory is the science that studies the relationship between geometry and the motion of machine parts and the forces causing such movement. It consists of synthesis and analysis.

© Springer International Publishing Switzerland 2016
A. Simón Mata et al., *Fundamentals of Machine Theory and Mechanisms*,
Mechanisms and Machine Science 40, DOI 10.1007/978-3-319-31970-4_1

Fig. 1.1 Areas to be studied
in rigid body mechanics

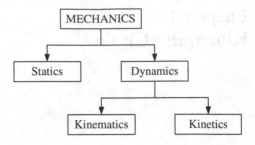

Synthesis or design refers to the creative process through which a model or
pattern can be generated, so that it satisfies a certain need while complying with
certain kinematic and dynamic constraints that define the problem.

Analysis refers to the study of dynamic behavior of a certain mechanism in order
to determine if it is suitable for its purpose.

1.2   Definitions

In this part we will describe the basic definitions that will help us to predict the
kinematic behavior of a given mechanism.

1.2.1   Degrees of Freedom (DOF)

We define degree of freedom (DOF) of a mechanical system as the number of
independent parameters that unambiguously define its position in space at every
instant. For example:

- We will consider a point moving on a plane surface (a flat two-dimensional
  space): two parameters will be needed to define its location. A rod moving on a
  plane will need three parameters to define its position: two coordinates (x, y) to
  locate one of its points and a third parameter, θ, that defines its angular position
  (Fig. 1.2).

Fig. 1.2   A rod moving on a
plane: coordinates x and
y define the location of one of
its points while angle θ
defines its angular position

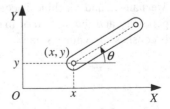

- The same rod in space will be unambiguously defined by six parameters: three for the location of any of its points and three for its angular position with respect to the coordinate planes.

1.2.2 Links and Kinematic Pairs

A link is a rigid body with two or more joints to other rigid bodies. We will call such joints joining elements. For kinematic analysis purposes, we consider links rigid bodies with no mass, as we study their movement without considering forces.

The number of kinematic links on a mechanism does not depend on the number or function of its parts but on their movement. Therefore, if two linked parts move jointly they are considered one single mechanical link.

Links are named binary (two joints), ternary (three joints) or quaternary (four joints) (Fig. 1.3).

A kinematic pair consists of a set of two links whose relative motion is constricted by their joint (pairing element). Depending on the relative planar motion between these two links, we distinguish prismatic pairs, when only sliding motion is allowed, and hinge or turning pairs, when only relative rotation between both elements is allowed.

1.2.3 Classification of Kinematic Pairs

According to the number of degrees of freedom, kinematic pairs are classified in:

- Pairs with one DOF: Resulting relative motion between the two links is limited to one degree of freedom. See the prismatic and rotating pairs represented in Fig. 1.4.

Fig. 1.3 a Binary link, **b** ternary link, **c** quaternary link

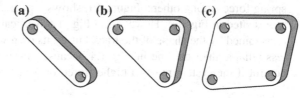

Fig. 1.4 a Prismatic pair. **b** Hinge pair

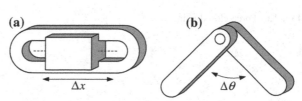

Fig. 1.5 Pair with two DOF

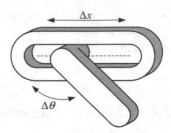

- Pairs with two DOF: Fig. 1.5 shows an example of this type of kinematic pair where both links can simultaneously have a rotating and sliding motion on a path defined along one of the connected links.
- Pairs with three DOF: They are only used in spatial mechanisms. For instance, a pair consisting of two links joined by a spherical pair will have three degrees of freedom as the joint allows rotation about three axes in a spatial frame system.

According to the way contact between elements is made, pairs can be classified as:

- Higher-pairs: They are joints with point or line contact. Pairs with more than one degree of freedom are usually higher pairs. However, spherical pairs are an exception. In the pair shown in Fig. 1.5, contact between links is made through a contact line created along a plane surface and a cylindrical one.
- Lower-pairs: They are joints with surface contact. Pairs in Fig. 1.4 are lower pairs as relative motion is transmitted through surface contact between two plain surfaces in a prismatic pair and between two cylindrical surfaces in a hinge pair.

Depending on the nature of the constraint, pairs can be classified as:

- Self-closed pairs: Links are mechanically joined due to their shape. In pairs shown in Figs. 1.4 and 1.5, relative motion between links is constrained by their geometry. Movement is defined as long as there is no break.
- Force-closed pairs: An external force is required to connect both links. For example, consider a slide over a surface. External forces could be gravity or a spring force among others. Figure 1.6 shows a prismatic pair similar to the one represented in Fig. 1.4. However, in Fig. 1.6, vertical motion of the slide is not constrained by the shape of the track but by its own weight. This kind of pair is less reliable than the one in Fig. 1.4 because link movement might get out of control if the system meets a higher force than the one carrying out the joint.

Fig. 1.6 Force-closed pair

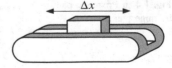

1.2.4 Kinematic Chains, Mechanisms, Kinematic Skeletons and Machines

A kinematic chain consists of a system of kinematic pairs, in other words, a set of links coupled by kinematic pairs that constrain their relative motion. Examples of kinematic chains are the four-bar chain and the slider-crank chain shown in Fig. 1.7.

A mechanism is defined as a kinematic chain in which one link is attached to the reference frame, so that a controlled output motion is generated in response to an input motion.

A kinematic skeleton is a sketch that represents a mechanism in a simplified way, so that it facilitates its kinematic analysis. Figure 1.8 shows kinematic skeletons for four-bar and slider-crank mechanisms.

A machine is a set of mechanisms joined in order to transmit forces and carry out work.

For instance, the slider-crank mechanism is used in different machines such as combustion engines and air compressors having different purposes. Figure 1.9 shows a reciprocating engine and its skeleton diagram.

Fig. 1.7 a Four-bar kinematic chain, **b** slider-crank kinematic chain

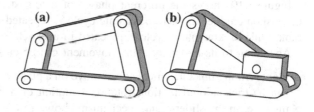

Fig. 1.8 a Four-bar mechanism and its skeleton diagram, **b** crank-shaft mechanism and its skeleton diagram

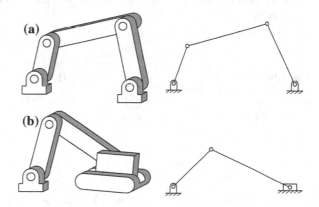

Fig. 1.9 Reciprocating
combustion engine and its
kinematic skeleton

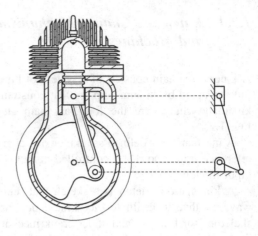

1.2.5   Link Movement

We define a kinematic cycle as the process in which a set of pieces of a machine
returns to its original position after going through all the intermediate positions it
can occupy. The time spent completing this cycle is called period.

Figure 1.10 shows the different phases of a four-stroke reciprocating internal
combustion engine. Its thermodynamic cycle is repeated in every 720°-turn of the
crank while its kinematic cycle is repeated in every 360°-turn.

According to its geometry, link movement can be classified as follows:

- Plane: All the links in a mechanism move on a plane or parallel ones. Movement
  can be translational motion, rotational motion or a combination of both. For
  instance, in the slider-crank mechanism shown in Fig. 1.10, the slider makes a
  translational movement, the crank rotates and the shaft carries out a combination
  of both.

Fig. 1.10 Four-stroke
thermodynamic cycle **a** power
b exhaust **c** intake
d compression

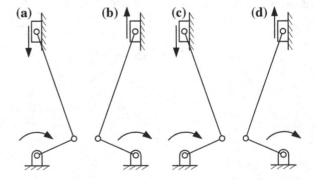

Fig. 1.11 Centrifugal
governor. The two spheres
describe a spatial motion

- Spatial: Movement happens in all three directions in space. Examples of this
  kind of movement can be those described by a threading screw (helical
  movement) or the one described by a centrifugal governor (spherical movement)
  in Fig. 1.11.

Depending on the freedom of movement, motion can be classified as:

- Free: Bodies are not physically joined to each other and external forces are
  applied. Motion depends on the forces applied to the body.
- Desmodromic: Bodies are physically joined and their motion is always defined.
  Movement does not depend on acting forces, but on the joints between links. An
  example of this kind of movement is found in four-bar or slider-crank mechanisms where all links will move with a controlled motion when one of them is
  moved and the location of every link can be determined knowing the location of
  just one.

According to motion continuity, movement can be classified as:

- Continuous: A link moves continuously when it does not stop on its way along
  its kinematic cycle. For example, the crank in four-bar and slider-crank
  mechanisms.

Fig. 1.12 Geneva wheel.
The follower moves
intermittently rotating 90° at
every full turn of the driver
wheel

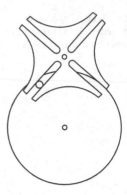

- Intermittent: A link has intermittent motion when it makes one or several stops along its kinematic cycle without changing direction. An example of this type of motion is given by the Geneva mechanism shown in Fig. 1.12. The follower will only move when the pin on the driver wheel enters one of the slots.
- Reciprocating: A link has reciprocating motion when it stops and reverses the direction of its motion. Take for instance the slider in a slider-crank mechanism.

1.3 Degrees of Freedom of Mechanisms

The number of degrees of freedom (*DOF*) of a mechanism depends on its number of links and the number and nature of its joining elements.

A link on a plane has three degrees of freedom. Therefore, two links will have six degrees of freedom and a system of N free links (not joined together) will have $3N$ degrees of freedom.

If we connect two links by a joining element, one or more degrees of freedom will be removed depending on the nature of the joint.

Moreover, when a link is fixed to the reference frame, it stays fixed and its three degrees of freedom are removed. Furthermore, a mechanism has already been defined as a kinematic chain with one fixed link, so all mechanisms will have one link with no degrees of freedom.

This reasoning leads to Grübler's equation (Eq. 1.1), which for the N links case is:

$$DOF = 3N - 2J - 3G \tag{1.1}$$

where:

- *DOF* is the number of degrees of freedom.
- *N* represents the number of links.

- $J = \sum J_i$ where for each i kinematic pair the value for J_i is:
    - $J_i = 1$ for every pair with one degree of freedom.
    - $J_i = 1/2$ for every pair with two degrees of freedom.
- G is the number of fixed links.

In a mechanism number of fixed links G will always be equal to one as all the parts fixed to the reference frame are the same link. Consequently, Grübler's equation becomes (Eq. 1.2):

$$DOF = 3(N - 1) - 2J \qquad (1.2)$$

Finally, we can also use Kutzbach's modification (Eq. 1.3) for Grübler's equation:

$$DOF = 3(N - 1) - 2J_1 - J_2 \qquad (1.3)$$

where:

- J_1 is the number of kinematic pairs with one degree of freedom.
- J_2 is the number of kinematic pairs with two degrees of freedom.

For a correct use of Kutzbach's equation, when more than two links are joined with one joining element, we will have as many kinematic pairs as links joined minus one. Figure 1.13 shows a set of three links (2, 3 and 4) joined by a single rotating constraint. This set is formed by two kinematic pairs. If we take link 3 as the reference and we join it to links 2 and 4, we will have pairs 2–3 and 3–4. Any of the three links can be regarded as the reference, but two kinematic pairs will be obtained in any case.

1.3.1 Fixed Mechanisms. Structures

When the DOF of a mechanism is zero, movement is impeded and it becomes a structure.

Fig. 1.13 Multiple joint with three links and two kinematic pairs

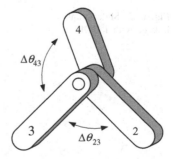

Fig. 1.14 Three-bar truss
with 0 DOF

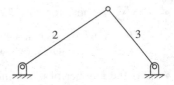

The mechanism shown in Fig. 1.14 presents three links forming three pairs with one degree of freedom. Equation (1.4) gives us the number of degrees of freedom of the mechanism:

$$DOF = 3(N - 1) - 2J_1 - J_2 = 3 \cdot (3 - 1) - 2 \cdot 3 = 0 \qquad (1.4)$$

It is a fixed mechanism (structure) called three-bar truss.

When a system has a negative DOF, it means that the number of restrictions is redundant and it becomes an overconstrained or preloaded structure.

1.3.2   Mechanisms with One DOF

When the DOF of a mechanism has a positive value, links move in relation to each other. If the number of DOF is one, the motion of all the links in the mechanism can be determined provided we know how one of the links moves. Accordingly, this kind of mechanism is used the most as only movement for one of the links has to be defined to control the output.

The slider-crank mechanism in Fig. 1.15 has four links with four pairs with one DOF. The same way as in the last example, this mechanism has no pairs with two DOF. Hence, Kutzbach's formula (Eq. 1.5) will be expressed as:

$$DOF = 3(N - 1) - 2J_1 - J_2 = 3 \cdot (4 - 1) - 2 \cdot 4 = 1 \qquad (1.5)$$

In the case of the four-bar linkage (Fig. 1.16), the number of links of the mechanism is four and the number of pairs with one degree of freedom is also four.

Fig. 1.15 Slider-crank
linkage with 1 DOF

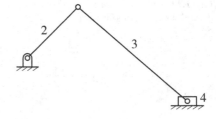

This mechanism does not have any pairs with two DOF either. Consequently, the number of degrees of freedom in the mechanism (Eq. 1.6) is:

$$DOF = 3(N-1) - 2J_1 - J_2 = 3 \cdot (4-1) - 2 \cdot 4 = 1 \qquad (1.6)$$

Another example of a mechanism with one degree of freedom is the one known as Withworth's quick-return mechanism shown in Fig. 1.17a. This mechanism has six links and seven pairs with one DOF. Therefore the number of degrees of freedom of the mechanism (Eq. 1.7) is:

$$DOF = 3(N-1) - 2J_1 - J_2 = 3 \cdot (6-1) - 2 \cdot 7 = 1 \qquad (1.7)$$

This mechanism can be modified removing one link and adding a pair with two degrees of freedom where rotation and linear motion are allowed. As seen in Fig. 1.17b, this new mechanism consists of five links that form five pairs with one DOF and one pair with two DOF. Kuzbach's equation for this mechanism (Eq. 1.8) is:

$$DOF = 3(N-1) - 2J_1 - J_2 = 3 \cdot (5-1) - 2 \cdot 5 - 1 = 1 \qquad (1.8)$$

It is verified that if pairs 4–5 and 5–6, each with one degree of freedom, are replaced by just one pair with two degrees of freedom connecting links 4 and 5, the number of degrees of freedom in the mechanism does not change.

Fig. 1.16 Four-bar linkage with 1 DOF

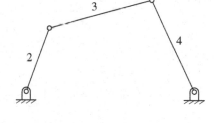

Fig. 1.17 a Withworth's quick-return mechanism **b** withworth's quick-return mechanism where one link has been removed and pairs 4–5 and 5–6 have been replaced by one pair with two *DOF*

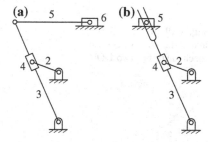

1.3.3 Mechanisms with More Than One DOF

This kind of mechanism is employed less than the latter as, in order to obtain a controlled output motion, the designer needs to control as many links as the number of degrees of freedom the system has.

In Fig. 1.18, an example of this type of mechanism can be found. It consists of seven links connected to each other forming eight pairs with one degree of freedom each. This time there are not any pairs with two DOF. The number of degrees of freedom of the mechanism is (Eq. 1.9) therefore given by:

$$DOF = 3(N-1) - 2J_1 - J_2 = 3 \cdot (4-1) - 2 \cdot 3 - 1 = 2 \qquad (1.9)$$

Having two degrees of freedom, the motion of two links, such as links 2 and 7, has to be defined in order to determine the movement of the rest.

Finally, Fig. 1.19 shows another example of a mechanism with two DOF. This mechanism is constituted by four links forming three pairs with one DOF and one pair (pair 3–4) with two DOF as wheel 3 rotates and slides over wheel 4. Kuthbach's equation for this mechanism Eq. (1.10) shows the following result:

$$DOF = 3(N-1) - 2J_1 - J_2 = 3 \cdot (4-1) - 2 \cdot 3 - 1 = 2 \qquad (1.10)$$

Fig. 1.18 Mechanism with two DOF

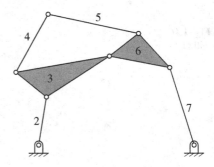

Fig. 1.19 If wheel 3 rotates and slides over wheel 4 the mechanism has two DOF

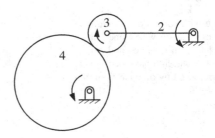

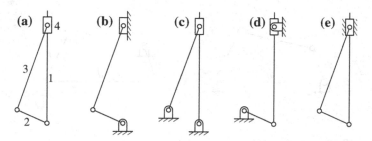

Fig. 1.20 Slider-crank kinematic chain **a** and its four specific inversions **b, c, d, e**

1.4   Kinematic Inversions

Kinematic inversions are different mechanisms obtained by defining different links as the reference frame in a kinematic chain. A mechanism has, therefore, a maximum of as many different inversions as the number of links it has.

Not all kinematic inversions present different movements of the links. Inversions with specifically different movements are called specific inversions. To illustrate this, Fig. 1.20 shows all four different inversions for the slider-crank kinematic chain.

Inversion 1 (Fig. 1.20b) takes link 1 as the frame, allowing pure translational motion for the slider. This inversion gives way to a slider-crank (also called engine mechanism).

In inversion 2 (Fig. 1.20c) link 2 is fixed and the slider has a combination of translational and rotational motion. It gives way to slider mechanisms that will be studied further on in this book.

Inversion 3 (Fig. 1.20d) fixes link 3 as the frame, which results in a reciprocating rotational motion of link 4.

Last, in inversion 4 (Fig. 1.20e) we block link 4. This inversion is used in manually driven mechanisms such as those used in water pumps.

1.5   Grashof's Criterion

This criterion predicts kinematic behavior of the inversions of a four-bar linkage based on the dimensions of the links. Grashof's criterion states that for at least one link of a four-bar mechanism to be able to fully rotate, it is necessary that the sum of the length of the longest and shortest link is less than or equal to the sum of the

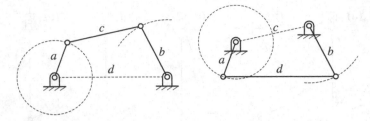

Fig. 1.21 Two non-specific rocker-crank inversions

length of the remaining two links (Eq. 1.11). In other words, in the mechanism in Fig. 1.21 one of the links will revolve if link lengths a, b, c and d comply with:

$$a+d \leq c+b \tag{1.11}$$

where $a \leq b \leq c \leq d$.

Otherwise, no links will be able to fully rotate with respect to the frame. This criterion is independent of link order and the reference frame.

Different generated movements in a four-bar mechanism, which complies with Grashof's criterion, can be seen in Figs. 1.21, 1.22, 1.23.

Fig. 1.22 Double-crank inversion of a Grashof chain

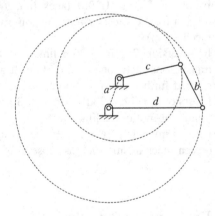

Fig. 1.23 Double-rocker inversion of a Grashof chain

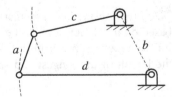

1.6   Mechanical Advantage

Mechanical advantage is defined as the ratio between the output torque and the input torque. In a four-bar mechanism it is directly proportional to the sine of angle γ and inversely proportional to the sine of angle β (Fig. 1.24). Mechanical advantage varies, consequently, in every position of the mechanism.

In Fig. 1.25, angle β equals zero, thus, $\sin \beta = 0$ and, consequently, the mechanical advantage is ∞. With a small input torque on link 2, we can overcome any resistant torque on link 4 no matter how big it is.

In the position in which $\beta = 180°$, shown in Fig. 1.26, the same happens as in the previous case, $\sin \beta = 0$. Note that if we change the input and output links, in other words, link 4 is now input and link 2 becomes output, then β and γ change too, being $\gamma = 0$ for these two positions and the mechanical advantage being zero.

Finally, in Fig. 1.27, the mechanism is shown in a position in which γ is rather small, so mechanical advantage decreases. In general, mechanisms with $\gamma < 45°$ should not be used as the mechanical advantage is too small.

Fig. 1.24 Four-bar linkage and parameters that define its mechanical advantage depending on the position of the links

Fig. 1.25 Dead-point. Mechanical advantage is infinite

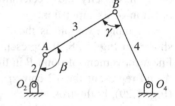

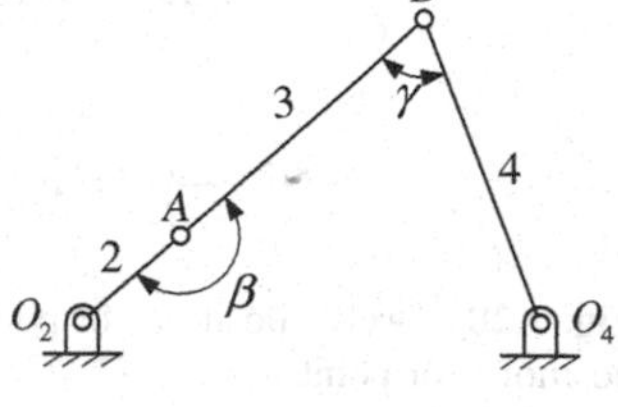

Fig. 1.26 Mechanical advantage is infinite when link 2 is the input and link 4 the output and it is zero when the input link is 4 and the output link is 2

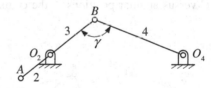

Fig. 1.27 Low mechanical
advantage when link 2 is the
input

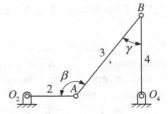

Dead-point positions are those in which two links become aligned resulting in a mechanical advantage of 0 or ∞ depending on the situation of the input link. Figures 1.25 and 1.26 show two dead-point configurations of a four-bar linkage.

1.7   Kinematic Curves

Kinematic curves are a graphic representation of two variables depending on each other in a Cartesian coordinate system. The variables represented are usually position, velocity and acceleration of a point or a link in a mechanism related to the position of its input link.

A simple example is the displacement curve for the slider-crank mechanism shown in Fig. 1.28. It represents the circular motion of link 2 (crank) versus the linear movement of point B in the piston. We take 12 equidistant positions of point A and represent the 12 corresponding positions for point B. As seen in the curve (Fig. 1.29), both strokes of the piston are symmetrical.

Another example of a kinematic curve is the one for the quick-return mechanism used in shaper machines (Fig. 1.30). It shows linear displacements of point C versus angular positions of the crank (Fig. 1.31).

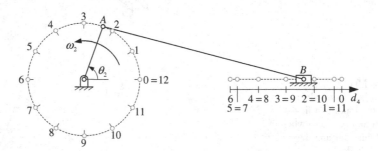

Fig. 1.28 Twelve positions of point A during one revolution of the crank and the correspondent positions for point B

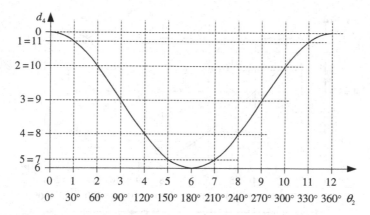

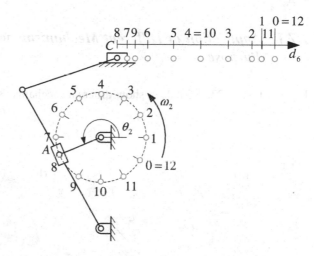

Fig. 1.29 Kinematic curve of slider displacement versus crank rotation of the slider-crank linkage in Fig. 1.28

Fig. 1.30 Twelve positions of point A of the crank during a complete kinematic cycle of withworth's

The graph shows that the working stroke of the tool (link 6) is longer than the return stroke. This is because these mechanisms are designed to minimize the time the tool spends returning to the starting point and beginning a new working cycle. In other words, while the driving link rotates at uniform speed, the working stroke is slow because the tool cannot exceed a certain speed above which it would not work properly. The return stroke is made as quickly as possible as it does not carry out any work.

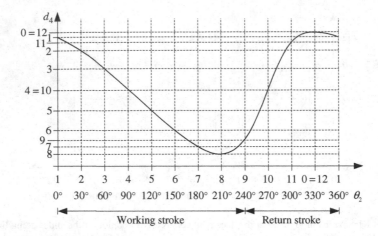

Fig. 1.31 Kinematic graph of slider displacement versus crank rotation in the quick-return mechanism in Fig. 1.30

1.7.1   Application of Different Mechanisms with Different Purposes

Figures 1.32, 1.33, 1.34, 1.35 show different mechanisms with different purposes.

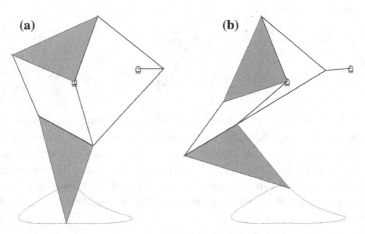

Fig. 1.32 Theo Jansen mechanism in two different positions **a** and **b**. Wheels in vehicles can be replaced with two Theo Jansen mechanisms. To do so, the lowest point has to touch the floor following a straight line along, at least, 180° rotation of the input link

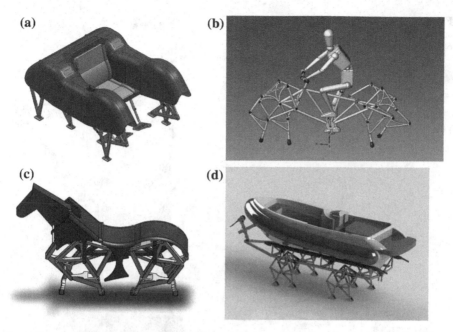

Fig. 1.33 Application of the Theo Jansen mechanism to vehicles without wheels developed by the Engineering Department of the University of Malaga. **a** Leg chair, **b** bicycle, **c** toy horse, **d** boat hand trailer

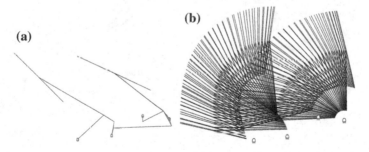

Fig. 1.34 **a** Kinematic skeleton of a eight-bar mechanism used in windshield wipers **b** area covered by the wipers along their cyclic movement

Fig. 1.35 Mechanism: robot
arm for laparoscopy
operations

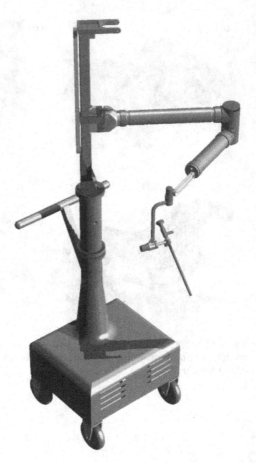

Chapter 2
Kinematic Analysis of Mechanisms. Relative Velocity and Acceleration. Instant Centers of Rotation

Abstract Kinematic analysis of a mechanism consists of calculating position, velocity and acceleration of any of its points or links. To carry out such an analysis, we have to know linkage dimensions as well as position, velocity and acceleration of as many points or links as degrees of freedom the linkage has. We will point out two different methods to calculate velocity of a point or link in a mechanism: the relative velocity method and the instant center of rotation method.

2.1 Velocity in Mechanisms

We will point out two different methods to calculate velocity of a point or link in a mechanism: the relative velocity method and the instant center of rotation method. However, before getting into the explanation of these methods, we will introduce the basic concepts for their development.

2.1.1 Position, Displacement and Velocity of a Point

To analyze motion in a system, we have to define its position and displacement previously. The movement of a point is a series of displacements in time, along successive positions.

2.1.1.1 Position of a Point

The position of a point is defined according to a reference frame. The coordinate system in a plane can be Cartesian or polar (Fig. 2.1).

© Springer International Publishing Switzerland 2016
A. Simón Mata et al., *Fundamentals of Machine Theory and Mechanisms*,
Mechanisms and Machine Science 40, DOI 10.1007/978-3-319-31970-4_2

Fig. 2.1 a Cartesian and polar coordinates of point P in a plane. **b** Polar coordinates of the same point

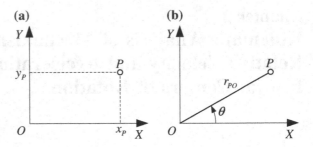

In any coordinate system, we have to define the following:

- Origin of coordinates: starting point from where measurements start.
- Axis of coordinates: established directions to measure distances and angles.
- Unit system: units to quantify distances.

If a polar coordinate system is used, the position of a point is defined by a vector called $\mathbf{r}_{PO}$ connecting the origin of coordinates O with the mentioned point. If O is a point on the frame this vector gives us the absolute position of point P and we will call it $\mathbf{r}_P$.

In most practical situations, an absolute reference system, considered stationary, is used. The stationary system coordinates do not depend on time. The absolute position of a point is defined as its position seen from this absolute reference system. If the reference system moves with respect to a stationary system, the position of the point is considered a relative position.

Anyway, this choice is not fundamental in kinematics as the movements to be studied will be relative. Take, for example, the suspension of a car where movements might refer to the car body, without considering whether the car is moving or not. Movements in the suspension system can be regarded as absolute motion with respect to the car body.

2.1.1.2   Displacement of a Point

When a point changes its position, a displacement takes place. If at instant t the point is at position P and at instant $t + \Delta t$, the point is at P', displacement during Δt is defined as the vector that measures the change in position (Eq. 2.1):

$$\Delta \mathbf{r} = \mathbf{r}_{P'} - \mathbf{r}_P \tag{2.1}$$

Displacement is a vector that connects point P at instant t with point P' at instant $t + \Delta t$ and does not depend on the path followed by the point but on the initial and final positions (Fig. 2.2).

Fig. 2.2 Displacement of
point P in a plane during
instant Δt

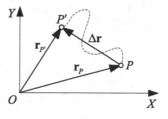

Fig. 2.3 Displacement of
point P in a plane during
instant dt close to zero

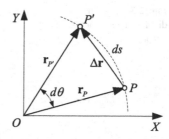

2.1.1.3 Velocity of a Point

The ratio between point displacement and time spent carrying it out is referred to as
average velocity of that point. Therefore, average velocity is a vector of magnitude
$\Delta r / \Delta t$ and has the same direction as displacement vector $\Delta \mathbf{r}$. If the time during
which displacement takes place is close to zero, the velocity of the point is called
instant velocity, or simply velocity (Eq. 2.2):

$$\mathbf{v} = \lim_{\Delta t \to 0} \frac{\Delta \mathbf{r}}{\Delta t} = \frac{d\mathbf{r}}{dt} \tag{2.2}$$

The instant velocity vector magnitude is dr/dt. In an infinitesimal position
change, the direction of the displacement vector coincides with the trajectory. When
O is the instantaneous center of the trajectory of point P, we can express the instant
velocity magnitude as Eq. (2.3):

$$v_P = \frac{dr}{dt} = \frac{ds}{dt} = \frac{d\theta}{dt} \cdot r_P = \omega \cdot r_P \tag{2.3}$$

The direction of this velocity is the same as $d\mathbf{r}$ which, at the same time, is
tangent to the motion trajectory of point P (Fig. 2.3).

2.1.2 *Position, Displacement and Angular Velocity*
of a Rigid Body

Any movement of a rigid body can be considered a combination of two motions: the
displacement of a point in the rigid body and its rotation with respect to the point.

Fig. 2.4 Angular position of
a rigid body θ_{AB}

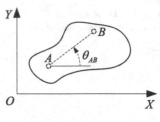

Fig. 2.5 Angular
displacement of a rigid body
$\Delta\theta_{AB}$

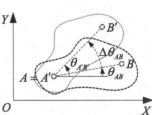

In the last section, we defined the displacement of a point, so the next subject to be studied is the rotation of a rigid body.

2.1.2.1 Angular Position of a Rigid Body

To define the angular position of a rigid body, we just need to know the angle formed by the axis of the coordinate system and reference line AB (Fig. 2.4).

2.1.2.2 Angular Displacement of a Rigid Body

When a rigid body changes its angular position from θ_{AB} to $\theta_{A'B'}$, angular displacement $\Delta\theta_{AB}$ takes place (Fig. 2.5).

$$\theta_{A'B'} = \theta_{AB} + \Delta\theta_{AB} \tag{2.4}$$

The angular displacement of a rigid body, $\Delta\theta_{AB}$, does not depend on the trajectory followed but on the initial and final angular position (Eq. 2.4).

2.1.2.3 Angular Velocity of a Rigid Body

We define the angular velocity of a rigid body as the ratio between angular displacement and its duration. If this time is, dt close to zero, this velocity is called instant angular velocity or simply angular velocity (Eq. 2.5).

$$\omega_{AB} = \frac{d\theta_{AB}}{dt} \tag{2.5}$$

2.1.3 Relative Velocity Method

In this section we will develop the relative velocity method that will allow calculating linear and angular velocities of points and links in a mechanism.

2.1.3.1 Relative Velocity Between Two Points

Let A be a point that travels from position A to position A' during time interval Δt and let B be a point that moves from position B to position B' in the same time interval (Fig. 2.6).

Absolute displacements of points A and B are given by vectors $\Delta\mathbf{r}_A$ and $\Delta\mathbf{r}_B$. Relative displacement of point B with respect to A is given by vector $\Delta\mathbf{r}_{BA}$, so it verifies (Eq. 2.6):

$$\Delta\mathbf{r}_B = \Delta\mathbf{r}_A + \Delta\mathbf{r}_{BA} \tag{2.6}$$

In other words, we can consider that point B moves to position B' with displacement equal to the one for point A to reach point B'' followed by another displacement, from point B'' to point B'. The latter coincides with vector $\Delta\mathbf{r}_{BA}$ for relative displacement. We can assert the same for the displacement of point A (Eq. 2.7), hence:

$$\Delta\mathbf{r}_A = \Delta\mathbf{r}_B + \Delta\mathbf{r}_{AB} \tag{2.7}$$

Evidently $\Delta\mathbf{r}_{BA}$ and $\Delta\mathbf{r}_{AB}$ are two vectors with the same magnitude but opposite directions.

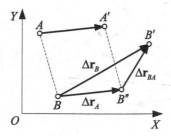

Fig. 2.6 Absolute displacements of points A and B, $\Delta\mathbf{r}_A$ and $\Delta\mathbf{r}_B$, and relative displacement of point B with respect to A, $\Delta\mathbf{r}_{BA}$

If we regard these as infinitesimal displacements and relate them to time dt, the time it took them to take place, we obtain the value of the relative velocities by Eq. (2.8):

$$\frac{d\mathbf{r}_B}{dt} = \frac{d\mathbf{r}_A}{dt} + \frac{d\mathbf{r}_{BA}}{dt} \Rightarrow \mathbf{v}_B = \mathbf{v}_A + \mathbf{v}_{BA} \tag{2.8}$$

Therefore, the velocity of a point can be determined by the velocity of another point and their relative velocity.

As we have mentioned before, relative displacements $\Delta\mathbf{r}_{BA}$ and $\Delta\mathbf{r}_{AB}$ have opposite directions. Therefore relative velocities $\mathbf{v}_{BA}$ and $\mathbf{v}_{AB}$ will be two vectors with the same magnitude that also have opposite directions (Eq. 2.9).

$$\mathbf{v}_{BA} = -\mathbf{v}_{AB} \tag{2.9}$$

2.1.3.2   Relative Velocity Between Two Points of the Same Link

Let AB be a reference line on a body that changes its position to $A'B'$ during time interval Δt.

As studied in the previous section, the vector equation for the displacement of point B is Eq. (2.6) (Fig. 2.7a). In the case of A and B belonging to the same body, distance $\overline{AB}$ does not change, so the only possible relative movement between A and B is a rotation of radius $\overline{AB}$. This way, relative displacement will always be a rotation of point B about point A (Fig. 2.7b).

If we divide these displacements by the time interval in which they happened, we obtain Eq. (2.11):

$$\frac{\Delta\mathbf{r}_B}{\Delta t} = \frac{\Delta\mathbf{r}_A}{\Delta t} + \frac{\Delta\mathbf{r}_{BA}}{\Delta t} \Rightarrow \mathbf{V}_B = \mathbf{V}_A + \mathbf{V}_{BA} \tag{2.10}$$

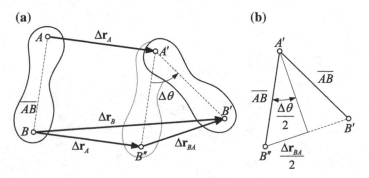

Fig. 2.7   a Relative displacement of point B with respect to point A (both being part of the same link). **b** Rotation of point B about point A

Fig. 2.8 Relative velocity of
point B with respect to point
A (both being part of the same
link)

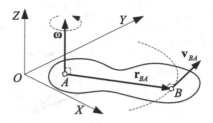

The value of $\mathbf{V}_{BA}$ (average relative velocity of point B with respect to point A) can be determined by using Eq. (2.11):

$$V_{BA} = \frac{\Delta r_{BA}}{\Delta t} = \frac{2\sin(\Delta\theta/2)}{\Delta t} \tag{2.11}$$

where $\Delta\theta$ is the angular displacement of body AB (Fig. 2.7b). If all displacements take place during an infinitesimal period of time, dt, then average velocities become instant velocities (Eq. 2.12):

$$\mathbf{v}_B = \mathbf{v}_A + \mathbf{v}_{BA} \tag{2.12}$$

This way, the velocity of point B can be obtained by adding relative instant velocity $\mathbf{v}_{BA}$ to the velocity of point A.

To obtain the magnitude of relative instant velocity $\mathbf{v}_{BA}$ in Eq. (2.13), we have to consider that the time during which displacement takes place is close to zero in Eq. (2.11)

$$v_{BA} = \lim_{\Delta t \to 0} \frac{\Delta r_{BA}}{\Delta t} = \frac{2\sin(d\theta/2)}{dt}\overline{AB} \simeq \frac{d\theta}{dt}\overline{AB} = \omega\overline{AB} \tag{2.13}$$

Therefore, any point on a rigid body, such as B, moves relatively to any other point on the same body, such as A, with velocity $\mathbf{v}_{BA}$, which can be expressed as a vector of magnitude equal to the product of the angular velocity of the body multiplied by the distance between both points (Eq. 2.14). Its direction is given by the angular velocity of the body, perpendicular to the straight line connecting both points (Fig. 2.8).

$$\mathbf{v}_{BA} = \boldsymbol{\omega} \wedge \mathbf{r}_{BA} \tag{2.14}$$

2.1.3.3 Application of the Relative Velocity Method to One Link

Equation (2.13) is the basis for the relative velocity method. It is a vector equation that allows us to calculate two algebraic unknowns such as one magnitude and one direction, two magnitudes or two directions.

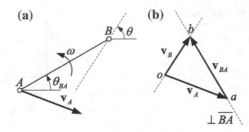

Fig. 2.9 **a** Calculation of the point B velocity magnitude knowing its direction and vector $\mathbf{v}_A$. **b** Velocity diagram

Figure 2.9a shows points A and B of a link moving at unknown angular velocity ω. Suppose that we know the velocity of point A, $\mathbf{v}_A$, and the velocity direction of point B. To calculate the velocity magnitude of point B, we use Eq. (2.13). Studying every parameter in the equation:

- $\mathbf{v}_A$ is a vector defined as $\mathbf{v}_A = v_{A_x}\hat{\mathbf{i}} + v_{A_y}\hat{\mathbf{j}}$ with known magnitude and direction.
- $\mathbf{v}_B$ is a vector with known direction and unknown magnitude. Assuming it is moving upward to the left (Fig. 2.9a), the direction of this vector will be given by angle θ and it will be defined as Eq. (2.15):

$$\mathbf{v}_B = v_B \cos\theta\hat{\mathbf{i}} + v_B \sin\theta\hat{\mathbf{j}} \tag{2.15}$$

- $\mathbf{v}_{BA}$ is a vector of unknown magnitude due to the fact that we do not know the angular velocity value, ω, of the rigid body. Its direction is perpendicular to segment line $\overline{AB}$ (Fig. 2.9b). Therefore, it can be obtained in Eq. (2.16):

$$\mathbf{v}_{BA} = \boldsymbol{\omega} \wedge \mathbf{r}_{BA} = \begin{vmatrix} \hat{\mathbf{i}} & \hat{\mathbf{j}} & \hat{\mathbf{k}} \\ 0 & 0 & \omega \\ r_{BA_x} & r_{BA_y} & 0 \end{vmatrix} = \begin{vmatrix} \hat{\mathbf{i}} & \hat{\mathbf{j}} & \hat{\mathbf{k}} \\ 0 & 0 & \omega \\ \overline{BA}\cos\theta_{BA} & \overline{BA}\sin\theta_{BA} & 0 \end{vmatrix} \tag{2.16}$$

$$= -r_{BA_y}\omega\hat{\mathbf{i}} + r_{BA_x}\omega\hat{\mathbf{j}} = \overline{BA}\omega(-\sin\theta_{BA}\hat{\mathbf{i}} + \cos\theta_{BA}\hat{\mathbf{j}})$$

where $r_{BA_x} = \overline{AB}\cos\theta_{AB}$ and $r_{BA_y} = \overline{AB}\sin\theta_{AB}$.

If we plug the velocity vectors into Eq. (2.13), we obtain Eq. (2.17):

$$v_B \cos\theta\hat{\mathbf{i}} + v_B \sin\theta\hat{\mathbf{j}} = v_{A_x}\hat{\mathbf{i}} + v_{A_y}\hat{\mathbf{j}} - r_{BA_x}\omega\hat{\mathbf{i}} + r_{BA_y}\omega\hat{\mathbf{j}}$$
$$= v_{A_x}\hat{\mathbf{i}} + v_{A_y}\hat{\mathbf{j}} - \overline{BA}\omega\sin\theta_{BA}\hat{\mathbf{i}} + \overline{BA}\omega\cos\theta_{BA}\hat{\mathbf{j}} \tag{2.17}$$

If we break the velocity vectors in the equation into their components, two algebraic equations are obtained (Eq. 2.18):

$$\left.\begin{aligned} v_B \cos \theta &= v_{A_x} - \overline{BA}\omega \sin \theta_{BA} \\ v_B \sin \theta &= v_{A_y} + \overline{BA}\omega \cos \theta_{BA} \end{aligned}\right\} \qquad (2.18)$$

We get two equations where the $\mathbf{v}_B$ magnitude and angular velocity ω are the unknowns, so the problem is completely defined. Once the magnitudes have been calculated by solving the system of Eq. (2.18), we obtain the rotation direction of ω and the direction of $\mathbf{v}_B$ depending on the + or − magnitude sign. In the example in Fig. 2.9a, the values obtained from Eq. (2.18) are positive for angular velocity ω as well as for the velocity magnitude of point B, v_B. This means that both have same directions from the ones that were assumed to write the equations. Therefore, point B moves upward right and the body rotates counterclockwise.

Equation (2.13) can also be solved graphically using a velocity diagram (Fig. 2.9b). Starting from point o (velocity pole), a straight line equal to the value of known velocity $\mathbf{v}_A$ is drawn using a scale factor. The velocity polygon is closed drawing the known direction of $\mathbf{v}_B$ from the pole and velocity direction $\mathbf{v}_{BA}$ (perpendicular to $\overline{AB}$) from the end point of $\mathbf{v}_A$. The intersection of these two directions defines the end points of vectors $\mathbf{v}_{BA}$ and $\mathbf{v}_B$. Measuring their length and using the scale factor, we obtain their magnitudes.

2.1.3.4 Calculation of Velocities in a Four-Bar Mechanism

Figure 2.10 represents a four-bar linkage in which we know the dimensions of all the links: $\overline{O_2A}$, $\overline{AB}$, $\overline{O_4B}$ and $\overline{O_2O_4}$. This mechanism has one degree of freedom, which means that the position and velocity of any point on any link can be determined from the position and velocity of one link. Assume that we know θ_2 and ω_2 and that we want to find the values of θ_3, θ_4, ω_3 and ω_4. To calculate θ_3 and θ_4, we can simply draw a scale diagram of the linkage at position θ_2 (Fig. 2.10) or solve the necessary trigonometric equations (Appendix A).

Once the link positions are obtained, we can start determining the velocities. First, velocity $\mathbf{v}_A$ will be calculated:

Fig. 2.10 Four-bar linkage where all the link dimensions are know as well as the position and velocity of link 2

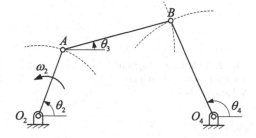

- The horizontal and vertical components of $\mathbf{v}_A$ are given by the expression (Eq. 2.19):

$$\mathbf{v}_A = \boldsymbol{\omega}_2 \wedge \mathbf{r}_{AO_2} = \begin{vmatrix} \hat{\mathbf{i}} & \hat{\mathbf{j}} & \hat{\mathbf{k}} \\ 0 & 0 & \omega_2 \\ \overline{AO_2}\cos\theta_2 & \overline{AO_2}\sin\theta_2 & 0 \end{vmatrix} \quad (2.19)$$

As the direction of ω_2 is counterclockwise (Fig. 2.10), its value in the previous equation will be negative. Point A describes a rotational motion about O_2 with a radius of $r_2 = \overline{O_2A}$ and an angular velocity of ω_2, so the direction of $\mathbf{v}_A$ will be perpendicular to $\overline{O_2A}$ to the left according to the rotation of link 2 (Fig. 2.11).
- Once $\mathbf{v}_A$ is known, we can obtain $\mathbf{v}_B$ with the following expression (Eq. 2.20):

$$\mathbf{v}_B = \boldsymbol{\omega}_4 \wedge \mathbf{r}_{BO_4} = \begin{vmatrix} \hat{\mathbf{i}} & \hat{\mathbf{j}} & \hat{\mathbf{k}} \\ 0 & 0 & \omega_4 \\ \overline{BO_4}\cos\theta_4 & \overline{BO_4}\sin\theta_4 & 0 \end{vmatrix} \quad (2.20)$$

Since point B rotates about steady point O_4 with a radius of $\overline{BO_4}$ and an angular velocity of ω_4, we cannot calculate the magnitude of $\mathbf{v}_B$ due to the fact that ω_4 is unknown. The direction of the linear velocity of point B has to be perpendicular to turning radius $\overline{BO_4}$ (Eq. 2.21). We can use the relative velocity method to find the magnitude of velocity $\mathbf{v}_B$:

$$\mathbf{v}_B = \mathbf{v}_A + \mathbf{v}_{BA} \quad (2.21)$$

Vector $\mathbf{v}_A$ as well as the direction of vector $\mathbf{v}_B$ are known in vector equation (2.21). We will now study vector $\mathbf{v}_{BA}$.
- The horizontal and vertical components of the point B relative velocity considering its rotation about point A are (Eq. 2.22):

$$\mathbf{v}_{BA} = \boldsymbol{\omega}_3 \wedge \mathbf{r}_{BA} = \begin{vmatrix} \hat{\mathbf{i}} & \hat{\mathbf{j}} & \hat{\mathbf{k}} \\ 0 & 0 & \omega_3 \\ \overline{BA}\cos\theta_3 & \overline{BA}\sin\theta_3 & 0 \end{vmatrix} \quad (2.22)$$

Fig. 2.11 Velocity diagram for the given position and velocity of link 2

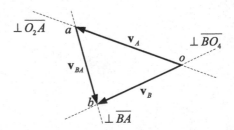

Since the angular velocity of link 3 is unknown, we cannot calculate the magnitude of $\mathbf{v}_{BA}$. The direction of $\mathbf{v}_{BA}$ is known since the relative velocity of a point that rotates about another is always perpendicular to the radius joining them. In this case, the direction will be perpendicular to $\overline{BA}$.

This way we confirm that Eq. (2.21) has two unknowns. In (Fig. 2.11) this equation is solved graphically to calculate the $\mathbf{v}_{BA}$ and $\mathbf{v}_B$ magnitudes the same way as in Fig. 2.9.

If we want to solve Eq. (2.21) mathematically, the unknowns are the ω_3 and ω_4 magnitudes. To obtain these values, we have to solve the vector equation (Eq. 2.23):

$$\begin{vmatrix} \hat{\mathbf{i}} & \hat{\mathbf{j}} & \hat{\mathbf{k}} \\ 0 & 0 & \omega_4 \\ \overline{BO_4}\cos\theta_4 & \overline{BO_4}\sin\theta_4 & 0 \end{vmatrix} = \begin{vmatrix} \hat{\mathbf{i}} & \hat{\mathbf{j}} & \hat{\mathbf{k}} \\ 0 & 0 & \omega_2 \\ \overline{AO_2}\cos\theta_2 & \overline{AO_2}\sin\theta_2 & 0 \end{vmatrix} \\ + \begin{vmatrix} \hat{\mathbf{i}} & \hat{\mathbf{j}} & \hat{\mathbf{k}} \\ 0 & 0 & \omega_3 \\ \overline{BA}\cos\theta_3 & \overline{BA}\sin\theta_3 & 0 \end{vmatrix} \qquad (2.23)$$

By developing and separating components, we obtain two algebraic equations (Eq. 2.24) where we can clear ω_3 and ω_4.

$$\left. \begin{aligned} \overline{BO_4}\omega_4\sin\theta_4 &= \overline{AO_2}\omega_2\sin\theta_2 + \overline{BA}\omega_3\sin\theta_3 \\ \overline{BO_4}\omega_4\cos\theta_4 &= \overline{AO_2}\omega_2\cos\theta_2 + \overline{BA}\omega_3\cos\theta_3 \end{aligned} \right\} \qquad (2.24)$$

Once the angular velocities are obtained, we can represent velocities $\mathbf{v}_A$, $\mathbf{v}_B$ and $\mathbf{v}_{BA}$ according to their components (Fig. 2.11).

Assume that we add point C to link 3 in the previous mechanism as shown in Fig. 2.12a and that we want to calculate its velocity. In this case, the value of angle θ'_3 is already known since angle β is a given value of the problem. Hence, $\theta'_3 = 360° - (\beta - \theta_3)$.

To obtain the velocity of point C once ω_3 has been determined, we make use of vector equation $\mathbf{v}_C = \mathbf{v}_A + \mathbf{v}_{CA}$, $\mathbf{v}_{CA}$ where is perpendicular to $\overline{CA}$ and its value is:

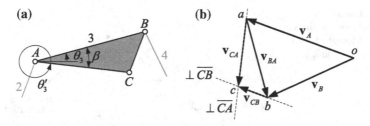

Fig. 2.12 a Four-bar linkage with new point C on link 3. b Velocity diagram

$$\mathbf{v}_{CA} = \boldsymbol{\omega}_3 \wedge \mathbf{r}_{CA} = \begin{vmatrix} \hat{\mathbf{i}} & \hat{\mathbf{j}} & \hat{\mathbf{k}} \\ 0 & 0 & \omega_3 \\ \overline{CA}\cos\theta_3' & \overline{CA}\sin\theta_3' & 0 \end{vmatrix} \tag{2.25}$$

Vector $\mathbf{v}_{CA}$ is obtained directly from Eq. (2.25) since angular velocity ω_3 is already known. $\mathbf{v}_C$ can be calculated adding the two known vectors, $\mathbf{v}_A$ and $\mathbf{v}_{CA}$. The velocity of point C can also be calculated based on the velocity of point B by using Eq. (2.26)

$$\mathbf{v}_C = \mathbf{v}_B + \mathbf{v}_{CB} \tag{2.26}$$

Figure 2.12b shows the calculation of $\mathbf{v}_C$ graphically.

2.1.3.5 Velocity Calculation in a Crankshaft Linkage

To calculate link velocities in a crankshaft linkage such as the one in Fig. 2.13, we start by calculating the positions of links 3 and 4. We consider that dimensions $\overline{O_2A}$ and $\overline{AB}$ are already known as well as the direction of the piston trajectory line and its distance r_{B_y} to O_2. If we draw a scale diagram of the linkage, the positions of links 3 and 4 are determined for a given position of link 2. We can also obtain their position by solving the following trigonometric equations (Eq. 2.27) (Appendix A):

$$\left. \begin{array}{l} \mu = \arcsin\dfrac{\overline{O_2A}\sin\theta_2 + r_{B_y}}{\overline{AB}} \\ r_{B_x} = \overline{O_2A}\cos\theta_2 + \overline{AB}\cos\mu \\ \theta_3 = 360° - \mu \end{array} \right\} \tag{2.27}$$

From this point, the calculation of the velocity of point A is the same as the one previously done for the four-bar linkage (Eq. 2.28).

$$\mathbf{v}_A = \boldsymbol{\omega}_2 \wedge \mathbf{r}_2 = \begin{vmatrix} \hat{\mathbf{i}} & \hat{\mathbf{j}} & \hat{\mathbf{k}} \\ 0 & 0 & \omega_2 \\ r_2\cos\theta_2 & r_2\sin\theta_2 & 0 \end{vmatrix} \tag{2.28}$$

Fig. 2.13 Crankshaft linkage: positions of links 3 and 4 are determined for a given position of link 2

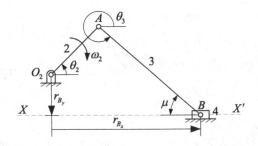

Since point A is rotating with respect to steady point O_2, the direction of velocity $\mathbf{v}_A$ is perpendicular to $\overline{O_2A}$ and it points in the same direction as the angular velocity of link 2, that is, ω_2.

We will now study velocity $\mathbf{v}_B$:

- The magnitude of velocity $\mathbf{v}_B$ is unknown. As the trajectory of point B moves along a straight line, its turning radius is infinite and its angular velocity is zero. Therefore, we cannot determine its velocity magnitude in terms of its angular velocity and turning radius.
- The direction of $\mathbf{v}_B$ is the same as the trajectory of the piston, XX'. Consequently, velocity $\mathbf{v}_B$ can be written as Eq. (2.29):

$$\mathbf{v}_B = v_B \hat{\mathbf{i}} \qquad (2.29)$$

To calculate $\mathbf{v}_B$ we need to make use of the relative velocity method (Eq. 2.30):

$$\mathbf{v}_B = \mathbf{v}_A + \mathbf{v}_{BA} \qquad (2.30)$$

The magnitude and direction of vector $\mathbf{v}_{BA}$ are given by Eq. (2.31):

$$\mathbf{v}_{BA} = \boldsymbol{\omega}_3 \wedge \mathbf{r}_{BA} = \begin{vmatrix} \hat{\mathbf{i}} & \hat{\mathbf{j}} & \hat{\mathbf{k}} \\ 0 & 0 & \omega_3 \\ \overline{BA}\cos\theta_3 & \overline{BA}\sin\theta_3 & 0 \end{vmatrix} \qquad (2.31)$$

Plugging the results into velocity equation (2.30), we obtain Eq. (2.32):

$$v_B\hat{\mathbf{i}} = \begin{vmatrix} \hat{\mathbf{i}} & \hat{\mathbf{j}} & \hat{\mathbf{k}} \\ 0 & 0 & \omega_2 \\ \overline{AO_2}\cos\theta_2 & \overline{AO_2}\sin\theta_2 & 0 \end{vmatrix} + \begin{vmatrix} \hat{\mathbf{i}} & \hat{\mathbf{j}} & \hat{\mathbf{k}} \\ 0 & 0 & \omega_3 \\ \overline{BA}\cos\theta_3 & \overline{BA}\sin\theta_3 & 0 \end{vmatrix} \qquad (2.32)$$

Breaking it into its components, we define the following equation system (Eq. 2.33):

$$\left. \begin{array}{l} v_B = -\overline{AO_2}\omega_2\sin\theta_2 - \overline{BA}\omega_3\sin\theta_3 \\ 0 = \overline{AO_2}\omega_2\cos\theta_2 + \overline{BA}\omega_3\cos\theta_3 \end{array} \right\} \qquad (2.33)$$

From which the magnitude of velocity $\mathbf{v}_B$ and angular velocity ω_3 are obtained. Once these velocities are known, we can represent them as shown in Fig. 2.14.

2.1.3.6   Velocity Analysis in a Slider Linkage

To analyze the slider linkage in Fig. 2.15a, we will start by calculating the position of links 3 and 4. As in previous examples, we know the length and position of link

Fig. 2.14 Calculation of velocities in a crankshaft linkage

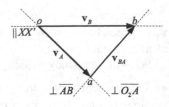

Fig. 2.15 Slider linkage: **a** positions of links 3 and 4 are determined for a given position of link 2, **b** calculation of velocities in a slider linkage

(a)   **(b)**

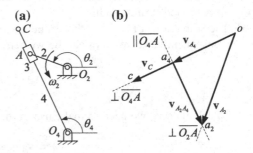

2 as well as the distance between both steady supports $\overline{O_2O_4}$. The position of link 4 is graphically determined by the line that joins O_4 and A. If we use trigonometry for our analysis, the equations needed are (Eq. 2.34) (Appendix A):

$$\left.\begin{array}{l} \overline{O_4A} = \sqrt{\overline{O_2O_4}^2 + \overline{O_2A}^2 - 2\,\overline{O_2O_4}\,\overline{O_2A}\cos(270° - \theta_2)} \\[2mm] \theta_4 = \arccos\frac{\overline{O_2A}\cos\theta_2}{\overline{O_4A}} \end{array}\right\} \qquad (2.34)$$

In the diagram, let A be a point that belongs to links 2 and 3 as in previous examples for the four-bar and crank-shaft linkages. It is not necessary to distinguish A_2 and A_3 as they are actually the same point. However, there is another point, A_4 in link 4, which coincides with A_2 at the instant represented in Fig. 2.15a. Nonetheless, point A_4 rotates about steady point O_4 while A_2 rotates about O_2. Due to this, they follow different trajectories at different velocities.

The velocity of point A_2 is perpendicular to $\overline{O_2A}$ and its magnitude and direction are represented by Eq. (2.35):

$$\mathbf{v}_{A_2} = \boldsymbol{\omega}_2 \wedge \mathbf{r}_2 = \begin{vmatrix} \hat{\mathbf{i}} & \hat{\mathbf{j}} & \hat{\mathbf{k}} \\ 0 & 0 & \omega_2 \\ r_2\cos\theta_2 & r_2\sin\theta_2 & 0 \end{vmatrix} \qquad (2.35)$$

The velocity of A_4 is perpendicular to $\overline{O_4A}$ and its magnitude is unknown because it depends on the angular velocity of link 4. Since point A_4 belongs to link 4 and it is rotating about steady point O_4, its velocity is represented by Eq. (2.36):

$$\mathbf{v}_{A_4} = \omega_4 \wedge \mathbf{r}_{A_4 O_4} = \begin{vmatrix} \hat{\mathbf{i}} & \hat{\mathbf{j}} & \hat{\mathbf{k}} \\ 0 & 0 & \omega_4 \\ \overline{A_4 O_4} \cos \theta_4 & \overline{A_4 O_4} \sin \theta_4 & 0 \end{vmatrix} \qquad (2.36)$$

To calculate $\mathbf{v}_{A_4}$, we will make use of the relative velocity method (Eq. 2.37):

$$\mathbf{v}_{A_2} = \mathbf{v}_{A_4} + \mathbf{v}_{A_2 A_4} \qquad (2.37)$$

To calculate the velocities and solve this vector equation, we have to study vector $\mathbf{v}_{A_2 A_4}$ first:

- The magnitude of $\mathbf{v}_{A_2 A_4}$ is unknown and represents the velocity at which link 3 slides over link 4.
- The direction of $\mathbf{v}_{A_2 A_4}$ coincides with direction $\overline{O_4 A}$. Therefore, this velocity is represented by Eq. (2.38):

$$\mathbf{v}_{A_2 A_4} = v_{A_2 A_4} \cos \theta_4 \hat{\mathbf{i}} + v_{A_2 A_4} \sin \theta_4 \hat{\mathbf{j}} \qquad (2.38)$$

Using Eq. (2.37), we obtain Eq. (2.39) with two algebraic unknowns (angular velocity ω_4 and the magnitude of velocity $v_{A_2 A_4}$):

$$\begin{vmatrix} \hat{\mathbf{i}} & \hat{\mathbf{j}} & \hat{\mathbf{k}} \\ 0 & 0 & \omega_2 \\ \overline{AO_2} \cos \theta_2 & \overline{AO_2} \sin \theta_2 & 0 \end{vmatrix} = \begin{vmatrix} \hat{\mathbf{i}} & \hat{\mathbf{j}} & \hat{\mathbf{k}} \\ 0 & 0 & \omega_4 \\ \overline{O_4 A} \cos \theta_4 & \overline{O_4 A} \sin \theta_4 & 0 \end{vmatrix} \\ + v_{A_2 A_4} \cos \theta_4 \hat{\mathbf{i}} + v_{A_2 A_4} \sin \theta_4 \hat{\mathbf{j}} \qquad (2.39)$$

This can be solved by breaking the equation into its components (Eq. 2.40):

$$\left. \begin{array}{l} -\overline{AO_2}\omega_2 \sin \theta_2 = -\overline{O_4 A}\omega_4 \sin \theta_4 + v_{A_2 A_4} \cos \theta_4 \\ \overline{AO_2}\omega_2 \cos \theta_2 = \overline{O_4 A}\omega_4 \cos \theta_4 + v_{A_2 A_4} \sin \theta_4 \end{array} \right\} \qquad (2.40)$$

Once the velocities have been obtained, we can represent them in the polygon shown on Fig. 2.15b.

2.1.3.7 Velocity Images

In the velocity polygon shown in Fig. 2.16b, the sides of triangle $\triangle abc$ are perpendicular to those of triangle $\triangle ABC$ of the linkage in Fig. 2.16a. The reason for this is that relative velocities are always perpendicular to their radius and, consequently, triangles $\triangle abc$ and $\triangle ABC$ are similar, with a scale ratio that depends on ω_3. The velocity image of link 3 is a triangle similar to the link, rotated 90° in the direction of ω_3.

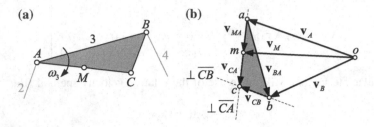

Fig. 2.16 a Four bar linkage with coupler point C. **b** Triangle $\triangle abc$ in the velocity diagram in grey represents the velocity image of link 3

Every side or link has its image in the velocity polygon. This way $\overline{ab}$, $\overline{bc}$ and $\overline{ac}$ are the images of $\overline{AB}$, $\overline{BC}$ and $\overline{AC}$ respectively. Vector $\overrightarrow{oa} = \mathbf{v}_A$ starting at pole o is the image of $\overline{O_2A}$ and vector $\overrightarrow{ob} = \mathbf{v}_B$ is the image of $\overline{O_4B}$. Moreover, the image of the frame link is pole o with null velocity. We can verify that velocities departing from o are always absolute velocities while velocities departing from any other point are relative ones.

If we add point M to link 3 of the linkage in Fig. 2.16a, we can obtain its velocity in the velocity polygon by looking for its image. We can verify that distance $\overline{am}$ in the velocity diagram is given by Eq. (2.41):

$$\frac{\overline{ab}}{\overline{AB}} = \frac{\overline{am}}{\overline{AM}} \Longrightarrow \overline{am} = \overline{AM}\frac{\overline{ab}}{\overline{AB}} \tag{2.41}$$

In conclusion, once the image of the velocity of a link has been obtained, it is very simple to calculate the velocity of any point in it. Finding the image of the point in the velocity polygon is enough. The vector that joins pole o with the image of a point represents its absolute velocity.

2.1.3.8 Application to Superior Pairs

This method can be applied to cams or geared teeth. In Fig. 2.17a, let link 2 be the driving element and link 3 the follower. Angular velocity ω_2 of the driving link is known.

In the considered instant, link 2 transmits movement to link 3 in point A. However, we have to distinguish between point A of link 2 (A_2) and point A of link 3 (A_3). These two points have different velocities and, consequently, there will be a relative velocity $\mathbf{v}_{A_3A_2}$ between them. We know that the vector sum in Eq. (2.42) has to be met:

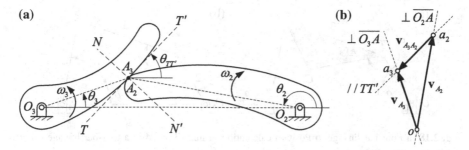

Fig. 2.17 **a** Superior pair linkage. **b** Calculation of velocities in a superior pair

$$\mathbf{v}_{A_3} = \mathbf{v}_{A_2} + \mathbf{v}_{A_3A_2} \tag{2.42}$$

The velocity of point A_2 is perpendicular to its turning radius $\overline{O_2A}$ while the velocity of point A_3 is perpendicular to $\overline{O_3A}$. To calculate these two velocities, we can use Eq. (2.43) in which ω_3 is unknown:

$$\begin{aligned} \mathbf{v}_{A_2} &= \boldsymbol{\omega}_2 \wedge \mathbf{r}_{AO_2} \\ \mathbf{v}_{A_3} &= \boldsymbol{\omega}_3 \wedge \mathbf{r}_{AO_3} \end{aligned} \tag{2.43}$$

Relative velocity $\mathbf{v}_{A_3A_2}$ of point A_3 relative to A_2 has an unknown magnitude. To find it, we need to determine the direction of vector $\mathbf{v}_{A_3A_2}$. Since the links are rigid, there is no relative motion in direction $\overline{NN'}$ due to physical constraints. Hence, relative motion happens at point A along the tangential line to the surface. This way, the direction of $\mathbf{v}_{A_3A_2}$ will coincide with tangential line TT' (Eq. 2.44):

$$\mathbf{v}_{A_3A_2} = v_{A_3A_2} \cos\theta_{TT'}\hat{\mathbf{i}} + v_{A_3A_2} \sin\theta_{TT'}\hat{\mathbf{j}} \tag{2.44}$$

Angular velocity ω_3 and linear velocity $\mathbf{v}_{A_3A_2}$ can be determined by rewriting Eq. (2.42) using the two velocity components of each vector (Eq. 2.45):

$$\left. \begin{aligned} -\overline{O_3A}\omega_3 \sin\theta_3 &= \overline{O_2A}\omega_2 \sin\theta_2 + v_{A_3A_2} \cos\theta_{TT'} \\ \overline{O_3A}\omega_3 \cos\theta_3 &= \overline{O_2A}\omega_2 \cos\theta_2 + v_{A_3A_2} \sin\theta_{TT'} \end{aligned} \right\} \tag{2.45}$$

Example 1 Determine velocities $\mathbf{v}_B$ and $\mathbf{v}_C$ of the four-bar mechanism in Fig. 2.18a. Its dimensions are: $\overline{O_2O_4} = 15\,\text{cm}$, $\overline{O_2A} = 6\,\text{cm}$, $\overline{AB} = 11\,\text{cm}$, $\overline{O_4B} = 9\,\text{cm}$, $\overline{AC} = 8\,\text{cm}$ and $\widehat{BAC} = 30°$. The input angle is $\theta_2 = 60°$ and the angular velocity of the driving link is $\omega_2 = -20\,\text{rad/s}$ (clockwise direction).

Angles θ_3, θ_4 and θ_3' can be obtained by applying the trigonometric method (Eqs. 2.46–2.51) developed in Appendix A where angles β, ϕ and δ are represented in Fig. 2.18b.

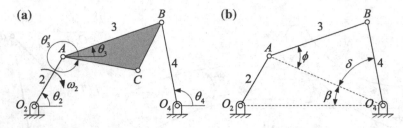

Fig. 2.18 a Four bar linkage. **b** Position calculation of links 3 and 4 in a four-bar linkage using the trigonometric method

$$\overline{O_4 A} = \sqrt{15^2 + 6^2 - 2 \cdot 15 \cdot 6 \cdot \cos 60°} = 13.08 \, \text{cm} \qquad (2.46)$$

$$\beta = \arcsin\left(\frac{6}{13.08} \sin 60°\right) = 23.41° \qquad (2.47)$$

$$\phi = \arccos\left(\frac{11^2 + 13.08^2 - 9^2}{2 \cdot 11 \cdot 13.08}\right) = 42.81° \qquad (2.48)$$

$$\delta = \arcsin\left(\frac{11}{9} \sin 42.81°\right) = 56.17° \qquad (2.49)$$

$$\begin{aligned} \theta_3 &= \phi - \beta = 19.4° \\ \theta_4 &= 180° - (\beta + \delta) = 100.42° \end{aligned} \qquad (2.50)$$

$$\theta_3' = 360° - \left(\widehat{BAC} - \theta_3\right) = 349.4° \qquad (2.51)$$

To calculate the velocity of point B, we will apply the relative velocity method. We start by analyzing velocities v_A, v_B and v_{BA} (Eqs. 2.52–2.54).

$$\mathbf{v}_A = \boldsymbol{\omega}_2 \wedge \mathbf{r}_{AO_2} = \begin{vmatrix} \hat{\mathbf{i}} & \hat{\mathbf{j}} & \hat{\mathbf{k}} \\ 0 & 0 & -20 \\ 6\cos 60° & 6\sin 60° & 0 \end{vmatrix} = 103.9\hat{\mathbf{i}} - 60\hat{\mathbf{j}} \qquad (2.52)$$

Operating with these components, we calculate its magnitude and direction:

$$\mathbf{v}_A = 120 \, \text{cm/s} \angle 330°$$

$$\mathbf{v}_{BA} = \boldsymbol{\omega}_3 \wedge \mathbf{r}_{BA} = \begin{vmatrix} \hat{\mathbf{i}} & \hat{\mathbf{j}} & \hat{\mathbf{k}} \\ 0 & 0 & \omega_3 \\ 11\cos 19.4° & 11\sin 19.4° & 0 \end{vmatrix} = -3.56\omega_3\hat{\mathbf{i}} + 10.38\omega_3\hat{\mathbf{j}}$$

$$(2.53)$$

$$v_B = \omega_4 \wedge r_{BO_4} = \begin{vmatrix} \hat{i} & \hat{j} & \hat{k} \\ 0 & 0 & \omega_4 \\ 9\cos 100.4° & 9\sin 100.4° & 0 \end{vmatrix} = -8.85\omega_4 \hat{i} + -1.62\omega_4 \hat{j}$$

$$(2.54)$$

To calculate ω_3 and ω_4 we use the relative velocity (Eq. 2.55):

$$v_B = v_A + v_{BA} \tag{2.55}$$

Clearing the components, we obtain Eq. (2.56):

$$\left. \begin{array}{l} -8.85\omega_4 = 103.9 - 3.65\omega_3 \\ -1.62\omega_4 = -60 + 10.38\omega_3 \end{array} \right\} \tag{2.56}$$

From which the following values for angular velocity $\omega_3 = 7.16$ rad/s clockwise and $\omega_4 = -8.78$ rad/s counterclockwise can be worked out. Operating with these values in Eqs. (2.53) and (2.54), we obtain velocities v_B and v_{BA}:

$$v_B = 77.75\hat{i} + 14.28\hat{j} = 79.1 \text{ cm/s} \angle 10.4°$$

$$v_{BA} = -26.13\hat{i} + 74.32\hat{j} = 78.78 \text{ cm/s} \angle 109.4°$$

To calculate the velocity of point C, we apply the relative velocity equation, $v_C = v_A + v_{CA}$, where v_A is already known and v_{CA} is given by Eq. (2.57):

$$v_{CA} = \omega_3 \wedge r_{CA} = \begin{vmatrix} \hat{i} & \hat{j} & \hat{k} \\ 0 & 0 & 7.16 \\ 8\cos 349.4° & 8\sin 349.4° & 0 \end{vmatrix} = 10.53\hat{i} + 56.3\hat{j} \quad (2.57)$$

$$v_{CA} = 57.28 \text{ cm/s} \angle 79.4°$$

Using these values in the relative velocity equation, we obtain:

$$v_C = 114.4\hat{i} - 3.7\hat{j} = 114.46 \text{ cm/s} \angle 358.1°$$

Example 2 Calculate velocity v_B in the crank-shaft linkage shown in Fig. 2.19. Consider the dimensions to be as follows: $\overline{O_2A} = 3$ cm, $\overline{AB} = 7$ cm and $y = 1.5$ cm. The trajectory followed by the piston is horizontal. The input angle is $\theta_2 = 60°$ and link 2 moves with angular velocity $\omega_2 = -10$ rad/s (clockwise).

We start by solving the position problem (Eqs. 2.58–2.60) using the trigonometric method (Fig. 2.19):

Fig. 2.19 Calculation of the position of the links for a given input angle in a crank-shaft linkage

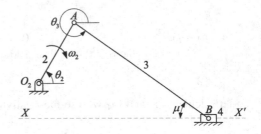

$$\mu = \arcsin \frac{3 \sin 60° + 1.5}{7} = 35.8° \tag{2.58}$$

$$x_B = 3 \cos 60° + 7 \cos 35.8° = 7.1 \, \text{cm} \tag{2.59}$$

$$\theta_3 = 360° - 35.8 = 324.2° \tag{2.60}$$

The velocity of point B is obtained from relative velocity (Eq. 2.61):

$$\mathbf{v}_B = \mathbf{v}_A + \mathbf{v}_{BA} \tag{2.61}$$

where $\mathbf{v}_A$, $\mathbf{v}_B$ and $\mathbf{v}_{BA}$ are given by Eqs. (2.62)–(2.64):

$$\mathbf{v}_A = \boldsymbol{\omega}_2 \wedge \mathbf{r}_{AO_2} = \begin{vmatrix} \hat{\mathbf{i}} & \hat{\mathbf{j}} & \hat{\mathbf{k}} \\ 0 & 0 & -10 \\ 3 \cos 60° & 3 \sin 60° & 0 \end{vmatrix} = 25.98\hat{\mathbf{i}} - 15\hat{\mathbf{j}} \tag{2.62}$$

$$\mathbf{v}_A = 30 \, \text{cm/s} \, \angle 330°$$

$$\mathbf{v}_{BA} = \boldsymbol{\omega}_3 \wedge \mathbf{r}_{BA} = \begin{vmatrix} \hat{\mathbf{i}} & \hat{\mathbf{j}} & \hat{\mathbf{k}} \\ 0 & 0 & \omega_3 \\ 7 \cos 324.2° & 7 \sin 324.2° & 0 \end{vmatrix} = 4.09\omega_3\hat{\mathbf{i}} + 5.68\omega_3\hat{\mathbf{j}}$$

$$\tag{2.63}$$

$$\mathbf{v}_B = v_B\hat{\mathbf{i}} \tag{2.64}$$

Using these values in the relative velocity (Eq. 2.61) we obtain Eq. (2.65):

$$\left. \begin{aligned} v_B &= 25.98 + 4.09\omega_3 \\ 0 &= -15 + 5.68\omega_3 \end{aligned} \right\} \tag{2.65}$$

Ultimately, resulting in the following values for angular and linear velocities $\omega_3 = 2.64 \, \text{rad/s}$ counterclockwise and $v_B = 36.78 \, \text{cm/s}$. Thus, the velocities will be:

Fig. 2.20 Position and
velocity calculation of the
links in a slider linkage. The
unknowns are θ_4, $\overline{O_4A}$, ω_4
and $v_{A_2A_4}$

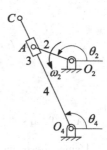

$$v_B = 36.78\hat{i} = 36.78\,\text{cm/s}\,\angle 0°$$

$$v_{BA} = 10.79\hat{i} + 15\hat{j} = 18.48\,\text{cm/s}\,\angle 54.24°$$

Example 3 Calculate velocity v_C of the slider linkage in Fig. 2.20 when the
dimensions of the links are $\overline{O_2A} = 3\,\text{cm}$, $\overline{O_2O_4} = 5\,\text{cm}$, $\overline{O_4C} = 9\,\text{cm}$ and the input
angle is $\theta_2 = 160°$. The input link moves counterclockwise with angular velocity
$\omega_2 = 10\,\text{rad/s}$.

The position problem can easily be solved using the trigonometric method
(Eqs. 2.66 and 2.67):

$$\overline{O_4A} = \sqrt{5^2 + 3^2 - 2 \cdot 5 \cdot 3 \cos(270° - 160°)} = 6.65\,\text{cm} \qquad (2.66)$$

$$\theta_4 = \arccos\frac{3\cos 160°}{6.65} = 115.08° \qquad (2.67)$$

In order to calculate the velocity of point C, we first have to calculate the
velocity of point A_4 which temporarily coincides with A_2 at the time instant considered while being part of link 4. We can relate v_{A_2} and v_{A_4} with relative velocity
(Eq. 2.68):

$$v_{A_2} = v_{A_4} + v_{A_2A_4} \qquad (2.68)$$

where v_{A_2}, v_{A_4} and $v_{A_2A_4}$ are given by Eqs. (2.69)–(2.71).

$$v_{A_2} = \omega_2 \wedge r_{AO_2} = \begin{vmatrix} \hat{i} & \hat{j} & \hat{k} \\ 0 & 0 & 10 \\ 3\cos 160° & 3\sin 160° & 0 \end{vmatrix} = -10.26\hat{i} - 28.19\hat{j} \qquad (2.69)$$

$$v_{A_2} = 30\,\text{cm/s}\,\angle 250°$$

$$\mathbf{v}_{A_4} = \boldsymbol{\omega}_4 \wedge \mathbf{r}_{AO_4} = \begin{vmatrix} \hat{\mathbf{i}} & \hat{\mathbf{j}} & \hat{\mathbf{k}} \\ 0 & 0 & \omega_4 \\ 6.65\cos 115.08^\circ & 6.65\sin 115.08^\circ & 0 \end{vmatrix} \tag{2.70}$$

$$= -6.02\omega_4\hat{\mathbf{i}} - 2.82\omega_4\hat{\mathbf{j}}$$

$$\mathbf{v}_{A_2A_4} = v_{A_2A_4}\cos 115.08^\circ\hat{\mathbf{i}} + v_{A_2A_4}\sin 115.08^\circ\hat{\mathbf{j}} = -0.42v_{A_2A_4}\hat{\mathbf{i}} + 0.91v_{A_2A_4}\hat{\mathbf{j}} \tag{2.71}$$

Using these values in Eq. (2.68) and clearing the components, we obtain Eq. (2.72):

$$\left.\begin{array}{l} -10.26 = -6.02\omega_4 - 0.42v_{A_2A_4} \\ -28.19 = -2.82\omega_4 + 0.9v_{A_2A_4} \end{array}\right\} \tag{2.72}$$

We calculate the values of angular velocity $\omega_4 = -3.19$ rad/s clockwise and the magnitude of $v_{A_2A_4} = -21.32$ cm/s. The negative sign indicates that the angle of $\mathbf{v}_{A_2A_4}$ is not θ_4 but $\theta_4 + 180^\circ$. Consequently, the velocity values are Eqs. (2.73) and (2.74):

$$\mathbf{v}_{A_4} = -19.2\hat{\mathbf{i}} - 9\hat{\mathbf{j}} = 21.2\,\text{cm/s}\,\angle 205.1^\circ \tag{2.73}$$

$$\mathbf{v}_{A_2A_4} = 8.95\hat{\mathbf{i}} - 19.19\hat{\mathbf{j}} = 21.32\,\text{cm/s}\,\angle 295^\circ \tag{2.74}$$

To calculate the velocity of point C, we make use of Eq. (2.75):

$$\mathbf{v}_C = \boldsymbol{\omega}_4 \wedge \mathbf{r}_{CO_4} = \begin{vmatrix} \hat{\mathbf{i}} & \hat{\mathbf{j}} & \hat{\mathbf{k}} \\ 0 & 0 & 3.19 \\ 9\cos 115.08^\circ & 9\sin 115.08^\circ & 0 \end{vmatrix} = -26\hat{\mathbf{i}} - 12.17\hat{\mathbf{j}} \tag{2.75}$$

$$\mathbf{v}_C = 28.7\,\text{cm/s}\,\angle 205.1^\circ$$

Example 4 In the mixing machine in Fig. 2.21a, calculate the velocity of extreme point C of the spatula knowing that the motor of the mixer moves counterclockwise with an angular velocity of 95.5 rpm and $\theta_2 = 0^\circ$. The dimensions in the drawing are in centimeters.

The kinematic skeleton of the mixing machine is shown in Fig. 2.21b. To determine the position of the linkage, we have to calculate the value of angle θ_3 and distance $\overline{O_4A}$. To do so, we apply Eq. (2.76):

$$\left.\begin{array}{l} \mu = \arctan\dfrac{r_2}{r_1} = \arctan\dfrac{7}{11} = 32.47^\circ \\[2mm] \overline{O_4A} = \sqrt{r_1^2 + r_2^2} = \sqrt{11^2 + 7^2} = 13.04\,\text{cm} \\[2mm] \theta_3 = 90^\circ + \mu = 122.47^\circ \end{array}\right\} \tag{2.76}$$

Fig. 2.21 a Mixing machine.
b Kinematic skeleton

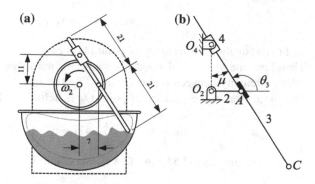

Before starting the calculation of velocities, we have to convert the given input velocity from rpm into rad/s (Eq. 2.77):

$$\omega_2 = 95.5 \text{ rpm} \frac{2\pi \text{ rad}}{60 \text{ s}} = 10 \text{ rad/s} \tag{2.77}$$

To calculate the velocity of point C, we first have to solve the velocity of point O_3, which coincides with the position of point O_4 at the instant considered while still being part of link 3. Since points O_3 and A belong to link 3, we can use the relative velocity method to relate their velocities (Eq. 2.78):

$$\mathbf{v}_{O_3} = \mathbf{v}_A + \mathbf{v}_{O_3 A} \tag{2.78}$$

where $\mathbf{v}_A$ and $\mathbf{v}_{O_3 A}$ are given by Eqs. (2.79) and (2.80):

$$\mathbf{v}_A = \boldsymbol{\omega}_2 \wedge \mathbf{r}_{AO_2} = \begin{vmatrix} \hat{\mathbf{i}} & \hat{\mathbf{j}} & \hat{\mathbf{k}} \\ 0 & 0 & 10 \\ 7\cos 0° & 7\sin 0° & 0 \end{vmatrix} = 70\hat{\mathbf{j}} \tag{2.79}$$

$$\mathbf{v}_A = 70 \text{ cm/s} \angle 90°$$

$$\mathbf{v}_{O_3 A} = \boldsymbol{\omega}_3 \wedge \mathbf{r}_{AO_3} = \begin{vmatrix} \hat{\mathbf{i}} & \hat{\mathbf{j}} & \hat{\mathbf{k}} \\ 0 & 0 & \omega_3 \\ \overline{AO_3}\cos 122.47° & \overline{AO_3}\sin 122.47° & 0 \end{vmatrix} = -11\omega_3\hat{\mathbf{i}} - 7\omega_3\hat{\mathbf{j}} \tag{2.80}$$

Using these values in Eq. (2.78), we obtain Eq. (2.81):

$$\mathbf{v}_{O_3} = (70\hat{\mathbf{j}}) + (-11\omega_3\hat{\mathbf{i}} - 7\omega_3\hat{\mathbf{j}}) \tag{2.81}$$

However, in Eq. (2.81) the direction as well as the magnitude of velocity $\mathbf{v}_{O_3}$ remain unknown. To obtain information on this velocity, we will relate the velocity of point O_3 with the velocity of point O_4 by using Eq. (2.82):

$$\mathbf{v}_{O_3} = \mathbf{v}_{O_4} + \mathbf{v}_{O_3 O_4} \tag{2.82}$$

In this equation, the velocity of point O_4 is zero, $\mathbf{v}_{O_4} = 0$, since it is a fixed point. Therefore, the velocity of point O_3 has the same magnitude and direction as the relative velocity between points O_3 and O_4. The direction of this velocity is given by link 3. Hence, the velocity of point O_3 is defined as Eq. (2.83):

$$\mathbf{v}_{O_3} = v_{O_3} \cos 122.47 \hat{\mathbf{i}} + v_{O_3} \sin 122.47 \hat{\mathbf{j}} \tag{2.83}$$

Evening out Eqs. (2.81) and (2.83), we obtain Eq. (2.84):

$$v_{O_3} \cos 122.47 \hat{\mathbf{i}} + v_{O_3} \sin 122.47 \hat{\mathbf{j}} = 70 \hat{\mathbf{j}} + (-11\omega_3 \hat{\mathbf{i}} - 7\omega_3 \hat{\mathbf{j}}) \tag{2.84}$$

By separating the components, we obtain Eq. (2.85):

$$\left. \begin{array}{l} v_{O_3} \cos 122.47 = -11\omega_3 \\ v_{O_3} \sin 122.47 = 70 + 7\omega_3 \end{array} \right\} \tag{2.85}$$

Solving Eq. (2.85), we obtain the values for angular velocity $\omega_3 = 2.88$ rad/s clockwise and the velocity magnitude of point O_3, $v_{O_3} = 58.71$ cm/s. This way, the vector velocity of point O_3 is defined by Eq. (2.86):

$$\mathbf{v}_{O_3} = -31.52 \hat{\mathbf{i}} + 49.53 \hat{\mathbf{j}} = 58.71 \text{ cm/s} \angle 122.47° \tag{2.86}$$

Eventually, in order to calculate the velocity of point C, we apply velocity (Eq. 2.87):

$$\mathbf{v}_C = \mathbf{v}_A + \mathbf{v}_{CA} \tag{2.87}$$

where relative velocity between points C and A is Eq. (2.88):

$$\mathbf{v}_{CA} = \boldsymbol{\omega}_3 \wedge \mathbf{r}_{CA} = \begin{vmatrix} \hat{\mathbf{i}} & \hat{\mathbf{j}} & \hat{\mathbf{k}} \\ 0 & 0 & 2.88 \\ 21\cos 302.47° & 21\sin 302.47° & 0 \end{vmatrix} = 51.03 \hat{\mathbf{i}} + 32.45 \hat{\mathbf{j}} \tag{2.88}$$

Operating with the known values in Eq. (2.87), we obtain the vector velocity of point C:

$$\mathbf{v}_C = 51.03 \hat{\mathbf{i}} + 102.45 \hat{\mathbf{j}} = 114.4 \text{ cm/s} \angle 63.52° $$

Once all the velocities are defined, we can represent them in the velocity polygon (Fig. 2.21c).

2.1.4 Instant Center of Rotation Method

Any planar displacement of a rigid body can be considered a rotation about a point. This point is called instantaneous center or instant center of rotation (I.C.R.).

2.1.4.1 Instant Center of Rotation of a Rigid Body

Let a rigid body move from position AB to position $A'B'$ (Fig. 2.22). Position change could be due to a pure rotation of triangle $\triangle OAB$ about O, intersection point of the bisectors of segments AA' and BB'. We can obtain the displacement of points A and B by using their distance to center O and the angular displacement of the body, $\Delta\theta$ (Eq. 2.89).

$$\left.\begin{array}{l} \Delta r_A = \overline{AA'} = 2\overline{OA}\sin\frac{\Delta\theta}{2} \\ \Delta r_B = \overline{BB'} = 2\overline{OB}\sin\frac{\Delta\theta}{2} \end{array}\right\} \tag{2.89}$$

Considering the time to be infinitesimal, we can consider the body to be rotating about O, the instant rotation center. Displacements will be Eq. (2.90):

$$\left.\begin{array}{l} dr_A = 2\overline{OA}\sin\frac{d\theta}{2} = \overline{OA}d\theta \\ dr_B = 2\overline{OB}\sin\frac{d\theta}{2} = \overline{OB}d\theta \end{array}\right\} \tag{2.90}$$

Dividing both displacements by the time spent, dt, we find the instant velocities of points A and B. Their directions are perpendicular to radius $\overline{OA}$ and $\overline{OB}$ respectively and their magnitudes are Eq. (2.91):

$$\left.\begin{array}{l} v_A = \overline{OA}\frac{d\theta}{dt} = \overline{OA}\omega \\ v_B = \overline{OB}\frac{d\theta}{dt} = \overline{OB}\omega \end{array}\right\} \tag{2.91}$$

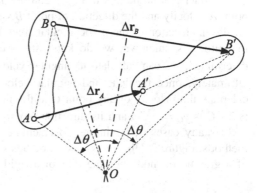

Fig. 2.22 A planar movement of the rigid body AB can be considered a rotation about point O

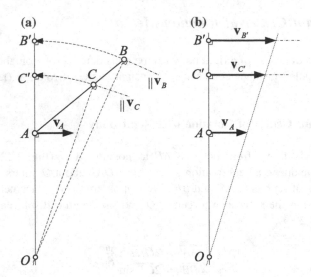

Fig. 2.23 Graphical calculation of direction (**a**) and magnitude (**b**) of the velocities of points B and C knowing $\mathbf{v}_A$ and the direction of $\mathbf{v}_B$

This way, it is verified that, at a certain instant of time, point O is the rotation center of points A and B. The magnitude of the velocity of any point in the body will be Eq. (2.92):

$$v = R\omega \tag{2.92}$$

where:

- R is the instant rotation radius of the point (distance from the point to O).
- ω is the angular velocity of the body measured in radians per second.

Velocity of every point in a link will have direction perpendicular to its instant rotation radius. Thus, if we know the direction of the velocities of two points of a link, we can find the ICR of the link on the intersection of two perpendicular lines to both velocities.

Consider that in the link in Fig. 2.23a, we know the magnitude and direction of point A velocity and the direction of point B velocity. The ICR of the link has to be on the intersection of the perpendicular lines to $\mathbf{v}_A$ and $\mathbf{v}_B$; even though the latter magnitude is unknown, we do know its direction. Once the ICR of the link is determined, we can calculate its angular velocity (Eq. 2.91) and so $\omega = v_A/\overline{OA}$. Ultimately, once the ICR and angular velocity of the link are known, we can calculate the velocity of any point C in the link. The magnitude of the velocity of point C is $v_C = \overline{OC}\omega$ and its direction is perpendicular to $\overline{OC}$.

In many cases, it is simpler to calculate velocity magnitudes with graphical methods. Figure 2.23b shows how velocities $\mathbf{v}_B$ and $\mathbf{v}_C$ can be calculated by means of a graphic method once the ICR of a rigid body and the velocity of one of its

Fig. 2.24 The ICR of a body moving on a plane with pure translation is placed at the infinite

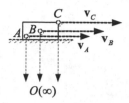

points (in this case v_A) are known. If we fold up points B and C over line OA, it must be verified that the triangles with their sides formed by each velocity and the rotation radius of each point are similar (Eq. 2.93), since:

$$\frac{v_A}{\overline{OA}} = \frac{v_B}{\overline{OB'}} = \frac{v_C}{\overline{OC'}} = \omega \qquad (2.93)$$

In the case of a body moving on a plane with no angular velocity (pure translation), its ICR is placed at the infinite since all points of the body have the same velocity and the perpendicular lines to such velocities intersect at the infinite (Fig. 2.24).

2.1.4.2 Instant Center of Rotation of a Pair of Links

So far, we have looked at the ICR of a link relative to a stationary reference system. However, we can define the ICR of a pair of links, not taking into account if one of them is fixed or not. This ICR between the two links is the point one link rotates about with respect to the other.

In Fig. 2.25, point I_{23} is the ICR of link 2 relative to link 3. In other words, link 2 rotates about this point relative to link 3. There is one point of each link that coincides in position with this ICR. If we consider that link 3 is moving, these two points move at the same absolute velocity, that is, null relative velocity. This is the only couple of points - one of each link - that has zero relative velocity at the instant studied.

To help us understand the ICR concept of a pair of links, we are going to calculate the ones corresponding to a four-bar linkage. Notice that in the linkage in Fig. 2.26, there is one ICR for every two links. To know the number of ICRs in a linkage, we have to establish all possible combinations of the number of links

Fig. 2.25 The ICR between links 2 and 3 is the point link 2 rotates about with respect to link 3 or vice versa

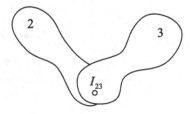

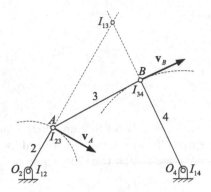

Fig. 2.26 ICR I_{13} is on the intersection point of two lines perpendicular to the velocity vectors of points A and B of link 3 with respect to link 1

taking two at a time since I_{ij} is the same IRC as I_{ji}. Therefore, the number of IRCs is given by Eq. (2.94):

$$NICRs = \frac{N(N-1)}{2} = 6 \tag{2.94}$$

where:

- *NICRs* is the number of ICRs
- N is the number of links.

The obvious ones are I_{12}, I_{23}, I_{34} and I_{14} since every couple of links is joined by a hinge, which is the rotating point of one link relative to another. Remember that the velocity of any point in the link has to be perpendicular to its instant rotation radius. In consequence, considering that points A and B are part of link 3, I_{13} is on the intersection of two lines perpendicular to the velocity vectors of points A and B (Fig. 2.26).

ICR I_{24} is obtained the same way but considering the inversion shown in Fig. 2.27. As in kinematic inversions, relative motion between links is maintained.

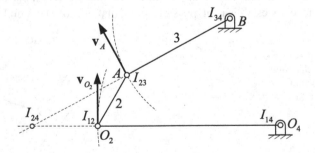

Fig. 2.27 ICR I_{24} is on the intersection of two lines perpendicular to the velocity vectors of points A and O_2 of link 2 with respect to link 4

Fig. 2.28 The velocity
vector of ICR I_{23} has to be
perpendicular to ICR I_{12} and
to ICR I_{13}. Therefore, it has to
be located on the straight line
defined by ICR I_{12} and ICR I_{13}

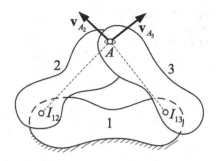

2.1.4.3 Kennedy's Theorem

Also known as the Three Centers Theorem, it is used to find the ICR of a linkage
without having to look into its kinematic inversions as we did in the last example.
Kennedy's Theorem states that all three ICRs of three links with planar motion have
to be aligned on a straight line.

In order to demonstrate this theorem, first note Fig. 2.28 representing a set of
three links (1, 2, 3) that have relative motion. Links 2 and 3 are joined to link 1
making two rotating pairs. Therefore, ICRs I_{12} and I_{13} are easy to locate.

Links 2 and 3 are not physically joined. However, as previously studied in this
chapter, there is a point link 2 rotates about, relative to link 3, at a given instant.
This point is ICR I_{23}. Initially, we do not know where to locate it, so we are going
to assume it coincides with point A.

In this case, point A would act as a hinge that joins links 2 and 3. In other words,
we could consider it as a point that is part of links 2 and 3 at the same time. If we
consider it to be a point of link 2, its velocity with respect to link 1 has to be
perpendicular to the rotating radius $\overline{I_{12}A}$ (Fig. 2.28). However, if we consider it to
be part of link 3, it has to rotate about I_{13} with a radius of $\overline{I_{13}A}$.

This gives us different directions for the velocity vectors of points A_2 and A_3,
which means that there is a relative velocity between them. Therefore, point
A cannot be ICR I_{23}. If the velocity of ICR I_{23} has to have the same direction when
calculated as a point of link 2 and a point of link 3, ICR I_{23} has to be located on the
straight line defined by I_{12} and I_{13}.

This rule is known as Kennedy's Theorem, which says that the three relative
ICRs of any three links have to be located on a straight line. This law is valid for
any set of three links that has relative planar motion, even if none of them is the
ground link (frame).

2.1.4.4 Locating the ICRs of a Linkage

To locate the ICRs of the links in a linkage, we will apply the following rules:

1. Identify the ones corresponding to rotating kinematic pairs. The ICR is the point
   that identifies the axis of the pair (hinge).

Fig. 2.29 Instant Centers of
Rotation of links 1, 2, 3 and 4
in a slider-crank linkage

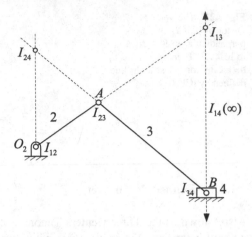

2. In sliding pairs, the ICR is on the curvature center of the path followed by the slide.
3. The rest of ICRs can be obtained by means of the application of Kennedy's Theorem to sets of three links in the linkage.

Example 5 Find the Instant Centers of Rotation of all the links in the slider-crank linkage in Fig. 2.29.

First, we identify ICRs I_{12}, I_{23}, I_{34}, and I_{14} that correspond to the four kinematic pairs in the linkage. Notice that ICR I_{14} is located at the infinite as the slider path is a straight line.

Next, we apply Kennedy's Theorem to links 1, 2 and 3. According to this theorem ICRs I_{13}, I_{23} and I_{14} have to be aligned. The same way, if we take links 1, 3 and 4, ICRs I_{13}, I_{34} and I_{14} also have to be aligned. By drawing the two straight lines, we find the position of ICR I_{13}.

To find ICR I_{24}, we proceed the same way applying Kennedy's Theorem to links 1, 2, 4 on one side and 2, 3, 4 on the other.

Example 6 Find the Instant Centers of Rotation of the links of the four-bar linkage in Fig. 2.30a.

To help us to locate all ICRs we are going to make use of a polygon formed by as many sides as there are links in the linkage to analyze. In this case, we use a four-sided polygon. Next, we number the vertex from 1 to 4 (Fig. 2.30b). Every side or diagonal of the polygon represents an ICR. In this case, the sides represent I_{12}, I_{23}, I_{34} and I_{14}. Both diagonals represent ICRs I_{13} and I_{24}. We will trace those sides or diagonals representing known ICRs with a solid line and the unknown ones with a dotted line.

In the example in Fig. 2.30a, ICRs I_{12}, I_{23}, I_{34} and I_{14} are known while ICRs I_{24} and I_{13} are unknown. In order to find ICR I_{24} we apply Kennedy's Theorem making use of the polygon. To find the two ICRs that are aligned with ICR I_{24}, we define a

Fig. 2.30 a Instant Centers
of Rotation of links 1, 2, 3
and 4 in a four-bar linkage.
b Polygon to analyze all ICRs

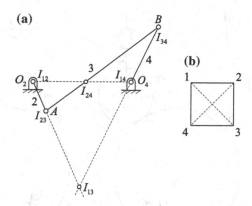

triangle in the polygon, where two sides represent already known ICRs (for instance, I_{23} and I_{34}) and a third side representing the unknown ICR (in this case I_{24}). We will repeat the operation with ICRs I_{12}, I_{14} and I_{24}. We find ICR I_{24} on the intersection of lines $I_{23}I_{24}$ and $I_{12}I_{14}$. To find ICR I_{13}, we define triangles I_{12}, I_{23}, I_{13} and I_{14}, I_{34}, I_{13}. ICR I_{13} is on the intersection of lines $I_{12}I_{23}$ and $I_{14}I_{34}$.

Example 7 Find the Instant Centers of Rotation of the links in the slider linkage in Fig. 2.31.

In the example in Fig. 2.31a, ICRs I_{12}, I_{23}, I_{34} and I_{14} are known while ICRs I_{24} and I_{13} are unknown. In order to find these ICRs we apply Kennedy's Theorem making use of the polygon (Fig. 2.31b) the same way we did in the last example.

Example 8 Find the all the Instant Centers of Rotation in the mechanism in Fig. 2.32a. Link 2 is an eccentric wheel that rotates about O_2 transmitting a rolling motion without slipping to link 3, which is a roller joined at point A to link 4 in straight motion inside a vertical guide.

Fig. 2.31 a Instant Centers
of Rotation of links 1, 2, 3
and 4 in a slider linkage.
b Polygon to analyze all ICRs

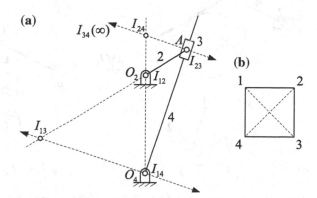

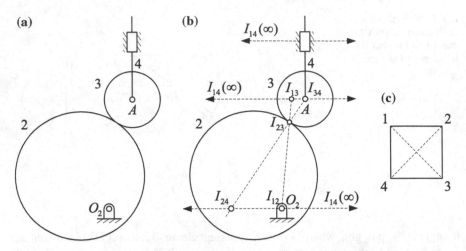

Fig. 2.32 **a** Instant Centers of Rotation of links 1, 2, 3 and 4 in a mechanism with two wheels and a slider. **b** ICRs. **c** Polygon to analyze all ICRs

The known ICRs are I_{12}, I_{23}, I_{34} and I_{14}. ICR I_{13} is on the intersection of $I_{12}I_{23}$ and $I_{14}I_{34}$ and ICR I_{24} is on the intersection of lines $I_{23}I_{34}$ and $I_{12}I_{14}$.

Example 9 Find the ICRs of the links in the five-bar linkage shown in Fig. 2.33a. Link 2 rolls and slips over link 3.

The known ICRs are I_{12}, I_{13}, I_{15}, I_{34}, I_{35} and I_{45}. ICR I_{23} is on the intersection of $I_{12}I_{13}$ and a line perpendicular to the contours of links 2 and 3 at the contact point (Fig. 2.33b). The rest of the ICRs can easily be found by applying Kennedy's Theorem making use of the polygon the same way we did in the previous examples (Fig. 2.33c).

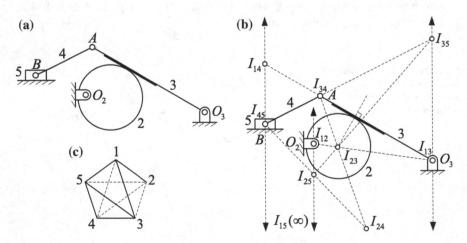

Fig. 2.33 **a** Mechanism with 5 links, **b** ICRs of all links in the linkage, **c** polygon helping to apply Kennedy's Theorem

2.1.4.5 Calculating Velocities with ICRs

We have already studied the relative velocity method for the calculation of point velocity in a linkage. Although it is a simple method to apply, it has one inconvenience. In order to calculate the velocity of one link, we need to calculate the velocities of all the links that connect it to the input link.

Calculating velocity by using instantaneous centers of rotation allows us to directly calculate the velocity of any point in a linkage without having to first calculate the velocities of other points.

Figure 2.34 shows a six-bar linkage in which the velocity of point A is already known. To calculate the velocity of point D by means of the relative velocity method, we first have to calculate the velocities of points B and C.

With the ICR method, it is not necessary to calculate the velocity of a point that physically joins the links. By calculating the relative ICR of two links, we can consider that we know the velocity of a point that is equally part of both links.

It is important to stress that the ICR behaves as if it were part of both links simultaneously and, consequently, its velocity is the same, no matter which link we look at to find it.

The process to calculate velocity is as follows:

4. We identify the following links:

   - The link the point with known velocity belongs to (in this example point A).
   - The link to which the point with unknown velocity belongs (point D).
   - The frame link.

   In the example of Fig. 2.34, the link with known velocity is link 2, the one with unknown velocity is link 6 and link 1 is the frame.
5. We identify all three relative ICRs of the mentioned links (I_{12}, I_{16} and I_{26} in the example) which are aligned according to Kennedy's Theorem.
6. We calculate the velocity of the ICR between the two non-fixed links v_{26}, considering that the ICR is a point that belongs to the link with known velocity. In this case, I_{26} will be considered part of link 2 and it will revolve about I_{12}.
7. We consider ICR I_{26} a point in the link with unknown velocity (link 6 in this example). Knowing the velocity of a point in this link, v_{26}, and its center of rotation, I_{16}, the velocity of any other point in the same link can easily be calculated.

This problem is solved in Example 13 of this chapter.

Fig. 2.34 Six-bar linkage with known velocity of point A

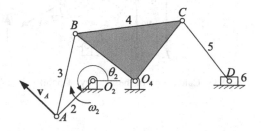

2.1.4.6 Application of ICRs to a Four-Bar Linkage

Figure 2.35 shows a four-bar linkage in which the velocity vector of point A, $\mathbf{v}_A$, is known and the velocity of point B, $\mathbf{v}_B$, is the one to be calculated. The steps to be followed are:

8. We identify the link the point of known velocity belongs to (in this example link 2). We also have to identify the link the point with unknown velocity belongs to (link 4), and the frame (link 1).
9. We locate the three ICRs between these three links: I_{12}, I_{14} and I_{24}. The straight line they form will be used as a folding line for points A and B.
1. We obtain velocity magnitude v_{24} as if I_{24} was part of link 2. Figure 2.35 shows the graphic calculation of this velocity making use of $\mathbf{v}_A$. See the analytical calculation in Eqs. (2.95)–(2.97).

$$v_{I_{24}} = \overline{I_{12}I_{24}}\omega_2 \tag{2.95}$$

$$v_A = \overline{I_{12}I_{23}}\omega_2 \tag{2.96}$$

Dividing and clearing $v_{I_{24}}$:

$$v_{I_{24}} = \frac{\overline{I_{12}I_{24}}}{\overline{I_{12}I_{23}}} v_A \tag{2.97}$$

2. ICR I_{24} is now considered a point on link 4. The velocity of point B is graphically obtained by drawing two similar triangles: the first one defined by sides $\overline{I_{12}I_{24}}$ and and the second one by sides $\overline{I_{14}B}$ (Fig. 2.35). It can also be obtained analytically in Eqs. (2.98)–(2.100):

$$v_{I_{24}} = \overline{I_{14}I_{24}}\omega_4 \tag{2.98}$$

$$v_B = \overline{I_{14}I_{34}}\omega_4 \tag{2.99}$$

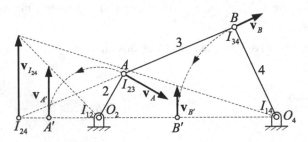

Fig. 2.35 Calculation of the velocity of point B in a four-bar linkage with the ICR method

Fig. 2.36 Velocity
calculation of point B in a
crank-shaft linkage using the
ICR method

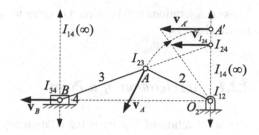

Dividing and clearing

$$v_B = \frac{\overline{I_{14}I_{34}}}{\overline{I_{14}I_{24}}} v_{I_{24}} \qquad\qquad (2.100)$$

If the angular velocity of link 4 is required, it can easily be calculated using the ω_2 value in Eqs. (2.101)–(2.103):

$$\omega_2 = \frac{v_{I_{24}}}{\overline{I_{12}I_{24}}} \qquad\qquad (2.101)$$

$$\omega_4 = \frac{v_{I_{24}}}{\overline{I_{14}I_{24}}} \qquad\qquad (2.102)$$

$$\omega_4 = \frac{\overline{I_{12}I_{24}}}{\overline{I_{14}I_{24}}} \omega_2 \qquad\qquad (2.103)$$

2.1.4.7 Application of the ICR Method to a Crank-shaft Linkage

We assume velocity vector $\mathbf{v}_A$ of point A to be known and we want to calculate $\mathbf{v}_B$ for point B (Fig. 2.36).

10. The link with known velocity is link 2. We want to find the velocity of link 4, while link 1 is fixed.
11. We locate the three ICRs related to these links: I_{12}, I_{24} and I_{14}.
12. We calculate velocity of ICR I_{24}, regarded as a point of link 2.
13. We consider ICR I_{24} as part of link 4. Note that all the points in link 4 have the same velocity. Consequently, if we know velocity $v_{I_{24}}$, we already know the velocity of point B: $\mathbf{v}_B = \mathbf{v}_{I_{24}}$.

2.2 Accelerations in Mechanisms

In this section we will start by defining the components of the linear acceleration of a point. Then we will develop the relative acceleration method that will allow us to calculate the linear and angular accelerations of all points and links in a mechanism.

These accelerations will be needed in order to continue with the dynamic analysis in future chapters.

2.2.1 Acceleration of a Point

The acceleration of a point is the relationship between the change of its velocity vector and time.

Point A moves from position A to A' along a curve during time Δt and changes its velocity vector from v_A to $v_{A'}$ (Fig. 2.37a). Vector Δv measures this velocity change (Fig. 2.37b).

The $\Delta v / \Delta t$ ratio, that is to say, the variation of velocity divided by the time it takes for that change to happen, is the average acceleration. When the time considered is infinitesimal, then, $\Delta v / \Delta t$ becomes dv/dt and this is called instantaneous acceleration or just acceleration.

From Fig. 2.37b we deduce that $\Delta v = \Delta v_1 + \Delta v_2$, where, since the magnitude of vector v_A is equal $\overline{om} = \overline{on}$, we can assert that:

- Δv_1 represents the change in direction of velocity v_A, thus $v_A + \Delta v_1 = v_{A'} - \Delta v_2$ is a vector with the same direction as $v_{A'}$ and the magnitude of v_A.
- Δv_2 represents the change in magnitude (magnitude change) of the velocity of point A when it switches from one position to another. Its magnitude is the difference between the magnitudes of vectors v_A and $v_{A'}$.

Relating these changes in velocity and the time it took for them to happen, we obtain average acceleration vector A of point A (Eq. 2.104) when it moves from point A to A'.

$$\mathbf{A}_A = \frac{\Delta v}{\Delta t} = \frac{\Delta v_1}{\Delta t} + \frac{\Delta v_2}{\Delta t} \tag{2.104}$$

This average acceleration has two components. One is only responsible for the change in direction $(\Delta v_1 / \Delta t)$, and the other is responsible for the change in velocity

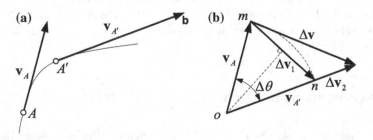

Fig. 2.37 a Change of point A velocity while changing its position from A to A' following a curve in Δt time. **b** Velocity change vector

magnitude ($\Delta v_2 / \Delta t$). In Fig. 2.37b we can calculate the magnitudes of $\Delta \mathbf{v}_1$ and $\Delta \mathbf{v}_2$ (Eqs. 2.108 and 2.109):

$$\Delta v_1 = 2v_A \sin \frac{\Delta \theta}{2} \tag{2.105}$$

$$\Delta v_2 = v_2 - v_1 \tag{2.106}$$

The directions of $\Delta \mathbf{v}_1$ and $\Delta \mathbf{v}_2$ in the limit as Δt approaches zero are respectively perpendicular and parallel to velocity vector $\mathbf{v}_A$, that is, normal and tangential to the trajectory at point A. These vectors are called normal and tangential accelerations, $\mathbf{a}_A^n$ and $\mathbf{a}_A^t$. The acceleration vector can be obtained by adding these two components (Eq. 2.107):

$$\mathbf{a}_A = \mathbf{a}_A^n + \mathbf{a}_A^t \tag{2.107}$$

The magnitudes of these components can be calculated as follows in Eqs. (2.108) and (2.109):

$$a_A^n = \lim_{\Delta t \to 0} \left(2v_A \frac{\sin \Delta \theta / 2}{\Delta t} \right) \simeq v_A \frac{d\theta}{dt} = v_A \omega = R\omega^2 = \frac{v_A^2}{R} \tag{2.108}$$

$$a_A^t = \lim_{\Delta t \to 0} \frac{v_{A'} - v_A}{\Delta t} = \frac{dv_A}{dt} = R\frac{d\omega}{dt} + \omega \frac{dR}{dt} = R\alpha + \omega \frac{dR}{dt} \tag{2.109}$$

where:

- v is the velocity of point A.
- R is the trajectory radius at point A.
- ω is the angular velocity of the radius.
- α is the angular acceleration of the radius.
- dR/dt is the radius variation with respect to time.

To sum up, acceleration of a point A can be broken into two components:

- The first one is called normal acceleration, $\mathbf{a}_A^n$. Its direction is normal to the trajectory followed by point A and it points towards the trajectory center (Fig. 2.38). This component is responsible for the change in velocity direction and its magnitude is Eq. (2.110):

$$a_A^n = R\omega^2 = \frac{v_A^2}{R} \tag{2.110}$$

- The second component, known as tangential acceleration, $\mathbf{a}_A^t$, has a direction tangential to the trajectory, that is, the same as the velocity vector of point A. It can point towards the same side as the velocity or towards the opposite one;

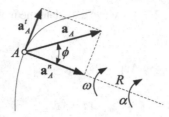

Fig. 2.38 The acceleration of a point has a normal component that points towards the center of the trajectory and a tangential component whose direction is tangential to the trajectory

it depends on whether the velocity magnitude increases or decreases. Tangential acceleration is responsible for the change in magnitude of the velocity vector and its value is Eq. (2.111):

$$a_A^t = R\alpha + \omega \frac{dR}{dt} \qquad (2.111)$$

If the trajectory radius is constant, dR/dt is zero and the value of the tangential acceleration is $a_A^t = R\alpha$.

The magnitude of the acceleration can be determined by the magnitudes of its normal and tangential components. Equation (2.112) will be applied:

$$a_A = \sqrt{(a_A^n)^2 + (a_A^t)^2} \qquad (2.112)$$

Finally, the angle formed by the acceleration vector and the normal direction to the trajectory is defined by Eq. (2.113):

$$\phi = \arctan \frac{a_A^t}{a_A^n} = \arctan \frac{\alpha}{\omega^2} \qquad (2.113)$$

Equation (2.113) is only valid when the radius is constant.

2.2.2 Relative Acceleration of Two Points

The relative acceleration of point A with respect to point B is the ratio between the change in their relative velocity vector and time.

Let us assume that point A moves from position A to A' in the same period of time it takes B to reach position B'. The velocities of points A and B are v_A and v_B and their change is given by vectors Δv_A and Δv_B (Fig. 2.39). This way, the new velocities will be Eqs. (2.114) and (2.115):

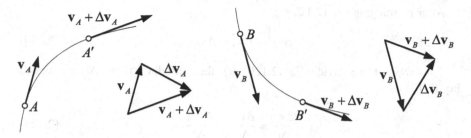

Fig. 2.39 Velocity change vectors $\Delta\mathbf{v}_A$ and $\Delta\mathbf{v}_B$ of points A and B when moving to new positions A' and B' respectively

$$\mathbf{v}_{A'} = \mathbf{v}_A + \Delta\mathbf{v}_A \tag{2.114}$$

$$\mathbf{v}_{B'} = \mathbf{v}_B + \Delta\mathbf{v}_B \tag{2.115}$$

On the other side, Eq. (2.116) that gives us the value of relative velocity $\mathbf{v}_{BA}$ between A and B is (Fig. 2.40a):

$$\mathbf{v}_{BA} = \mathbf{v}_B - \mathbf{v}_A \tag{2.116}$$

And between A' and B' it is Eq. (2.117) (Fig. 2.40b):

$$\mathbf{v}_{BA} + \Delta\mathbf{v}_{BA} = (\mathbf{v}_B + \Delta\mathbf{v}_B) - (\mathbf{v}_A + \Delta\mathbf{v}_A) \tag{2.117}$$

If we plug the value of relative velocity $\mathbf{v}_{BA}$ from Eq. (2.116) in Eq. (2.117), we obtain Eq. (2.118):

$$(\mathbf{v}_B - \mathbf{v}_A) + \Delta\mathbf{v}_{BA} = (\mathbf{v}_B + \Delta\mathbf{v}_B) - (\mathbf{v}_A + \Delta\mathbf{v}_A) \tag{2.118}$$

By simplifying the previous equation, we get Eq. (2.119):

$$\Delta\mathbf{v}_{BA} = \Delta\mathbf{v}_B - \Delta\mathbf{v}_A \tag{2.119}$$

Fig. 2.40 Relative velocity between **a** points A and B, **b** points A' and B'

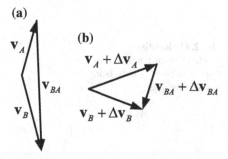

After rearranging Eq. (2.120):

$$\Delta \mathbf{v}_B = \Delta \mathbf{v}_A + \Delta \mathbf{v}_{BA} \tag{2.120}$$

This way, if we divide Eq. (2.120) by the period of time, Δt, we obtain Eq. (2.121):

$$\frac{\Delta \mathbf{v}_B}{\Delta t} = \frac{\Delta \mathbf{v}_A}{\Delta t} + \frac{\Delta \mathbf{v}_{BA}}{\Delta t} \tag{2.121}$$

Each one of the terms in equation (Eq. 2.121) is an average acceleration (Eq. 2.122):

$$\mathbf{A}_B = \mathbf{A}_A + \mathbf{A}_{BA} \tag{2.122}$$

When Δt approaches zero (dt), the average accelerations become instantaneous accelerations (Eq. 2.123):

$$\mathbf{a}_B = \mathbf{a}_A + \mathbf{a}_{BA} \tag{2.123}$$

Therefore, the acceleration vector of point B equals the sum of the acceleration vector of point A plus the relative acceleration vector of point B with respect to point A. The latter has a normal as well as a tangential component (Eq. 2.124):

$$\mathbf{a}_B = \mathbf{a}_A + \mathbf{a}_{BA}^n + \mathbf{a}_{BA}^t \tag{2.124}$$

2.2.3 Relative Acceleration of Two Points in the Same Rigid Body

As the distance between two points of a rigid body cannot change, relative motion between them is a rotation of one point about the other. In the example shown in Fig. 2.41, point B rotates about point A, both being part of a link that moves with angular velocity ω and angular acceleration α. The relative acceleration vector of point B with respect to point A can be broken into two components:

Fig. 2.41 Relative acceleration of point B with respect to point A both being in the same link

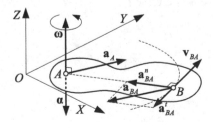

- The normal component, $\mathbf{a}^n_{BA}$, is always perpendicular to the relative velocity vector and it points towards the center of curvature of the trajectory. In this case, it points towards point A.
- The tangential component, $\mathbf{a}^t_{BA}$, has the same direction as the relative velocity vector. In the example shown in Fig. 2.41, as the direction of angular acceleration α opposes the direction of angular velocity ω, the tangential component points in the opposite direction to relative velocity $\mathbf{v}_{BA}$, which means that the magnitude of this velocity is decreasing.

These normal and tangential components of the relative acceleration of point B with respect to point A can be obtained with Eqs. (2.125) and (2.126):

$$\mathbf{a}^n_{BA} = \omega \wedge \mathbf{v}_{BA} = \omega \wedge \omega \wedge \mathbf{r}_{BA} \qquad (2.125)$$

$$\mathbf{a}^t_{BA} = \alpha \wedge \mathbf{r}_{BA} \qquad (2.126)$$

The angle formed by the relative acceleration vector and the normal direction to the trajectory is Eq. (2.127):

$$\phi = \arctan \frac{\mathbf{a}^t_{BA}}{\mathbf{a}^n_{BA}} = \arctan \frac{\alpha}{\omega^2} \qquad (2.127)$$

In other words, angle ϕ is independent from distance $\overline{AB}$. It only depends on the acceleration α and the angular velocity ω.

With these components, we can calculate the acceleration of point B based on the one of point A (Eq. 2.128):

$$\mathbf{a}_B = a_{A_x}\hat{\mathbf{i}} + a_{A_y}\hat{\mathbf{j}} + \omega \wedge \mathbf{v}_{BA} + \alpha \wedge \mathbf{r}_{BA} \qquad (2.128)$$

In the case of a rigid body that revolves about steady point O (Fig. 2.42) the absolute acceleration vector of point P (a generic point of the body) is the relative acceleration with respect to point O (Eq. 2.129):

$$\mathbf{a}_P = \mathbf{a}_O + \mathbf{a}^n_{PO} + \mathbf{a}^t_{PO} = \mathbf{a}^n_{PO} + \mathbf{a}^t_{PO} \qquad (2.129)$$

Fig. 2.42 Normal and tangential components of the acceleration of point P on a link that revolves about steady point O

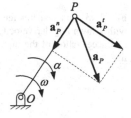

Where the normal and tangential components are Eqs. (2.130) and (2.131):

$$\mathbf{a}_{PO}^n = \boldsymbol{\omega} \wedge \mathbf{v}_{PO} = \boldsymbol{\omega} \wedge \boldsymbol{\omega} \wedge \mathbf{r}_{PO} \qquad (2.130)$$

$$\mathbf{a}_{PO}^t = \boldsymbol{\alpha} \wedge \mathbf{r}_{PO} \qquad (2.131)$$

2.2.4 Computing Acceleration in a Four-Bar Linkage

To apply the relative acceleration method, we will calculate the acceleration of points B and C in the linkage in Example 1 of this chapter (Fig. 2.43). We know angular velocity $\omega_2 = -20\,\text{rad/s}$ clockwise and angular acceleration $\alpha_2 = 150\,\text{rad/s}^2$ counterclockwise of the motor link as well as the geometrical data of the linkage. We will also make use of the following results obtained from the position and velocity analysis in Example 1:

$$\mathbf{v}_A = 103.9\hat{\mathbf{i}} - 60\hat{\mathbf{j}}$$

$$\theta_3 = 19.4° \qquad \omega_3 = 7.16\,\text{rad/s} \qquad \mathbf{v}_B = 77.75\hat{\mathbf{i}} + 14.28\hat{\mathbf{j}}$$
$$\theta_3' = 349.4° \qquad \qquad \mathbf{v}_{BA} = -26.13\hat{\mathbf{i}} + 74.32\hat{\mathbf{j}}$$
$$\theta_4 = 100.42° \qquad \omega_4 = -8.78\,\text{rad/s} \qquad \mathbf{v}_C = 114.4\hat{\mathbf{i}} - 3.7\hat{\mathbf{j}}$$

$$\mathbf{v}_{CA} = 10.53\hat{\mathbf{i}} + 56.3\hat{\mathbf{j}}$$

To solve the problem we will apply the vector equation that relates the accelerations of points B and A (Eq. 2.132):

$$\mathbf{a}_B = \mathbf{a}_A + \mathbf{a}_{BA} = (\mathbf{a}_B^n + \mathbf{a}_B^t) = (\mathbf{a}_A^n + \mathbf{a}_A^t) + (\mathbf{a}_{BA}^n + \mathbf{a}_{BA}^t) \qquad (2.132)$$

In general, normal components will be known, since they depend on velocity, while tangential components will be unknown as they depend on angular acceleration, α. We will start calculating the acceleration of point A. Then we will analyze the acceleration of point B with respect to point A. Next, we will study the acceleration of point B and, finally, we will obtain the acceleration of point C.

Fig. 2.43 Four-bar linkage with known angular velocity and acceleration of the input link

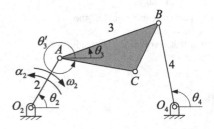

Acceleration Vector of Point A:

Point A rotates about steady point O_2, so the normal acceleration component is given by Eq. (2.133):

$$\mathbf{a}_A^n = \omega_2 \wedge \mathbf{v}_A = \begin{vmatrix} \hat{\mathbf{i}} & \hat{\mathbf{j}} & \hat{\mathbf{k}} \\ 0 & 0 & -20 \\ 103.9 & -60 & 0 \end{vmatrix} = -1200\hat{\mathbf{i}} - 2078\hat{\mathbf{j}} \qquad (2.133)$$

$$\mathbf{a}_A^n = 2400 \,\mathrm{cm/s^2} \,\angle 240°$$

We can verify that this vector has a perpendicular direction to $\mathbf{v}_A$ and it points towards the trajectory curvature center of point A, that is to say, towards O_2.

The tangential component of the acceleration vector of point A is given by Eq. (2.134):

$$\mathbf{a}_A^t = \alpha_2 \wedge \mathbf{r}_{AO_2} = \begin{vmatrix} \hat{\mathbf{i}} & \hat{\mathbf{j}} & \hat{\mathbf{k}} \\ 0 & 0 & 150 \\ 6\cos 60° & 6\sin 60° & 0 \end{vmatrix} = -779.42\hat{\mathbf{i}} + 450\hat{\mathbf{j}} \qquad (2.134)$$

$$\mathbf{a}_A^t = 900 \,\mathrm{cm/s^2} \,\angle 150°$$

Acceleration vector $\mathbf{a}_A^t$ is parallel to velocity vector $\mathbf{v}_A$ but in the opposite direction, as the direction of angular acceleration α_2 is opposite to the direction of angular velocity ω_2.

We can start drawing the acceleration polygon (Fig. 2.44) by tracing vectors $\mathbf{a}_A^n$ and $\mathbf{a}_A^t$. We define point a in the polygon at the end of vector acceleration $\mathbf{a}_A$. The acceleration image of link $\overline{O_2A}$ is $\overline{oa}$.

Relative Acceleration of Point B with Respect to Point A:

Since relative motion is a revolution of point B about point A, the normal component of the relative acceleration of B with respect to A is Eq. (2.135):

$$\mathbf{a}_{BA}^n = \omega_3 \wedge \mathbf{v}_{BA} = \begin{vmatrix} \hat{\mathbf{i}} & \hat{\mathbf{j}} & \hat{\mathbf{k}} \\ 0 & 0 & 7.16 \\ -26.13 & 74.32 & 0 \end{vmatrix} = -532.13\hat{\mathbf{i}} - 187.1\hat{\mathbf{j}} \qquad (2.135)$$

$$\mathbf{a}_{BA}^n = 564.06 \,\mathrm{cm/s^2} \,\angle 199.4°$$

The direction of this vector is perpendicular to velocity $\mathbf{v}_{BA}$ and it heads towards the trajectory center of point B, In other words, the direction is from B to A.

The tangential component of the relative acceleration vector of B with respect to A is expressed as Eq. (2.136):

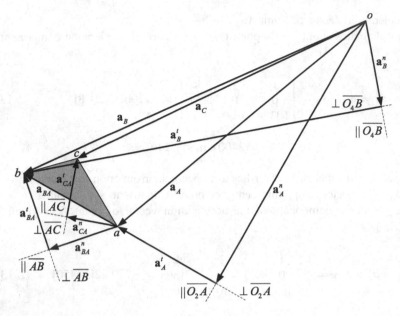

Fig. 2.44 Acceleration polygon of the four-bar linkage in Fig. 2.43

$$\mathbf{a}_{BA}^t = \alpha_3 \wedge \mathbf{r}_{BA} = \begin{vmatrix} \hat{\mathbf{i}} & \hat{\mathbf{j}} & \hat{\mathbf{k}} \\ 0 & 0 & \alpha_3 \\ 11\cos 19.4° & 11\sin 19.4° & 0 \end{vmatrix} = -3.65\alpha_3\hat{\mathbf{i}} + 10.38\alpha_3\hat{\mathbf{j}}$$

(2.136)

To calculate the value of $\mathbf{a}_{BA}^t$, we need to know α_3, which will be obtained in the next step.

Acceleration of Point B:

This point rotates about O_4, so the normal component of its acceleration (Eq. 2.137) will be:

$$\mathbf{a}_B^n = \omega_4 \wedge \mathbf{v}_B = \begin{vmatrix} \hat{\mathbf{i}} & \hat{\mathbf{j}} & \hat{\mathbf{k}} \\ 0 & 0 & -8.78 \\ 77.75 & 14.28 & 0 \end{vmatrix} = 125.37\hat{\mathbf{i}} - 682.64\hat{\mathbf{j}}$$ (2.137)

$$\mathbf{a}_B^n = 694.06\,\text{cm/s}^2 \angle 280.4°$$

This vector is perpendicular to $\mathbf{v}_B$ and heads towards the trajectory curvature center of point B. In other words, from B to O_4.

The tangential acceleration is defined by Eq. (2.138):

$$\mathbf{a}_B^t = \alpha_4 \wedge \mathbf{r}_{BO_4} = \begin{vmatrix} \hat{\mathbf{i}} & \hat{\mathbf{j}} & \hat{\mathbf{k}} \\ 0 & 0 & \alpha_4 \\ 9\cos 100.4° & 9\sin 100.4° & 0 \end{vmatrix} = -8.85\alpha_4\hat{\mathbf{i}} - 1.62\alpha_4\hat{\mathbf{j}}$$

$$(2.138)$$

This component depends on α_4, which is another unknown that we need to find. In order to determine it, we need to plug the obtained values in the acceleration vector (Eqs. 2.139 and 2.140):

$$\mathbf{a}_B = \mathbf{a}_A + \mathbf{a}_{BA} = (\mathbf{a}_A^n + \mathbf{a}_A^t) + (\mathbf{a}_{BA}^n + \mathbf{a}_{BA}^t) \tag{2.139}$$

$$\begin{aligned} (125.37\hat{\mathbf{i}} - 682.64\hat{\mathbf{j}}) + (-8.85\alpha_4\hat{\mathbf{i}} - 1.62\alpha_4\hat{\mathbf{j}}) \\ = (-1200\hat{\mathbf{i}} - 2078\hat{\mathbf{j}}) + (-779.42\hat{\mathbf{i}} + 450\hat{\mathbf{j}}) \\ + (-532.13\hat{\mathbf{i}} - 187.1\hat{\mathbf{j}}) + (-3.65\alpha_3\hat{\mathbf{i}} + 10.38\alpha_3\hat{\mathbf{j}}) \end{aligned} \tag{2.140}$$

Breaking Eq. (2.140) into its components, we obtain Eq. (2.141):

$$\left. \begin{aligned} 125.37 - 8.85\alpha_4 &= -1200 - 779.42 - 532.13 - 3.65\alpha_3 \\ -682.64 - 1.62\alpha_4 &= -2078 + 450 - 187.1 + 10.38\alpha_3 \end{aligned} \right\} \tag{2.141}$$

By solving the system, the angular accelerations are obtained: $\alpha_3 = 58.81$ rad/s^2 and $\alpha_4 = 322.21$ rad/s^2. They can be used in Eqs. (2.136)–(2.138) to calculate the values of the tangential components.

$$\mathbf{a}_{BA}^t = -214.65\hat{\mathbf{i}} + 610.48\hat{\mathbf{j}} = 647.08 \text{ cm/s}^2 \angle 109.37°$$

$$\mathbf{a}_B^t = -2851.55\hat{\mathbf{i}} - 522\hat{\mathbf{j}} = 2898.9 \text{ cm/s}^2 \angle 190.37°$$

Acceleration of Point C:
Finally, we can find the acceleration of point C (Eq. 2.142) by using the following vector equation:

$$\mathbf{a}_C = \mathbf{a}_A + \mathbf{a}_{CA} = \mathbf{a}_A + \mathbf{a}_{CA}^n + \mathbf{a}_{CA}^t \tag{2.142}$$

We already know acceleration $\mathbf{a}_A$ and we can calculate the components of the relative acceleration of point C with respect to point A (Eqs. 2.143 and 2.144):

$$\mathbf{a}_{CA}^n = \omega_3 \wedge \mathbf{v}_{CA} = \begin{vmatrix} \hat{\mathbf{i}} & \hat{\mathbf{j}} & \hat{\mathbf{k}} \\ 0 & 0 & 7.16 \\ 10.53 & 56.3 & 0 \end{vmatrix} = -403.11\hat{\mathbf{i}} + 75.39\hat{\mathbf{j}} \tag{2.143}$$

$$\mathbf{a}_{CA}^n = 410.1 \, \text{cm/s}^2 \, \angle 169.4°$$

$$\mathbf{a}_{CA}^t = \alpha_3 \wedge \mathbf{r}_{CA} = \begin{vmatrix} \hat{\mathbf{i}} & \hat{\mathbf{j}} & \hat{\mathbf{k}} \\ 0 & 0 & 58.81 \\ 8\cos 349.4° & 8\sin 349.4° & 0 \end{vmatrix} = 86.54\hat{\mathbf{i}} + 462.45\hat{\mathbf{j}}$$

$$(2.144)$$

$$\mathbf{a}_{CA}^t = 470.48 \, \text{cm/s}^2 \, \angle 79.4°$$

Hence, the acceleration of point C is Eq. (2.145):

$$\begin{aligned} \mathbf{a}_C &= (-1200\hat{\mathbf{i}} - 2078\hat{\mathbf{j}}) + (-779.42\hat{\mathbf{i}} + 450\hat{\mathbf{j}}) \\ &\quad + (-403.11\hat{\mathbf{i}} + 75.39\hat{\mathbf{j}}) + (86.54\hat{\mathbf{i}} + 462.45\hat{\mathbf{j}}) \qquad (2.145) \\ &= -2295.99\hat{\mathbf{i}} - 1090.16\hat{\mathbf{j}} \end{aligned}$$

$$\mathbf{a}_C = 2541.66 \, \text{cm/s}^2 \, \angle 205.4°$$

Once all the accelerations have been obtained, they can be represented in an acceleration polygon like the one shown in Fig. 2.44.

In the acceleration polygon in Fig. 2.44, triangle $\triangle abc$ is defined by the end points of the absolute acceleration vectors of points A, B and C. The same as in velocity analysis, triangle $\triangle abc$ in the polygon is similar to triangle $\triangle ABC$ in the mechanism (Eq. 2.146).

$$\frac{\overline{ab}}{\overline{AB}} = \frac{\overline{ac}}{\overline{AC}} = \frac{\overline{bc}}{\overline{BC}} \qquad (2.146)$$

2.2.4.1 Accelerations in a Slider-crank Linkage

Figure 2.45 shows the slider-crank mechanism whose velocity was calculated in Example 2 of Sect. 2.1.3.8. The crank rotates with constant angular velocity of $-10 \, \text{rad/s}$ clockwise. We want to find the acceleration of point B.

We will use the results obtained in Example 2:

$$\mathbf{v}_A = 25.98\hat{\mathbf{i}} - 15\hat{\mathbf{j}}$$

$$\theta_3 = 324.26 \quad \omega_3 = 2.64 \, \text{rad/s} \quad \mathbf{v}_{BA} = 10.79\hat{\mathbf{i}} + 15\hat{\mathbf{j}}$$

$$\mathbf{v}_B = 36.78\hat{\mathbf{i}}$$

To calculate the acceleration of point B (Eq. 2.147), we apply the following vector equation:

Fig. 2.45 Slider-crank linkage with constant angular velocity in link 2

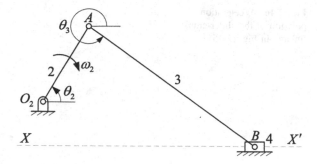

$$\mathbf{a}_B = \mathbf{a}_A + \mathbf{a}_{BA} = (\mathbf{a}_A^n + \mathbf{a}_A^t) + (\mathbf{a}_{BA}^n + \mathbf{a}_{BA}^t) \tag{2.147}$$

Acceleration Vector of Point A:

The normal component of acceleration $\mathbf{a}_A^n$ is given by Eq. (2.148):

$$\mathbf{a}_A^n = \omega_2 \wedge \mathbf{v}_A = \begin{vmatrix} \hat{\mathbf{i}} & \hat{\mathbf{j}} & \hat{\mathbf{k}} \\ 0 & 0 & -10 \\ 25.98 & -15 & 0 \end{vmatrix} = -150\hat{\mathbf{i}} - 259.8\hat{\mathbf{j}} \tag{2.148}$$

$$\mathbf{a}_A^n = 300\,\mathrm{cm/s^2}\,\angle 240°$$

This vector is perpendicular to velocity $\mathbf{v}_A$ and it heads towards the trajectory curvature center of point A, that is to say, from A to O_2.

The tangential component of the vector is zero since the angular velocity of link 2 is constant, that is, $\alpha_2 = 0$.

Relative Acceleration of Point B with Respect to Point A:

The normal component of the relative acceleration vector of point B with respect to point A is given by Eq. (2.149):

$$\mathbf{a}_{BA}^n = \omega_3 \wedge \mathbf{v}_{BA} = \begin{vmatrix} \hat{\mathbf{i}} & \hat{\mathbf{j}} & \hat{\mathbf{k}} \\ 0 & 0 & 2.64 \\ 10.79 & 15 & 0 \end{vmatrix} = -39.6\hat{\mathbf{i}} + 28.49\hat{\mathbf{j}} \tag{2.149}$$

$$\mathbf{a}_{BA}^n = 48.78\,\mathrm{cm/s^2}\,\angle 144.27°$$

This vector is perpendicular to velocity $\mathbf{v}_{BA}$ and heads from B towards A.

The tangential component of the relative acceleration vector of point B with respect to point A is given by Eq. (2.150):

$$\mathbf{a}_{BA}^t = \alpha_3 \wedge \mathbf{r}_{BA} = \begin{vmatrix} \hat{\mathbf{i}} & \hat{\mathbf{j}} & \hat{\mathbf{k}} \\ 0 & 0 & \alpha_3 \\ 7\cos 324.26° & 7\sin 324.26° & 0 \end{vmatrix} = 4.09\alpha_3\hat{\mathbf{i}} + 5.68\alpha_3\hat{\mathbf{j}}$$

$$\tag{2.150}$$

Prior to calculating its value, we have to find α_3.

Fig. 2.46 Acceleration
polygon of the slider-crank
linkage in Fig. (2.45)

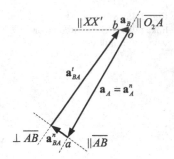

Acceleration of Point B:

We have to take into account that link 4 has a pure translational motion following the trajectory defined by line XX' (Fig. 2.45). The direction of the acceleration of point B coincides with the trajectory direction (Eq. 2.151):

$$\mathbf{a}_B = \mathbf{a}_B^t = a_B\hat{\mathbf{i}} \tag{2.151}$$

By substituting Eqs. (2.148)–(2.150) in Eq. (2.147) we obtain Eq. (2.152):

$$a_B\hat{\mathbf{i}} = (-150\hat{\mathbf{i}} - 259.8\hat{\mathbf{j}}) + (-39.6\hat{\mathbf{i}} + 28.49\hat{\mathbf{j}}) + (4.09\alpha_3\hat{\mathbf{i}} + 5.68\alpha_3\hat{\mathbf{j}}) \tag{2.152}$$

The following algebraic components are obtained if we break this acceleration vector into its components (Eq. 2.153):

$$\left.\begin{array}{l} a_B = -150 - 39.6 + 4.09\alpha_3 \\ 0 = -259.8 + 28.49 + 5.68\alpha_3 \end{array}\right\} \tag{2.153}$$

Based on these equations, we find angular acceleration, $\alpha_3 = 40.72 \text{ rad/s}^2$, and the magnitude of the linear acceleration at point B, $a_B = -23.04 \text{ cm/s}^2$. This way, the remaining acceleration value is Eq. (2.154):

$$\mathbf{a}_{BA}^t = 166.54\hat{\mathbf{i}} + 231.29\hat{\mathbf{j}} = 285 \text{ cm/s}^2 \angle 54.24° \tag{2.154}$$

$$\mathbf{a}_B = -23.04\hat{\mathbf{i}} = 23.04 \text{ cm/s}^2 \angle 180°$$

Once all the accelerations have been determined, we can represent them in an acceleration polygon as the one shown in Fig. 2.46.

2.2.5 The Coriolis Component of Acceleration

When a body moves along a trajectory defined over a rotating body, the acceleration of any point on the first body relative to a coinciding point on the second body will have, in addition to the normal and tangential components, a new one named the Coriolis acceleration. To demonstrate its value, we will use a simple example. Although a demonstration in a particular situation is not valid to demonstrate a generic situation, we will use this example because of its simplicity.

In Fig. 2.47, link 3 represents a slide that moves over straight line. Point P_3 of link 3 is above point P_2 of link 2. Therefore, the position of both points coincides at the instant represented in the figure.

The acceleration of point P_3 can be computed as Eq. (2.155):

$$\mathbf{a}_{P_3} = \mathbf{a}_{P_2} + \mathbf{a}_{P_3 P_2} \tag{2.155}$$

Relative acceleration $\mathbf{a}_{P_3 P_2}$ has, in addition to normal and tangential components studied so far, a new component called the Coriolis acceleration (Eq. 2.156):

$$\mathbf{a}_{P_3 P_2} = \mathbf{a}^n_{P_3 P_2} + \mathbf{a}^t_{P_3 P_2} + \mathbf{a}^c_{P_3 P_2} \tag{2.156}$$

To demonstrate the existence of this new component, see Fig. 2.48. It represents slider 3 moving with constant relative velocity $\mathbf{v}_{P_3 P_2} = \mathbf{v}$ over link 2, which, at the same time, rotates with constant angular velocity ω_2. P_3 is a point of slider 3 that moves along trajectory $O_2 F$ on body 2. P_2 is a point of link 2 that coincides with P_3 at the instant represented.

In a dt time interval, line $O_2 F$ rotates about O_2, angle $d\theta$, and moves to position $O_2 F'$. In the same period of time point P_2 moves to P'_2 and P_3 moves to P'_3. This last displacement can be regarded as the sum of displacements $\Delta P_2 P'_2$, $\Delta P'_2 B$ and $\Delta B P'_3$.

Displacement $\Delta P_2 P'_2$ takes place at constant velocity since $\overline{O_2 P_2}$ and ω_2 are constant. Movement $\Delta P'_2 B$ also takes place at constant velocity as $\mathbf{v}_{P_3 P_2}$ is constant.

Fig. 2.47 Link 3 moves along a trajectory defined over link 2 which rotates with angular velocity ω_2

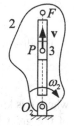

Fig. 2.48 Link 3 moves with constant relative velocity along a straight trajectory defined over link 2 which, at the same time, rotates with constant angular velocity ω_2

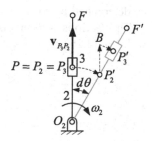

However, displacement $\Delta BP'_3$ is triggered by an acceleration. To obtain this acceleration, we start by calculating the length of arc $\overset{\frown}{BP'_3}$ (Eq. 2.157):

$$\overset{\frown}{BP'_3} = \overline{P'_2B}d\theta \tag{2.157}$$

But $\overline{P'_2B} = v_{P_3P_2}dt$ and $d\theta = \omega_2 dt$, which yields (Eq. 2.158):

$$\overset{\frown}{BP'_3} = \omega_2 v_{P_3P_2}dt^2 \tag{2.158}$$

The velocity of point P_3 is perpendicular to line O_2F and its magnitude is $\omega_2\overline{O_2P_3}$. Since ω_2 is constant and $\overline{O_2P_3}$ increases its value with a constant ratio, the magnitude of the velocity of point P_3, perpendicular to line O_2F, changes uniformly, that is, with constant acceleration.

In general, a displacement (ds) with constant acceleration (a) is defined by Eq. (2.159):

$$ds = \frac{1}{2}adt^2 \tag{2.159}$$

Then BP'_3 is expressed as Eq. (2.160):

$$\overset{\frown}{BP'_3} = \frac{1}{2}adt^2 \tag{2.160}$$

Evening out the two equations for arc $\overset{\frown}{BP'_3}$, we obtain Eq. (2.161):

$$\omega_2 v_{P_3P_2}dt^2 = \frac{1}{2}adt^2 \tag{2.161}$$

Finally, we clear the acceleration value Eq. (2.162):

$$a = 2\omega_2 v_{P_3P_2} \tag{2.162}$$

where a is known as the Coriolis component of the acceleration of point P_3 in honor of the great French mathematician of the XIX century. The Coriolis acceleration component is a vector perpendicular to the relative velocity vector. Its direction can be determined by rotating vector $\mathbf{v}_{P_3P_2}$ 90° in the direction of ω_2.

The Coriolis acceleration component vector can be obtained mathematically with Eq. (2.163):

$$\mathbf{a}^c_{P_3P_2} = 2\mathbf{\omega}_2 \wedge \mathbf{v}_{P_3P_2} \tag{2.163}$$

Fig. 2.49 Link 3 moves along a curved trajectory over link 2 while the latter is rotating with angular velocity ω_2

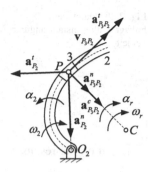

A general case of relative motion on a plane between two rigid bodies is shown in Fig. 2.49. P_2 is a point on body 2 and P_3 is a point on link 3, which moves along a curved trajectory over body 2 with its center in point C.

The absolute acceleration of point P_3 is Eq. (2.164):

$$\mathbf{a}_{P_3} = \mathbf{a}_{P_2} + \mathbf{a}_{P_3 P_2} \tag{2.164}$$

Or, expressed in terms of their intrinsic components (Eq. 2.165):

$$\mathbf{a}^n_{P_3} + \mathbf{a}^t_{P_3} = \mathbf{a}^n_{P_2} + \mathbf{a}^t_{P_2} + \mathbf{a}^n_{P_3 P_2} + \mathbf{a}^t_{P_3 P_2} + \mathbf{a}^c_{P_3 P_2} \tag{2.165}$$

where the Coriolis component is part of the relative acceleration of point P_3 with respect to P_2 and its value is given by Eq. (2.163).

The radius of the trajectory followed by P_3 over link 2 at the instant shown in Fig. 2.49 is $\overline{CP}$. Consequently, the normal and tangential acceleration vectors of P_3 with respect P_2 (Eqs. 2.166 and 2.167) are normal and tangential to the trajectory at the instant considered and their values are:

$$\mathbf{a}^n_{P_3 P_2} = \boldsymbol{\omega}_r \wedge \mathbf{v}_{P_3 P_2} \tag{2.166}$$

$$\mathbf{a}^t_{P_3 P_2} = \boldsymbol{\alpha}_r \wedge \mathbf{r}_{PC} \tag{2.167}$$

2.2.5.1 Accelerations in a Quick-return Mechanism

In the mechanism in Fig. 2.50, link 2 is the motor link. We want to calculate the velocity and acceleration of link 4. The information on the length of the links as well as angular velocity ω_2 are the same as in Example 3. Angular acceleration of the motor link is zero, that is, $\alpha_2 = 0$.

Fig. 2.50 Quick-return
linkage with constant angular
velocity

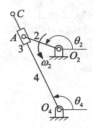

Based on the results obtained in Example 3:

$$\mathbf{v}_{A_2} = -10.26\hat{\mathbf{i}} - 28.19\hat{\mathbf{j}}$$

$$\mathbf{v}_{A_4} = -19.2\hat{\mathbf{i}} - 9\hat{\mathbf{j}}$$

$$\theta_4 = 115.08° \quad \omega_4 = 3.19 \text{ rad/s}^2$$

$$\mathbf{v}_{A_2A_4} = 8.95\hat{\mathbf{i}} - 19.19\hat{\mathbf{j}}$$

$$\mathbf{v}_C = -26\hat{\mathbf{i}} - 12.17\hat{\mathbf{j}}$$

To calculate the accelerations, we use Eq. (2.168):

$$\mathbf{a}_{A_2}^n + \mathbf{a}_{A_2}^t = \mathbf{a}_{A_4}^n + \mathbf{a}_{A_4}^t + \mathbf{a}_{A_2A_4}^n + \mathbf{a}_{A_2A_4}^t + \mathbf{a}_{A_2A_4}^c \qquad (2.168)$$

We study the value of each component, starting with the acceleration of point A_2 (Eqs. 2.169 and 2.170):

$$\mathbf{a}_{A_2}^n = \omega_2 \wedge \mathbf{v}_{A_2} = \begin{vmatrix} \hat{\mathbf{i}} & \hat{\mathbf{j}} & \hat{\mathbf{k}} \\ 0 & 0 & 10 \\ -10.26 & -28.19 & 0 \end{vmatrix} = 281.9\hat{\mathbf{i}} - 102.6\hat{\mathbf{j}} \qquad (2.169)$$

$$\mathbf{a}_{A_2}^n = 300 \text{ cm/s}^2 \angle 340°$$

$$\mathbf{a}_{A_2}^t = \alpha_2 \wedge \mathbf{r}_{AO_2} = 0 \qquad (2.170)$$

We continue with the acceleration of point A_4 (Eqs. 2.171 and 2.172):

$$\mathbf{a}_{A_4}^n = \omega_4 \wedge \mathbf{v}_{A_4} = \begin{vmatrix} \hat{\mathbf{i}} & \hat{\mathbf{j}} & \hat{\mathbf{k}} \\ 0 & 0 & 3.19 \\ -19.2 & -9 & 0 \end{vmatrix} = 28.71\hat{\mathbf{i}} - 61.23\hat{\mathbf{j}} \qquad (2.171)$$

$$\mathbf{a}_{A_4}^n = 67.63 \text{ cm/s}^2 \angle 295.08°$$

$$\mathbf{a}_{A_4}^t = \alpha_4 \wedge \mathbf{r}_{AO_4} = \begin{vmatrix} \hat{\mathbf{i}} & \hat{\mathbf{j}} & \hat{\mathbf{k}} \\ 0 & 0 & \alpha_4 \\ \overline{O_4A}\cos 115.08° & \overline{O_4A}\sin 115.08° & 0 \end{vmatrix} \qquad (2.172)$$

$$= -6.02\alpha_4\hat{\mathbf{i}} - 2.82\alpha_4\hat{\mathbf{j}}$$

We will now study the value of the acceleration of point A_2 relative to point A_4. Because the relative motion of point A_2 with respect to point A_4 follows a straight trajectory, the normal component is zero, $a_{A_2A_4}^t = 0$. The tangential component has the direction of link 4 (Eq. 2.173), consequently:

$$\mathbf{a}_{A_2A_4}^t = a_{A_2A_4}^t \cos 115.08°\hat{\mathbf{i}} + a_{A_2A_4}^t \sin 115.08°\hat{\mathbf{j}} \qquad (2.173)$$

Finally, we calculate the Coriolis component of the acceleration (Eq. 2.174):

$$\mathbf{a}_{A_2A_4}^c = 2\boldsymbol{\omega}_4 \wedge \mathbf{v}_{A_2A_4} = 2 \begin{vmatrix} \hat{\mathbf{i}} & \hat{\mathbf{j}} & \hat{\mathbf{k}} \\ 0 & 0 & 3.19 \\ 8.95 & -19.19 & 0 \end{vmatrix} = 122.43\hat{\mathbf{i}} + 57.1\hat{\mathbf{j}} \qquad (2.174)$$

In Eq. (2.175) we plug the calculated values in the vector equation (Eq. 2.168):

$$(281.9\hat{\mathbf{i}} - 102.6\hat{\mathbf{j}}) = (28.71\hat{\mathbf{i}} - 61.23\hat{\mathbf{j}}) + (-6.02\alpha_4\hat{\mathbf{i}} - 2.82\alpha_4\hat{\mathbf{j}})$$
$$+ (a_{A_2A_4}^t \cos 115.08°\hat{\mathbf{i}} + a_{A_2A_4}^t \sin 115.08°\hat{\mathbf{j}}) \qquad (2.175)$$
$$+ (122.43\hat{\mathbf{i}} + 57.1\hat{\mathbf{j}})$$

By separating the components, Eq. (2.176) is obtained:

$$\left. \begin{array}{l} 281.9 = 28.71 - 6.02\alpha_4 + a_{A_2A_4}^t \cos 115.08° + 122.43 \\ -102.6 = -61.23 \quad 2.82\alpha_4 + a_{A_2A_4}^t \sin 115.08° + 57.1 \end{array} \right\} \qquad (2.176)$$

This way, we find the value of angular acceleration $\alpha_4 = -11.56$ rad/s^2 and the magnitude of tangential relative acceleration $a_{A_2A_4}^t = -145.63$ cm/s^2. The accelerations then remain as follows:

$$\mathbf{a}_{A_4}^t = 69.59\hat{\mathbf{i}} + 32.6\hat{\mathbf{j}} = 76.85 \text{ cm/s}^2 \angle 25°$$

$$\mathbf{a}_{A_2A_4}^t = 61.73\hat{\mathbf{i}} - 131.9\hat{\mathbf{j}} = 145.63 \text{ cm/s}^2 \angle 295.08°$$

We proceed to calculating the acceleration of point C (Eqs. 2.177 and 2.178):

$$\mathbf{a}_C^n = \boldsymbol{\omega}_4 \wedge \mathbf{v}_C = \begin{vmatrix} \hat{\mathbf{i}} & \hat{\mathbf{j}} & \hat{\mathbf{k}} \\ 0 & 0 & 3.19 \\ -26 & -12.17 & 0 \end{vmatrix} = 38.82\hat{\mathbf{i}} - 82.94\hat{\mathbf{j}} \qquad (2.177)$$

$$\mathbf{a}_C^n = 91.57 \text{ cm/s}^2 \angle 295.08°$$

Fig. 2.51 Acceleration
polygon of the quick-return
linkage in Fig. 2.50

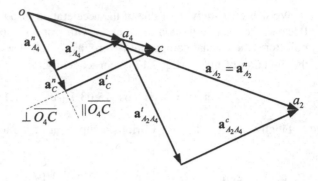

$$\mathbf{a}_C^t = \alpha_4 \wedge \mathbf{r}_{CO_4} = \begin{vmatrix} \hat{\mathbf{i}} & \hat{\mathbf{j}} & \hat{\mathbf{k}} \\ 0 & 0 & -11.56 \\ \overline{O_4 C} \cos 115.08° & \overline{O_4 C} \sin 115.08° & 0 \end{vmatrix}$$

$$= 94.23\hat{\mathbf{i}} + 44.1\hat{\mathbf{j}} \qquad\qquad (2.178)$$

$$\mathbf{a}_C^t = 104.04\,\mathrm{cm/s^2}\,\angle 25.08°$$

Once all the accelerations have been determined, they can be represented in an
acceleration polygon as in previous examples (Fig. 2.51).

2.3   Exercises with Their Solutions

In this section we will carry out the kinematic analysis of different mechanisms by
applying the methods developed in this chapter up to now.

Example 10 In the mixing machine in Example 4 (Fig. 2.52a), calculate the
velocity of point C (Fig. 2.52b) by means of the ICR method once the velocity of
point A is known. Use the relative acceleration method to calculate the acceleration
of point C knowing that the motor (link 2) moves at constant angular velocity
$\omega_2 = 10\,\mathrm{rad/s}$ counterclockwise, in other words, $\alpha_2 = 0$.

We start by calculating the velocity of point C by using the ICR method. Since
the point of known velocity, A, is part of link 3 and point C also belongs to link 3,
we only need to find the ICR of links 1 and 3. That is to say, I_{13}. This center is
shown in Fig. 2.53.

For velocity calculation purposes, we first have to find the velocity of point
A (Eq. 2.179) considered part of link 2. We know that $\omega_2 = 10\,\mathrm{rad/s}$ and distance
$\overline{I_{12}A} = 7\,\mathrm{cm}$:

Fig. 2.52 **a** Mixing machine.
b Kinematic skeleton

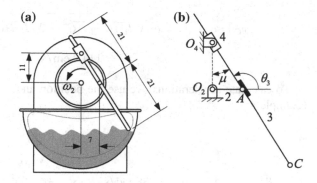

Fig. 2.53 Calculation of the
point *C* velocity using the
ICR method

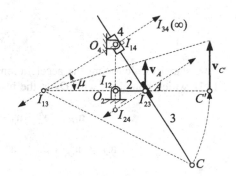

$$v_A = \omega_2 \overline{I_{12}A} = 70 \, \text{cm/s} \qquad (2.179)$$

This same velocity associated to link 3 (Eq. 2.180) is:

$$v_A = \omega_3 \overline{I_{13}I_{23}} \Rightarrow \omega_3 = \frac{v_A}{\overline{I_{13}I_{23}}} \qquad (2.180)$$

To determine the angular velocity of link 3, we need to calculate distance $\overline{I_{13}I_{23}}$
(Eq. 2.181). We will make use of values $\mu = 32.47°$ and $\overline{O_4A} = 13.04 \, \text{cm}$ obtained
in Example 4:

$$\overline{I_{13}I_{23}} = \frac{\overline{O_4A}}{\sin \mu} = 24.289 \, \text{cm} \qquad (2.181)$$

$$\omega_3 = \frac{v_A}{\overline{I_{13}I_{23}}} = 2.88 \, \text{rad/s} \qquad (2.182)$$

Finally, we can calculate the velocity of point *C* (Eq. 2.183) knowing that
$\theta_3 = 122.47°$ and $\overline{AC} = 21 \, \text{cm}$:

$$v_C = \omega_3 \overline{I_{13}C} = \omega_3 \sqrt{\overline{I_{13}I_{23}}^2 + \overline{AC}^2 - 2 \cdot \overline{I_{13}I_{23}} \cos \theta_3} \qquad (2.183)$$

$$v_C = 114.4 \, \text{cm/s}$$

To calculate accelerations, we use the position and velocity results obtained in Example 4:

$$\mathbf{v}_A = 70\hat{\mathbf{j}}$$

$$\mathbf{v}_{O_3} = \mathbf{v}_{O_3O_4} = -31.52\hat{\mathbf{i}} - 20.16\hat{\mathbf{j}}$$

$$\theta_3 = 122.47° \qquad \omega_3 = 288 \, \text{rad/s} \qquad \mathbf{v}_C = 51.03\hat{\mathbf{i}} + 102.45\hat{\mathbf{j}}$$

$$\mathbf{v}_{O_3A} = -31.68\hat{\mathbf{i}} - 20.16\hat{\mathbf{j}}$$

$$\mathbf{v}_{CA} = 51.03\hat{\mathbf{i}} + 32.45\hat{\mathbf{j}}$$

We start by calculating the acceleration of point O_3 on link 3 (Eqs. 2.184 and 2.185), which coincides with O_4 at the instant being studied.

$$\mathbf{a}_{O_3} = \mathbf{a}_A^n + \mathbf{a}_A^t + \mathbf{a}_{O_3A}^n + \mathbf{a}_{O_3A}^t \qquad (2.184)$$

$$\mathbf{a}_{O_3} = \mathbf{a}_{O_4}^n + \mathbf{a}_{O_4}^t + \mathbf{a}_{O_3O_4}^n + \mathbf{a}_{O_3O_4}^t + \mathbf{a}_{O_3O_4}^c \qquad (2.185)$$

In these equations, $\mathbf{a}_A^t = 0$ due to the fact that the angular acceleration of link 2 is zero. Acceleration $\mathbf{a}_{O_4} = 0$ because point O_4 is on the frame. Finally, acceleration $\mathbf{a}_{O_3O_4}^n = 0$ since the relative motion of point O_3 with respect to link 4 follows a straight path.

The rest of the acceleration components (Eqs. 2.186–2.190) have the following values:

$$\mathbf{a}_A^n = \boldsymbol{\omega}_2 \wedge \mathbf{v}_A = \begin{vmatrix} \hat{\mathbf{i}} & \hat{\mathbf{j}} & \hat{\mathbf{k}} \\ 0 & 0 & 10 \\ 0 & 70 & 0 \end{vmatrix} = -700\hat{\mathbf{i}} \qquad (2.186)$$

$$\mathbf{a}_A^n = 700 \, \text{cm/s}^2 \, \angle 180°$$

$$\mathbf{a}_{O_3A}^n = \boldsymbol{\omega}_3 \wedge \mathbf{v}_{O_3A} = \begin{vmatrix} \hat{\mathbf{i}} & \hat{\mathbf{j}} & \hat{\mathbf{k}} \\ 0 & 0 & 2.88 \\ -31.68 & -20.16 & 0 \end{vmatrix} = 58.06\hat{\mathbf{i}} - 91.24\hat{\mathbf{j}} \qquad (2.187)$$

$$\mathbf{a}_{O_3A}^n = 108.15 \, \text{cm/s}^2 \, \angle 302.47°$$

$$\mathbf{a}^t_{O_3A} = \alpha_3 \wedge \mathbf{r}_{O_3A} = \begin{vmatrix} \hat{\mathbf{i}} & \hat{\mathbf{j}} & \hat{\mathbf{k}} \\ 0 & 0 & \alpha_4 \\ \overline{AO_3} \cos 122.47° & \overline{AO_3} \sin 122.47° & 0 \end{vmatrix} = -11\alpha_3\hat{\mathbf{i}} - 7\alpha_3\hat{\mathbf{j}}$$

$$(2.188)$$

The relative acceleration component of will have the following values:

$$\mathbf{a}^t_{O_3O_4} = a^t_{O_3O_4} \cos 122.47°\hat{\mathbf{i}} + a^t_{O_3O_4} \sin 122.47°\hat{\mathbf{j}} \qquad (2.189)$$

$$\mathbf{a}^c_{O_3O_4} = 2\omega_3 \wedge \mathbf{v}_{O_3O_4} = 2\begin{vmatrix} \hat{\mathbf{i}} & \hat{\mathbf{j}} & \hat{\mathbf{k}} \\ 0 & 0 & 2.88 \\ -31.52 & 49.53 & 0 \end{vmatrix} = -285.29\hat{\mathbf{i}} - 181.56\hat{\mathbf{j}}$$

$$(2.190)$$

$$\mathbf{a}^c_{O_3O_4} = 338.16\,\text{cm/s}^2 \angle 212.47°$$

Evening out the two vector equations that define the value of the acceleration of point O_3 and introducing the calculated values, we obtain Eq. (2.191):

$$(-700\hat{\mathbf{i}}) + (58.06\hat{\mathbf{i}} - 91.24\hat{\mathbf{j}}) + (-11\alpha_3\hat{\mathbf{i}} - 7\alpha_3\hat{\mathbf{j}})$$
$$= a^t_{O_3O_4}(-0.54\hat{\mathbf{i}} + 0.843\hat{\mathbf{j}}) + (-285.29\hat{\mathbf{i}} - 181.56\hat{\mathbf{j}})$$

$$(2.191)$$

By breaking up the components, we obtain Eq. (2.192):

$$\left.\begin{array}{l} -700 + 58.06 - 11\alpha_3 = 0.54a^t_{O_3O_4} - 285.29 \\ -91.24 - 7\alpha_3 = 0.843a^t_{O_3O_4} - 181.56 \end{array}\right\} \qquad (2.192)$$

This way, the angular acceleration of link 3 can be found, $\alpha_3 = -19.3 \,\text{rad/s}^2$, as well as the magnitude of the relative tangential acceleration, $a^t_{O_3O_4} = 267.37\,\text{cm/s}^2$. Ultimately, we apply Eq. (2.193) to calculate the acceleration of point C:

$$\mathbf{a}_C = \mathbf{a}^n_A + \mathbf{a}^t_A + \mathbf{a}^n_{CA} + \mathbf{a}^t_{CA} \qquad (2.193)$$

Relative accelerations (Eqs. 2.194 and 2.195) can be worked out as follows:

$$\mathbf{a}^n_{CA} = \omega_3 \wedge \mathbf{v}_{CA} = \begin{vmatrix} \hat{\mathbf{i}} & \hat{\mathbf{j}} & \hat{\mathbf{k}} \\ 0 & 0 & 2.88 \\ 51.03 & 32.45 & 0 \end{vmatrix} = -93.46\hat{\mathbf{i}} + 146.97\hat{\mathbf{j}} \qquad (2.194)$$

$$\mathbf{a}^t_{CA} = \alpha_3 \wedge \mathbf{r}_{CA} = \begin{vmatrix} \hat{\mathbf{i}} & \hat{\mathbf{j}} & \hat{\mathbf{k}} \\ 0 & 0 & -19.3 \\ 11.27 & -17.71 & 0 \end{vmatrix} = -341.94\hat{\mathbf{i}} - 217.59\hat{\mathbf{j}} \qquad (2.195)$$

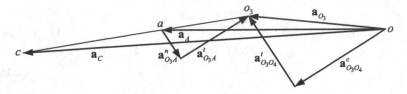

Fig. 2.54 Acceleration polygon of the mixing machine shown in Fig. (2.52)

Therefore, the acceleration of point C is Eq. (2.196):

$$\mathbf{a}_C = (-700\hat{\mathbf{i}}) + (-93.46\hat{\mathbf{i}} + 146.97\hat{\mathbf{j}}) + (-341.94\hat{\mathbf{i}} - 217.59\hat{\mathbf{j}}) \qquad (2.196)$$

$$\mathbf{a}_C = -1135.4\hat{\mathbf{i}} - 70.62\hat{\mathbf{j}} = 1137.6\,\text{cm}/\text{s}^2 \angle 183.56°$$

Once the acceleration problem has been solved, we can represent the vectors obtained and draw the acceleration polygon in Fig. 2.54:

Example 11 Figure 2.55a represents a mechanism that is part of a calculating machine that carries out multiplications and divisions. At the instant shown, knowing that $\overline{O_2A} = 1\,\text{cm}$, $\overline{O_2B} = 0.5\,\text{cm}$, input $x = 0.25\,\text{cm}$ and link 2 moves with constant angular velocity of $\omega_2 = 1\,\text{rad}/\text{s}$ counterclockwise, calculate:

1. Which constant value, k, the input has to be multiplied by to obtain output, y. That is to say, the value of constant k in equation $y = kx$.
2. The velocity vector of points A, B, C and D using the relative velocity method.
3. The acceleration vector of point C.
4. The velocity vector of point C using the ICR method. Use the velocity of point A calculated in question 2.

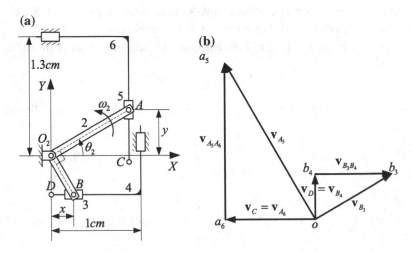

Fig. 2.55 Calculating machine (**a**) and its velocity polygon (**b**) for $x = 0.25$ cm and $\omega_2 = 1\,\text{rad}/\text{s}$

1. We start by calculating x and y (Eq. 2.197) in terms of θ_2.

$$x = \overline{O_2A} \sin \theta_2$$
$$y = \overline{O_2B} \sin \theta_2 \qquad (2.197)$$

Constant k will be given by Eq. (2.198):

$$k = \frac{y}{x} = \frac{\overline{O_2A}}{\overline{O_2B}} = 2 \qquad (2.198)$$

2. Angle θ_2 needs to be determined to calculate the velocity of point A:

$$y = kx = 2 \cdot 0.25 \,\text{cm} = 0.5 \,\text{cm} \Rightarrow \theta_2 = \arcsin \frac{0.5}{1} = 30°$$

We now define the relationship between the velocity of point A_5 and the velocity of point A_6 (Eq. 2.199), which coincides with it at that certain instant but belongs to link 6:

$$\mathbf{v}_{A_5} = \mathbf{v}_{A_6} + \mathbf{v}_{A_5A_6} \qquad (2.199)$$

This yields the following velocity values (Eqs. 2.200 and 2.201):

$$\mathbf{v}_{A_5} = \mathbf{v}_{A_2} = \omega_2 \wedge \mathbf{r}_{AO_2} = \begin{vmatrix} \hat{\mathbf{i}} & \hat{\mathbf{j}} & \hat{\mathbf{k}} \\ 0 & 0 & 1 \\ 1\cos 30° & 1\sin 30° & 0 \end{vmatrix} = -0.5\hat{\mathbf{i}} + 0.866\hat{\mathbf{j}} \quad (2.200)$$

$$\mathbf{v}_{A_5} = 1 \,\text{cm/s} \angle 120°$$

A_6 is a point on link 6 with straight horizontal motion. Relative motion of points A_5 and A_6 is also a straight movement but with vertical direction.

$$\mathbf{v}_{A_6} = v_{A_6}\hat{\mathbf{i}}$$
$$\mathbf{v}_{A_5A_6} = v_{A_5A_6}\hat{\mathbf{j}} \qquad (2.201)$$

We introduce these values in Eq. (2.199) and we obtain Eq. (2.202):

$$0.5\hat{\mathbf{i}} + 0.866\hat{\mathbf{j}} = v_{A_6}\hat{\mathbf{i}} + v_{A_5A_6}\hat{\mathbf{j}} \qquad (2.202)$$

The unknowns can easily be determined by separating the components.

$$v_{A_6} = -0.5\hat{i} = 0.5\,\text{cm/s}\,\angle 180°$$
$$v_{A_5 A_6} = 0.866\hat{j} = 0.866\,\text{cm/s}\,\angle 90°$$

Link 6 moves with straight translational motion. Therefore, all its points have the same velocity. Since point C belongs to this link, its velocity is:

$$v_C = v_{A_6} = -0.5\hat{i} = 0.5\,\text{cm/s}\,\angle 180°$$

To calculate the velocity of point D, we proceed in a similar way. We have to calculate the velocity of point B_4:

$$v_{B_3} = v_{B_4} + v_{B_3 B_4} \tag{2.203}$$

In Eq. (2.203), the velocities are given by Eqs. (2.204) and (2.205):

$$v_{B_3} = \omega_2 \wedge r_{BO_2} = \begin{vmatrix} \hat{i} & \hat{j} & \hat{k} \\ 0 & 0 & 1 \\ 0.5\cos 300° & 0.5\sin 300° & 0 \end{vmatrix} = 0.43\hat{i} + 0.25\hat{j} \tag{2.204}$$

$$v_{B_3} = 0.5\,\text{cm/s}\,\angle 30°$$

$$v_{B_4} = v_{B_4}\hat{j}$$
$$v_{B_3 B_4} = v_{B_3 B_4}\hat{i} \tag{2.205}$$

Plugging these values into Eq. (2.203), we obtain Eq. (2.206):

$$0.43\hat{i} + 0.25\hat{j} = v_{B_4}\hat{j} + v_{B_3 B_4}\hat{i} \tag{2.206}$$

From which we can calculate the following values:

$$v_{B_4} = 0.25\hat{j} = 0.25\,\text{cm/s}\,\angle 90°$$
$$v_{B_3 B_4} = 0.43\hat{i} = 0.43\,\text{cm/s}\,\angle 0°$$

Since point D belongs to link 4, which moves along a straight line, its velocity will be the same as the point B_4 velocity:

$$v_D = v_{B_4} = 0.25\hat{j} = 0.25\,\text{cm/s}\,\angle 90°$$

Once all the velocities have been determined, they can be represented in a velocity polygon (Fig. 2.55b).

3. To calculate the acceleration of point C, we start by defining Eq. (2.207), which relates the accelerations of point A_5 on link 5 with A_6 on link 6.

$$\mathbf{a}^n_{A_5} + \mathbf{a}^t_{A_5} = \mathbf{a}^n_{A_6} + \mathbf{a}^t_{A_6} + \mathbf{a}^n_{A_5A_6} + \mathbf{a}^t_{A_5A_6} + \mathbf{a}^c_{A_5A_6} \qquad (2.207)$$

The values of the acceleration components Eq. (2.208) in Eq. (2.207) are:

$$\mathbf{a}^n_{A_5} = \mathbf{a}^n_{A_2} = \boldsymbol{\omega}_2 \wedge \mathbf{v}_{A_5} = \begin{vmatrix} \hat{\mathbf{i}} & \hat{\mathbf{j}} & \hat{\mathbf{k}} \\ 0 & 0 & 1 \\ -0.5 & 0.866 & 0 \end{vmatrix} = -0.866\hat{\mathbf{i}} - 0.5\hat{\mathbf{j}} \qquad (2.208)$$

$$\mathbf{a}^t_{A_5} = \mathbf{a}^t_{A_2} = 0$$

$$\mathbf{a}^n_{A_6} = 0$$

$$\mathbf{a}^t_{A_6} = a^t_{A_6}\hat{\mathbf{i}}$$

$$\mathbf{a}^n_{A_5A_6} = 0$$

$$\mathbf{a}^t_{A_5A_6} = a^t_{A_5A_6}\hat{\mathbf{j}}$$

$$\mathbf{a}^c_{A_5A_6} = 0$$

Introducing these values in Eq. (2.207), we obtain Eq. (2.209):

$$-0.866\hat{\mathbf{i}} - 0.5\hat{\mathbf{j}} = a^t_{A_6}\hat{\mathbf{i}} + a^t_{A_5A_6}\hat{\mathbf{j}} \qquad (2.209)$$

Clearing the components, we obtain the values of the accelerations:

$$\mathbf{a}^t_{A_6} = -0.866\hat{\mathbf{i}} = 0.866\,\text{cm/s}^2 \angle 180°$$

$$\mathbf{a}^t_{A_5A_6} = -0.5\hat{\mathbf{i}} = 0.5\,\text{cm/s}^2 \angle 270°$$

Once more, as point C belongs to link 6, which moves with straight translational motion, the acceleration of point C is the same as the acceleration of point A_6:

$$\mathbf{a}_C = \mathbf{a}_{A_6} = -0.866\hat{\mathbf{i}} = 0.866\,\text{cm/s}^2 \angle 180°$$

Once the accelerations have been determined, they can be represented in an acceleration polygon (Fig. 2.56).

Fig. 2.56 Acceleration polygon of points A_2, A_5, A_6 and C in the mechanism shown in Fig. (2.55a)

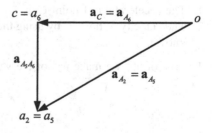

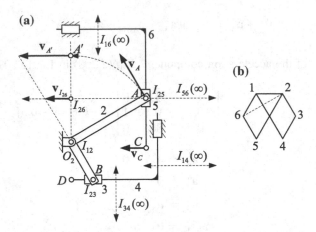

Fig. 2.57 **a** Calculation of the velocity of point C with the ICR method using velocity $\mathbf{v}_{A_2}$. **b** Polygon to analyze ICRs positions

4. To calculate the velocity of point C by means of the ICR method, we start by calculating the relative instantaneous centers of rotation of link 1 (frame), link 2 (the link point A belongs to) and link 6 (the link point C belongs to).

Once ICRs I_{12}, I_{16} and I_{26} have been obtained (Fig. 2.57a), we can calculate the velocity of I_{26}. At the considered instant, I_{26} is part of links 2 and 6 simultaneously. As a point on link 2, it rotates about O_2 and its velocity is Eq. (2.210):

$$v_A = \omega_2 \overline{I_{12}A}$$

$$v_{I_{26}} = \frac{v_A}{\overline{I_{12}A}} \overline{I_{12}I_{26}} = \frac{1}{1} 0.5 = 0.5 \, \text{cm/s} \tag{2.210}$$

Moreover, I_{26} also belongs to link 6, which moves with linear translational motion and all its points have the same velocity. Hence, $\mathbf{v}_C = \mathbf{v}_{I_{26}}$ (Fig. 2.57a).

Example 12 Figure 2.58 represents a Scotch Yoke mechanism used in the assembly line of a production chain. The lengths of the links are $\overline{O_2A} = 1$ m and $\overline{O_2B} = 1.2$ m. At the instant considered, angle $\theta_2 = 60°$ and angular velocity $\omega_2 = -1$ rad/s (clockwise). Knowing that link 2 moves with constant velocity, calculate:

1. The acceleration of point C.
2. The velocity of point C by using the ICR method when the velocity of point B is known.

1. First of all, we make use of the vector (Eq. 2.211) to calculate the velocities:

$$\mathbf{v}_{A_3} = \mathbf{v}_{A_4} + \mathbf{v}_{A_3A_4} \tag{2.211}$$

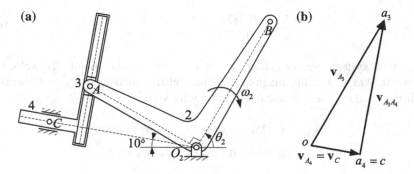

Fig. 2.58 a Scotch Yoke mechanism. **b** Velocity polygon

Next, we obtain the expressions of these components Eqs. (2.212)–(2.214). We start by calculating v_{A_3} (Eq. 2.212) knowing that $\overline{O_2A}$ forms an angle of $60° + 90° = 150°$ with the X-axis (Fig. 2.58a) and that $\omega_2 = -1$ rad/s clockwise.

$$\mathbf{v}_{A_3} = \mathbf{v}_{A_2} = \omega_2 \wedge \mathbf{r}_{AO_2} = \begin{vmatrix} \hat{\mathbf{i}} & \hat{\mathbf{j}} & \hat{\mathbf{k}} \\ 0 & 0 & -1 \\ 1\cos 150° & 1\sin 150° & 0 \end{vmatrix} = 0.5\hat{\mathbf{i}} + 0.866\hat{\mathbf{j}}$$

(2.212)

$$\mathbf{v}_{A_3} = \mathbf{v}_{A_2} = 1 \, \text{m/s} \angle 60°$$

The trajectory of link 4 forms an angle of $180° - 10° = 170°$ with the X-axis (Fig. 2.58a), thus:

$$\mathbf{v}_{A_4} = v_{A_4}\cos 170°\hat{\mathbf{i}} + v_{A_4}\sin 170°\hat{\mathbf{j}} = -0.985 v_{A_4}\hat{\mathbf{i}} + 0.174 v_{A_4}\hat{\mathbf{j}} \qquad (2.213)$$

The motion of A_3 with respect to A_4 follows a straight line that forms an angle of $80°$ with the X-axis, consequently:

$$\mathbf{v}_{A_3A_4} = v_{A_3A_4}\cos 80°\hat{\mathbf{i}} + v_{A_3A_4}\sin 80°\hat{\mathbf{j}} = 0.174 v_{A_3A_4}\hat{\mathbf{i}} + 0.985 v_{A_3A_4}\hat{\mathbf{j}} \qquad (2.214)$$

We plug these values into relative velocity (Eq. 2.211) and we obtain Eq. (2.215):

$$(0.5\hat{\mathbf{i}} + 0.866\hat{\mathbf{j}}) = (-0.985 v_{A_4}\hat{\mathbf{i}} + 0.174 v_{A_4}\hat{\mathbf{j}}) + (0.174 v_{A_3A_4}\hat{\mathbf{i}} + 0.985 v_{A_3A_4}\hat{\mathbf{j}})$$

(2.215)

By separating the vector components, we find the system of algebraic equations (Eq. 2.216):

$$\left.\begin{array}{l} 0.5 = -0.985v_{A_4} + 0.174v_{A_3A_4} \\ 0.866 = 0.174v_{A_4} + 0.985v_{A_3A_4} \end{array}\right\} \qquad (2.216)$$

In this system we can find the magnitude of the point A_4 velocity, $v_{A_4} = -0.34\,\text{m/s}$, and the magnitude of the relative velocity, $v_{A_3A_4} = 0.94\,\text{m/s}$. With these values, we can calculate the velocity vectors:

$$\mathbf{v}_{A_4} = 0.335\hat{\mathbf{i}} - 0.059\hat{\mathbf{j}} = 0.34\,\text{m/s}\,\angle 350°$$

$$\mathbf{v}_{A_3A_4} = 0.163\hat{\mathbf{i}} + 0.925\hat{\mathbf{j}} = 0.94\,\text{m/s}\,\angle 80°$$

Since all points on link 4 follow a straight path, the velocity of point C is the same as the one of point A_4:

$$\mathbf{v}_C = \mathbf{v}_{A_4} = 0.335\hat{\mathbf{i}} - 0.059\hat{\mathbf{j}} = 0.34\,\text{m/s}\,\angle 350°$$

Figure 2.58b shows the velocity polygon corresponding to these results. To calculate the acceleration, we apply Eq. (2.217):

$$\mathbf{a}_{A_3} = \mathbf{a}_{A_4} + \mathbf{a}_{A_3A_4} \qquad (2.217)$$

where the components of the accelerations (Eqs. 2.218–2.220) can be expressed as:

$$\mathbf{a}_{A_3}^n = \mathbf{a}_{A_2}^n = \omega_2 \wedge \mathbf{v}_{A_3} = \begin{vmatrix} \hat{\mathbf{i}} & \hat{\mathbf{j}} & \hat{\mathbf{k}} \\ 0 & 0 & -1 \\ 0.5 & 0.866 & 0 \end{vmatrix} = 0.866\hat{\mathbf{i}} - 0.5\hat{\mathbf{j}} \qquad (2.218)$$

$$\mathbf{a}_{A_3}^t = 0$$

$$\mathbf{a}_{A_4}^n = 0$$

$$\mathbf{a}_{A_4} = \mathbf{a}_{A_4}^t = a_{A_4}\cos 170°\hat{\mathbf{i}} + a_{A_4}\sin 170°\hat{\mathbf{j}} \qquad (2.219)$$

$$\mathbf{a}_{A_3A_4} = \mathbf{a}_{A_3A_4}^t = a_{A_3A_4}\cos 80°\hat{\mathbf{i}} + a_{A_3A_4}\sin 80°\hat{\mathbf{j}} \qquad (2.220)$$

We obtain Eq. (2.221) plugging these values into relative acceleration (Eq. 2.217):

$$(0.866\hat{\mathbf{i}} - 0.5\hat{\mathbf{j}}) = (-0.985a_{A_4}\hat{\mathbf{i}} + 0.174a_{A_4}\hat{\mathbf{j}}) + (0.174a_{A_3A_4}\hat{\mathbf{i}} + 0.985a_{A_3A_4}\hat{\mathbf{j}}) \qquad (2.221)$$

Separating the vector components (Eq. 2.222):

$$\left.\begin{array}{l} 0.866 = -0.985a_{A_4} + 0.174a_{A_3A_4} \\ -0.5 = 0.174a_{A_4} + 0.985a_{A_3A_4} \end{array}\right\} \qquad (2.222)$$

Finally, solving the algebraic equation system, we obtain:

$$\mathbf{a}_{A_4} = 0.926\hat{\mathbf{i}} - 0.163\hat{\mathbf{j}} = 0.94\,\text{m/s}^2\,\angle350°$$

$$\mathbf{a}_{A_3A_4} = -0.059\hat{\mathbf{i}} - 0.335\hat{\mathbf{j}} = 0.34\,\text{m/s}^2\,\angle260°$$

Since link 4 moves with translational motion without rotation, the acceleration of point C is the same as the one of point A_4:

$$\mathbf{a}_C = \mathbf{a}_{A_4} = 0.926\hat{\mathbf{i}} - 0.163\hat{\mathbf{j}} = 0.94\,\text{m/s}^2\,\angle350°$$

We can see these results represented in the acceleration polygon in Fig. 2.59.

1. We now calculate the velocity of point C using the ICR method. Since the known velocity belongs to a point in link 2 (point B) and the one we want to find corresponds to a point of link 4 (point C), the relative ICRs of links 1, 2 and 4 need to be determined. In other words, we need to determine ICRs I_{12}, I_{14} and I_{24}. Once the centers have been obtained (Fig. 2.60), we calculate the velocity of I_{24}. Point B rotates about point O_2 with the following velocity:

$$v_B = \omega_2\overline{I_{12}B} = 1.2\,\text{m/s}$$
$$v_{I_{24}} = \frac{v_B}{\overline{I_{12}B}}\overline{I_{12}I_{24}} = \frac{1.2}{1.2}\overline{O_2A}\sin20° = 0.34\,\text{m/s}$$

Since all points in link 4 follow a straight path, instantaneous center I_{24} yields the velocity of any point of link 4. The velocity of point C is:

$$v_C = v_{I_{24}} = 0.34\,\text{m/s}$$

Its direction is as shown in Fig. 2.60.

Example 13 Figure 2.61 represents a quick return mechanism that has the following dimensions: $\overline{O_2O_4} = 25\,\text{mm}$, $\overline{O_2A} = 30\,\text{mm}$, $\overline{AB} = 50\,\text{mm}$, $\overline{O_4B} = \overline{O_4C} = 45\,\text{mm}$ and $\overline{CD} = 50\,\text{mm}$. The angle between sides $\overline{O_4B}$ and $\overline{O_4C}$ of the triangle

Fig. 2.59 Acceleration polygon of points A_3, A_4 and C of the Scotch Yoke mechanism in Fig. (2.58a)

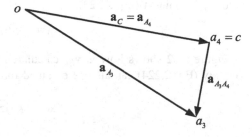

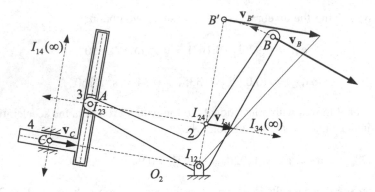

Fig. 2.60 Calculation of the velocity of point C with the ICR method using velocity $\mathbf{v}_B$

Fig. 2.61 Quick-return mechanism

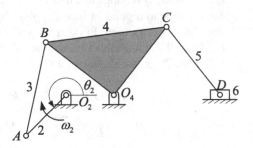

$\triangle BO_4C$ is 90°. Calculate the velocity of point D of the slider when the position and velocity of the input link are $\theta_2 = 225°$ and $\omega_2 = 80$ rpm clockwise. Use the ICR method.

Consider that point D, of unknown velocity, is part of link 6 (it could also be considered part of link 5) and that point A, of known velocity, is part of link 2 (it is also part of link 3). We have to find the relative ICRs of links 1, 2 and 6, that is, centers I_{12}, I_{16} and I_{26}.

Figure 2.62 shows the ICRs and the auxiliary lines used to find them. Before starting the calculation, we have to convert the velocity of the input link from rpm into rad/s (80 rpm = 8.38 rad/s).

In order to obtain the velocity of point D graphically, we need to know the velocity of point A (Eq. 2.223):

$$v_A = \omega_2 \overline{I_{12}A} \tag{2.223}$$

Figure 2.62 shows how is $\mathbf{v}_{I_{26}}$ calculated graphically using $\mathbf{v}_A$. However, this velocity (Eq. 2.224) can also be obtained mathematically:

$$v_{26} = \omega_2 \overline{I_{12}I_{26}} \tag{2.224}$$

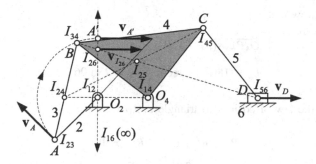

Fig. 2.62 Calculation of the velocity of point D with the ICR method knowing v_A

Since link 6 is moving with non-angular velocity, all the points in this link have the same velocity:

$$v_D = v_{I_{26}} = \omega_2 \overline{I_{12} I_{26}} = 211.8 \,\text{mm/s} \qquad (2.225)$$

Its direction is shown in Fig. 2.62.

Example 14 The mechanism shown in Fig. 2.63 represents the second inversion of a slider-crank mechanism where the slider follows a circular arc trajectory with radius $R = 24.4 \,\text{cm}$. The mechanism has the following dimensions: $\overline{O_2 P} = 10 \,\text{cm}$ and $\overline{O_2 O_4} = 14.85 \,\text{cm}$. Consider that link 2 rotates clockwise with a constant velocity of $\omega_2 = 12 \,\text{rad/s}$ clockwise and that $\theta_2 = 180°$ at the instant considered. Find the velocity and acceleration of link 4. Point Q of link 4 and point P of link 2 are superposed.

We start by solving the position problem using trigonometry (Eqs. 2.226–2.229). We apply the law of cosines to triangle $\triangle O_4 O_2 P$:

$$\overline{O_4 P} = \sqrt{\overline{O_2 O_4}^2 + \overline{O_2 P}^2 - 2\,\overline{O_2 O_4}\,\overline{O_2 P} \cos 90°} = 17.9 \,\text{cm} \qquad (2.226)$$

Fig. 2.63 Inverted slider-crank mechanism with the slider following a curved path. Point P on link 2 is coincident with point Q on link 4 at the studied instant

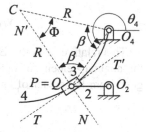

We also know that:

$$\overline{O_4P}\cos(\theta_4 - 180°) = \overline{O_2P}$$

$$\theta_4 = 180° + \arccos\frac{10}{17.9} = 236.06° \qquad (2.227)$$

Applying the law of cosines to triangle $\triangle O_4PC$:

$$\overline{O_4P}^2 = R^2 + R^2 - 2R^2 \cos \Phi$$

$$\Phi = \arccos\frac{2R^2 - \overline{O_4P}^2}{2R^2} = 43.04° \qquad (2.228)$$

$$2\beta + \Phi = 180° \Rightarrow \beta = \frac{180° - \Phi}{2} = 68.48° \qquad (2.229)$$

In Fig. 2.63 we can observe that the normal direction to link 4 on point Q is NN'. Its angle is defined by $\theta_{NN'} = \theta_4 - 180° + \beta = 124.54°$. The tangential direction to link 4 on point Q is defined by TT' and its angle is $\theta_{TT'} = \theta_4 - 180° + \beta - 90° = 34.54°$.

Applying the relative velocity method to points P and Q (Eq. 2.230):

$$\mathbf{v}_P = \mathbf{v}_Q + \mathbf{v}_{PQ} \qquad (2.230)$$

The absolute velocity of point P (Eq. 2.231), the extreme point of link 2, is:

$$\mathbf{v}_P = \boldsymbol{\omega}_2 \wedge \mathbf{r}_{PO_2} = \begin{vmatrix} \hat{\mathbf{i}} & \hat{\mathbf{j}} & \hat{\mathbf{k}} \\ 0 & 0 & -12 \\ 10\cos 180° & 10\cos 180° & 0 \end{vmatrix} = 120\hat{\mathbf{j}} \qquad (2.231)$$

$$\mathbf{v}_P = 120\,\text{cm/s} \angle 90°$$

The direction of velocity $\mathbf{v}_{PQ}$ is tangential to the trajectory followed by link 3 when it slides inside the guide rail of link 4 (Eq. 2.232). Nevertheless, its magnitude will be one of the unknowns of the problem.

$$\mathbf{v}_{PQ} = v_{PQ}\cos\theta_{TT'}\hat{\mathbf{i}} + v_{PQ}\sin\theta_{TT'}\hat{\mathbf{j}} \qquad (2.232)$$

This relative velocity (Eq. 2.233) can also be expressed as the vector product of the angular velocity of radius R associated to slider movement and the vector that goes from the center of curvature C to point P:

$$\mathbf{v}_{PQ} = \boldsymbol{\omega}_R \wedge \mathbf{r}_{PC} \qquad (2.233)$$

The velocity of point Q (Eq. 2.234) is:

$$\mathbf{v}_Q = \boldsymbol{\omega}_4 \wedge \mathbf{r}_{QO_4} = \begin{vmatrix} \hat{\mathbf{i}} & \hat{\mathbf{j}} & \hat{\mathbf{k}} \\ 0 & 0 & \omega_4 \\ 17.9\cos 236.06^\circ & 17.9\sin 236.06^\circ & 0 \end{vmatrix} \qquad (2.234)$$

$$= 14.85\omega_4\hat{\mathbf{i}} - 10\omega_4\hat{\mathbf{j}}$$

Substituting each vector in the relative velocity expression and separating its components, we obtain the following system of two equations and two unknowns (Eq. 2.235), ω_4 and v_{PQ}:

$$\left. \begin{array}{l} 0 = 14.85\omega_4 + 0.823v_{PQ} \\ 120 = -10\omega_4 + 0.567v_{PQ} \end{array} \right\} \qquad (2.235)$$

$$\omega_4 = -5.94\ \text{rad/s}$$
$$v_{PQ} = 106.96\ \text{cm/s}$$

Hence, velocities $\mathbf{v}_Q$ and $\mathbf{v}_{PQ}$ are:

$$\mathbf{v}_Q = -88.21\hat{\mathbf{i}} + 59.4\hat{\mathbf{j}} = 106.33\ \text{cm/s} \angle 146.06^\circ$$

$$\mathbf{v}_{PQ} = 88.1\hat{\mathbf{i}} + 60.64\hat{\mathbf{j}} = 106.96\ \text{cm/s} \angle 34.54^\circ$$

Once the velocities have been obtained, the velocity polygon can be drawn (Fig. 2.64).

Once vector $\mathbf{v}_{PQ}$ is defined, we have to calculate ω_R in Eq. (2.236). This value will be needed to solve the acceleration problem.

$$\mathbf{v}_{PQ} = \boldsymbol{\omega}_R \wedge \mathbf{r}_{PC} = \begin{vmatrix} \hat{\mathbf{i}} & \hat{\mathbf{j}} & \hat{\mathbf{k}} \\ 0 & 0 & \omega_R \\ R\cos(\theta_{NN'} + 180^\circ) & R\sin(\theta_{NN'} + 180^\circ) & 0 \end{vmatrix} \qquad (2.236)$$

$$88.1\hat{\mathbf{i}} + 60.64\hat{\mathbf{j}} = -R\omega_R \sin 304.54^\circ\hat{\mathbf{i}} + R\omega_R \cos 304.54^\circ\hat{\mathbf{j}}$$

Fig. 2.64 Velocity polygon of the mechanism shown in Fig. (2.63)

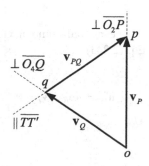

Clearing, we obtain:

$$\omega_R = \frac{-88.1}{R \sin 304.54°} = 4.38 \text{ rad/s}$$

Finally, we can study the acceleration problem. The angular velocity of link 2 is constant. Therefore, its angular acceleration (α_2) is null. Defining the vector expression of the relative acceleration (Eq. 2.237) between P and Q:

$$\mathbf{a}_P = \mathbf{a}_Q + \mathbf{a}_{PQ} \tag{2.237}$$

$$\mathbf{a}_P^n + \mathbf{a}_P^t = \mathbf{a}_Q^n + \mathbf{a}_Q^t + \mathbf{a}_{PQ}^n + \mathbf{a}_{PQ}^t + \mathbf{a}_{PQ}^c \tag{2.238}$$

And analyzing each vector in Eq. (2.238), we obtain Eqs. (2.239)–(2.245):

$$\mathbf{a}_P^n = \boldsymbol{\omega}_2 \wedge \mathbf{v}_P = \begin{vmatrix} \hat{\mathbf{i}} & \hat{\mathbf{j}} & \hat{\mathbf{k}} \\ 0 & 0 & -12 \\ 0 & 120 & 0 \end{vmatrix} = 1440\hat{\mathbf{i}} \tag{2.239}$$

$$\mathbf{a}_P^t = \boldsymbol{\alpha}_2 \wedge \mathbf{r}_{PO_2} = 0 \tag{2.240}$$

$$\mathbf{a}_P = 1440\hat{\mathbf{i}} = 1440 \text{ cm/s}^2 \angle 0°$$

The normal acceleration of point Q (Eq. 2.241) is:

$$\mathbf{a}_Q^n = \boldsymbol{\omega}_4 \wedge \mathbf{v}_Q = \begin{vmatrix} \hat{\mathbf{i}} & \hat{\mathbf{j}} & \hat{\mathbf{k}} \\ 0 & 0 & 5.94 \\ -88.21 & 59.4 & 0 \end{vmatrix} = 352.62\hat{\mathbf{i}} + 523.97\hat{\mathbf{j}} \tag{2.241}$$

$$\mathbf{a}_Q^n = 631.6 \text{ cm/s}^2 \angle 56.06°$$

The tangential acceleration of point Q (Eq. 2.242) is:

$$\mathbf{a}_Q^t = \boldsymbol{\alpha}_4 \wedge \mathbf{r}_{QO_4} = \begin{vmatrix} \hat{\mathbf{i}} & \hat{\mathbf{j}} & \hat{\mathbf{k}} \\ 0 & 0 & \alpha_4 \\ 17.9 \cos 236.06° & 17.9 \sin 236.06° & 0 \end{vmatrix} = 14.85\alpha_4\hat{\mathbf{i}} - 10\alpha_4\hat{\mathbf{j}}$$

$$\tag{2.242}$$

The normal component of the acceleration of point P relative to point Q is Eq. (2.243):

$$\mathbf{a}_{PQ}^n = \boldsymbol{\omega}_R \wedge \mathbf{v}_{PQ} = \begin{vmatrix} \hat{\mathbf{i}} & \hat{\mathbf{j}} & \hat{\mathbf{k}} \\ 0 & 0 & 4.38 \\ 88.1 & 60.64 & 0 \end{vmatrix} = -265.62\hat{\mathbf{i}} + 385.9\hat{\mathbf{j}} \tag{2.243}$$

$$\mathbf{a}_{PQ}^n = 468.1 \text{ cm/s}^2 \angle 124.54°$$

This vector is perpendicular to velocity $\mathbf{v}_{PQ}$ and it points towards the trajectory curvature center followed by link 3 when it slides along the link 4, that is to say, from P to C. The tangential component of the acceleration of P relative to Q (Eq. 2.244) is:

$$\mathbf{a}_{PQ}^t = \alpha_R \wedge \mathbf{r}_{PC} = \begin{vmatrix} \hat{\mathbf{i}} & \hat{\mathbf{j}} & \hat{\mathbf{k}} \\ 0 & 0 & \alpha_R \\ 24.4\cos 304.54° & 24.4\cos 304.54° & 0 \end{vmatrix}$$

$$= 20.1\alpha_R\hat{\mathbf{i}} + 13.83\alpha_R\hat{\mathbf{j}} \tag{2.244}$$

And the Coriolis component of the acceleration of P relative to Q (Eq. 2.245) is:

$$\mathbf{a}_{PQ}^c = 2\omega_4 \wedge \mathbf{v}_{PQ} = 2 \begin{vmatrix} \hat{\mathbf{i}} & \hat{\mathbf{j}} & \hat{\mathbf{k}} \\ 0 & 0 & -5.94 \\ 88.1 & 60.64 & 0 \end{vmatrix} = 720.45\hat{\mathbf{i}} - 1046.7\hat{\mathbf{j}} \tag{2.245}$$

$$\mathbf{a}_{PQ}^c = 1270.7 \text{ cm/s}^2 \angle 304.54°$$

Plugging these values into Eq. (2.238) and separating its components, we obtain the system of two equations and two unknowns, α_R and α_4, Eq. (2.246).

$$(1440\hat{\mathbf{i}}) = (352.62\hat{\mathbf{i}} + 523.97\hat{\mathbf{j}}) + (14.85\alpha_4\hat{\mathbf{i}} - 10\alpha_4\hat{\mathbf{j}})$$
$$+ (-265.62\hat{\mathbf{i}} + 385.9\hat{\mathbf{j}}) + (20.1\alpha_R\hat{\mathbf{i}} + 13.83\alpha_R\hat{\mathbf{j}})$$
$$+ (720.45\hat{\mathbf{i}} - 1046.7\hat{\mathbf{j}})$$

$$\left. \begin{matrix} 1440 = 352.62 + 14.85\alpha_4 - 265.62 + 20.1\alpha_R + 720.45 \\ 0 = 523.97 - 10\alpha_4 + 385.9 + 13.83\alpha_R - 1046.7 \end{matrix} \right\} \tag{2.246}$$

where:

$$\alpha_4 = 14.77 \text{ rad/s}^2$$
$$\alpha_R = 20.56 \text{ rad/s}^2$$

With these values we can calculate the tangential components of the acceleration of point Q relative to O_4 and to P.

$$\mathbf{a}_Q^t = 219.33\hat{\mathbf{i}} - 147.7\hat{\mathbf{j}} = 264.38 \text{ cm/s}^2 \angle 326.06°$$
$$\mathbf{a}_{PQ}^t = 413.26\hat{\mathbf{i}} + 284.34\hat{\mathbf{j}} = 501.7 \text{ cm/s}^2 \angle 34.54°$$

With these values, the acceleration polygon can be drawn (Fig. 2.65).

92   2 Kinematic Analysis of Mechanisms. Relative Velocity ...

Fig. 2.65 Acceleration
polygon of the mechanism
shown in Fig. (2.63)

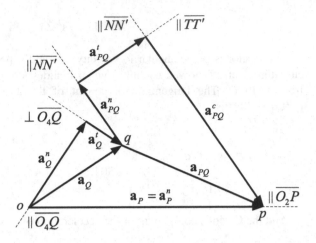

Example 15 Make a complete kinematic analysis of the shaft press mechanism in Fig. 2.66, provided that: $\theta_2 = 241°$ and $\omega_2 = -10 \, \text{rad/s}$ clockwise constant. Piston 8 follows a trajectory along a vertical line that passes through O_6. The dimensions of the mechanism are: $\overline{O_2A} = 1.4 \, \text{cm}$, $\overline{AB} = 3.5 \, \text{cm}$, $\overline{O_4B} = 1.4 \, \text{cm}$, $\mathbf{r}_{O_2} = (0,0) \, \text{cm}$, $\mathbf{r}_{O_4} = (-3.1, -1.6) \, \text{cm}$, $\overline{BC} = 3.7 \, \text{cm}$, $\overline{CD} = 5 \, \text{cm}$, $\overline{O_6D} = 1.2 \, \text{cm}$, $\mathbf{r}_{O_6} = (0, -2.8) \, \text{cm}$, $\overline{O_6E} = 1.5 \, \text{cm}$, $\psi = 67.5°$ and $\overline{EF} = 2.3 \, \text{cm}$. Use the relative velocity and acceleration methods.

Before we start to calculate velocities, it is necessary to solve the position problem. We will use the trigonometric method (Eqs. 2.247–2.266) developed in Appendix A for this purpose.

We start by studying the position of links 3 and 4 (Fig. 2.67).

Fig. 2.66 Shaft press
mechanism

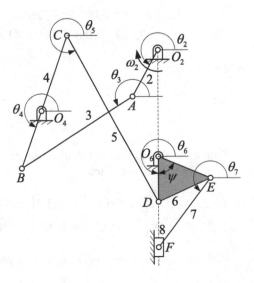

Fig. 2.67 Calculation of the position of links 3 and 4 with the trigonometric method

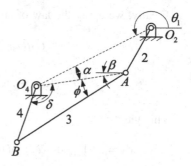

$$\overline{O_2O_4} = \sqrt{x_{O_4}^2 + y_{O_4}^2} = \sqrt{3.1^2 + 1.6^2} = 3.49\,\text{cm} \qquad (2.247)$$

$$\theta_1 = 180° + \arctan\frac{y_{O_4}}{x_{O_4}} = 207.3° \qquad (2.248)$$

The application of the law of cosines to triangle $\triangle O_2O_4A$ yields:

$$\overline{O_4A} = \sqrt{\overline{O_2O_4}^2 + \overline{O_2A}^2 - 2\,\overline{O_2O_4}\,\overline{O_2A}\cos(\theta_2 - \theta_1)} = 2.54\,\text{cm} \qquad (2.249)$$

We also apply the law of sines to the same triangle:

$$\overline{O_2A}\sin(\theta_2 - \theta_1) = \overline{O_4A}\sin\alpha \qquad (2.250)$$

$$\alpha = \arcsin\frac{1.4\sin 33.7°}{2.45} = 18.47°$$

Hence:

$$\beta = \theta_1 - 180° - \alpha = 27.3° - 18.47° = 8.83° \qquad (2.251)$$

Next, we apply the law of cosines to triangle $\triangle O_4AB$:

$$\overline{O_4B}^2 = \sqrt{\overline{O_4A}^2 + \overline{AB}^2 - 2\,\overline{O_4A}\,\overline{AB}\cos\phi} \qquad (2.252)$$

$$\phi = \arccos\frac{2.45^2 + 3.5^2 - 1.6^2}{2\cdot 2.45\cdot 3.5} = 23.81°$$

Thus:

$$\theta_3 = 180° + \beta + \phi = 180° + 8.83° + 23.81° = 212.64° \qquad (2.253)$$

Finally, we apply the law of sines to triangle $\triangle O_4AB$:

$$\overline{O_4B}\sin(180° - \delta) = \overline{AB}\sin\phi \qquad (2.254)$$

$$\delta = 180° - \arcsin\frac{3.2\sin 23.81°}{1.6} = 117.98°$$

This yields:

$$\theta_4 = 360° + \beta - \delta = 360° + 8.83° - 117.98 = 250.85° \qquad (2.255)$$

Next, we study links 5 and 6 (Fig. 2.68).

$$\overline{O_4O_6} = \sqrt{(x_{O_4} - x_{O_6})^2 + (y_{O_4} - y_{O_6})^2} = \sqrt{3.1^2 + 1.2^2} = 3.32\,\text{cm} \qquad (2.256)$$

$$\theta_1' = \arctan\frac{y_{O_4} - y_{O_6}}{x_{O_4} - x_{O_6}} = \arctan\frac{-1.2}{3.1} = 338.84° \qquad (2.257)$$

We apply the law of cosines to triangle $\triangle O_4O_6C$:

$$\overline{O_6C} = \sqrt{\overline{O_4C}^2 + \overline{O_4O_6}^2 - 2\,\overline{O_4O_6}\,\overline{O_4C}\cos(\theta_4 + 180° - \theta_{1'})} = 3.99\,\text{cm} \qquad (2.258)$$

We also apply the law of sines to the same triangle:

$$\overline{O_4C}\sin(\theta_4 + 180° - \theta_{1'}) = \overline{O_6C}\sin\alpha \qquad (2.259)$$

$$\alpha = \arcsin\frac{2.1\sin 92.01°}{3.99} = 31.735°$$

$$\beta = \alpha + 360° - \theta_{1'} = 31.735° + 360° - 338.84° = 52.895° \qquad (2.260)$$

Fig. 2.68 Variables defined to calculate the position of links 5 and 6 with the trigonometric method

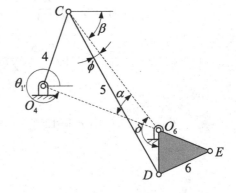

Next, we apply the law of cosines to triangle $\triangle O_6CD$:

$$\overline{O_6D}^2 = \overline{O_6C}^2 + \overline{CD}^2 - 2\,\overline{O_6C}\,\overline{CD}\cos\phi \qquad (2.261)$$

$$\phi = \arccos\frac{3.99^2 + 5^2 - 1.2^2}{2 \cdot 3.99 \cdot 5} = 8.32°$$

Thus:

$$\theta_5 = 360° - \beta - \phi = 360° - 52.895° - 8.32° = 298.8° \qquad (2.262)$$

The application of the law of sines on triangle $\triangle O_6CD$ yields:

$$\overline{O_6D}\sin(180° - \delta) = \overline{CD}\sin\phi \qquad (2.263)$$

$$\delta = 180° - \arcsin\frac{5\sin 8.32°}{1.2} = 142.92°$$

Thus:

$$\theta_6 = 180° - \beta + \delta = 180° - 52.895° + 142.92° = 270°$$

Finally, we will solve the position problem for links 7 and 8 (Fig. 2.69).

$$\overline{O_6E}\sin\psi = \overline{EF}\sin\mu \qquad (2.264)$$

$$\mu = \arcsin\frac{1.5\sin 67.5°}{2.3} = 37.05°$$

Hence:

$$\theta_7 = 270° - \mu = 232.95° \qquad (2.265)$$

Projecting $\overline{O_6E}$ and $\overline{EF}$ on the vertical axis, we obtain:

$$\overline{O_6E}\cos\psi + \overline{EF}\cos\mu = \overline{FO_{6y}} \qquad (2.266)$$

Fig. 2.69 Variables defined to calculate the position of links 7 and 8 using the trigonometric method

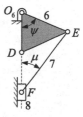

$$\overline{FO_{6y}} = 1.5\cos 67.5° + 2.3\cos 37.05° = 2.41\,\text{cm}$$

The position of point F with respect to O_2 will be given by Eq. (2.267):

$$\overline{FO_{2y}} = \overline{FO_{6y}} + \overline{O_6O_{2y}} = -2.41 - 2.8 = -5.21\,\text{cm} \qquad (2.267)$$

Once the position of each link has been obtained, we can solve the velocity problem. We start by analyzing the velocities of points A and B. The velocity of point A (Eq. 2.268), which is the end of the crank (link 2), is:

$$\mathbf{v}_A = \boldsymbol{\omega}_2 \wedge \mathbf{r}_{AO_2} = \begin{vmatrix} \hat{\mathbf{i}} & \hat{\mathbf{j}} & \hat{\mathbf{k}} \\ 0 & 0 & \omega_2 \\ \overline{O_2A}\cos\theta_2 & \overline{O_2A}\sin\theta_2 & 0 \end{vmatrix} = 12.24\hat{\mathbf{i}} - 6.79\hat{\mathbf{j}} \qquad (2.268)$$

$$\mathbf{v}_A = 14\,\text{cm/s} \angle 331°$$

Taking into account that point A is also part of link 3, we can calculate the velocity of any other point (point B) on the same link by means of the relative velocity (Eq. 2.269):

$$\mathbf{v}_B = \mathbf{v}_A + \mathbf{v}_{BA} \qquad (2.269)$$

The velocity of point B of link 4 (Eq. 2.270) is:

$$\mathbf{v}_B = \boldsymbol{\omega}_4 \wedge \mathbf{r}_{BO_4} = \begin{vmatrix} \hat{\mathbf{i}} & \hat{\mathbf{j}} & \hat{\mathbf{k}} \\ 0 & 0 & \omega_4 \\ \overline{O_4B}\cos\theta_4 & \overline{O_4B}\sin\theta_4 & 0 \end{vmatrix} = 1.511\omega_4\hat{\mathbf{i}} - 0.525\omega_4\hat{\mathbf{j}}$$

$$(2.270)$$

And the velocity of point B relative to point A (Eq. 2.271) is:

$$\mathbf{v}_{BA} = \boldsymbol{\omega}_3 \wedge \mathbf{r}_{BA} = \begin{vmatrix} \hat{\mathbf{i}} & \hat{\mathbf{j}} & \hat{\mathbf{k}} \\ 0 & 0 & \omega_3 \\ \overline{AB}\cos\theta_3 & \overline{AB}\sin\theta_3 & 0 \end{vmatrix} = 1.888\omega_3\hat{\mathbf{i}} - 2.947\omega_3\hat{\mathbf{j}} \qquad (2.271)$$

Substituting these values in the relative velocity (Eq. 2.269) and separating this equation into its components, we obtain Eq. (2.272):

$$(1.511\omega_4\hat{\mathbf{i}} - 0.525\omega_4\hat{\mathbf{j}}) = (12.24\hat{\mathbf{i}} - 6.79\hat{\mathbf{j}}) + (1.888\omega_3\hat{\mathbf{i}} - 2.947\omega_3\hat{\mathbf{j}})$$

$$\left.\begin{array}{l} 1.511\omega_4 = 12.24 + 1.888\omega_3 \\ -0.525\omega_4 = -6.79 - 2.947\omega_3 \end{array}\right\} \qquad (2.272)$$

The solution of Eq. (2.272) is:

$$\omega_3 = -1.11 \text{ rad/s}$$
$$\omega_4 = 6.71 \text{ rad/s}$$

Using the angular velocities of links 3 and 4 we can now calculate vectors $\mathbf{v}_B$, $\mathbf{v}_{BA}$ and $\mathbf{v}_C$ (Eq. 2.273):

$$\mathbf{v}_B = 10.17\hat{\mathbf{i}} - 3.55\hat{\mathbf{j}} = 10.77 \text{ cm/s} \angle 340.85°$$
$$\mathbf{v}_{BA} = -2.1\hat{\mathbf{i}} + 3.27\hat{\mathbf{j}} = 3.885 \text{ cm/s} \angle 122.6°$$

$$\mathbf{v}_C = \boldsymbol{\omega}_4 \wedge \mathbf{r}_{CO_4} = \begin{vmatrix} \hat{\mathbf{i}} & \hat{\mathbf{j}} & \hat{\mathbf{k}} \\ 0 & 0 & \omega_4 \\ \overline{O_4C}\cos(\theta_4+180°) & \overline{O_4C}\sin(\theta_4+180°) & 0 \end{vmatrix} \tag{2.273}$$
$$= -13.31\hat{\mathbf{i}} + 4.62\hat{\mathbf{j}}$$

$$\mathbf{v}_C = 14.1 \text{ cm/s} \angle 160.85°$$

We will continue the velocity analysis with points D and E. Assuming now that point C belongs to link 5, we can calculate the velocity of point D of the same link by means of the relative velocity (Eq. 2.274):

$$\mathbf{v}_D = \mathbf{v}_C + \mathbf{v}_{DC} \tag{2.274}$$

Hence, the velocity of point D of link 6 (Eq. 2.275) is:

$$\mathbf{v}_D = \boldsymbol{\omega}_6 \wedge \mathbf{r}_{DO_6} = \begin{vmatrix} \hat{\mathbf{i}} & \hat{\mathbf{j}} & \hat{\mathbf{k}} \\ 0 & 0 & \omega_6 \\ \overline{O_6D}\cos\theta_6 & \overline{O_6D}\sin\theta_6 & 0 \end{vmatrix} = -1.2\omega_6\hat{\mathbf{i}} \tag{2.275}$$

The relative velocity of point D relative to point C (Eq. 2.276) can be expressed as:

$$\mathbf{v}_{DC} = \boldsymbol{\omega}_5 \wedge \mathbf{r}_{DC} = \begin{vmatrix} \hat{\mathbf{i}} & \hat{\mathbf{j}} & \hat{\mathbf{k}} \\ 0 & 0 & \omega_5 \\ \overline{CD}\cos\theta_5 & \overline{CD}\sin\theta_5 & 0 \end{vmatrix} = -4.382\omega_5\hat{\mathbf{i}} - 2.409\omega_5\hat{\mathbf{j}}$$

$$(2.276)$$

Plugging these values into the relative velocity (Eq. 2.274) and separating the components, we obtain Eq. (2.277):

$$(-1.2\omega_6\hat{\mathbf{i}}) = (-13.31\hat{\mathbf{i}} + 4.62\hat{\mathbf{j}}) + (-4.382\omega_5\hat{\mathbf{i}} - 2.409\omega_5\hat{\mathbf{j}})$$

$$\left.\begin{array}{l} -1.2\omega_6 = -13.31 - 4.382\omega_5 \\ 0 = 4.62 - 2.409\omega_5 \end{array}\right\} \qquad (2.277)$$

where the values of the angular velocities can be obtained:

$$\omega_5 = -1.92 \text{ rad/s}$$
$$\omega_6 = -18.09 \text{ rad/s}$$

Therefore, the velocities of points D and E (Eq. 2.278) are:

$$\mathbf{v}_D = -21.7\hat{\mathbf{i}} = 21.71 \text{ cm/s} \angle 180°$$
$$\mathbf{v}_{DC} = -8.41\hat{\mathbf{i}} - 4.625\hat{\mathbf{j}} = 9.6 \text{ cm/s} \angle 208.8°$$

$$\mathbf{v}_E = \boldsymbol{\omega}_6 \wedge \mathbf{r}_{EO_6} = \begin{vmatrix} \hat{\mathbf{i}} & \hat{\mathbf{j}} & \hat{\mathbf{k}} \\ 0 & 0 & \omega_6 \\ \overline{O_6E}\cos(\theta_6 + \psi) & \overline{O_6E}\sin(\theta_6 + \psi) & 0 \end{vmatrix} \qquad (2.278)$$
$$= -10.38\hat{\mathbf{i}} - 25.07\hat{\mathbf{j}}$$

$$\mathbf{v}_E = 27.13 \text{ cm/s} \angle 247.5°$$

Finally, to calculate the velocity of point F we use Eq. (2.279):

$$\mathbf{v}_F = \mathbf{v}_E + \mathbf{v}_{FE} \qquad (2.279)$$

The velocity of point F of link 8 is a vector that has the direction of the Y-axis since the displacement of the piston follows a vertical trajectory (Eq. 2.280). Therefore, the velocity of point F can be expressed as:

$$\mathbf{v}_F = v_F\hat{\mathbf{j}} \qquad (2.280)$$

The velocity of F relative to E is given by Eq. (2.281):

$$\mathbf{v}_{FE} = \boldsymbol{\omega}_7 \wedge \mathbf{r}_{FE} = \begin{vmatrix} \hat{\mathbf{i}} & \hat{\mathbf{j}} & \hat{\mathbf{k}} \\ 0 & 0 & \omega_7 \\ \overline{EF}\cos\theta_7 & \overline{EF}\sin\theta_7 & 0 \end{vmatrix} = 1.835\omega_7\hat{\mathbf{i}} - 1.385\omega_7\hat{\mathbf{j}} \quad (2.281)$$

Plugging these values into Eq. (2.279) and separating its components, we obtain Eq. (2.282):

$$(v_F\hat{\mathbf{j}}) = (-10.38\hat{\mathbf{i}} - 25.07\hat{\mathbf{j}}) + (1.835\omega_7\hat{\mathbf{i}} - 1.385\omega_7\hat{\mathbf{j}})$$

$$\left. \begin{array}{l} 0 = -10.38 + 1.835\omega_7 \\ v_F = -25.07 - 1.385\omega_7 \end{array} \right\} \qquad (2.282)$$

where the unknowns can easily be calculated.

$$\omega_7 = 5.65 \text{ rad/s}$$
$$v_F = -32.91 \text{ cm/s}$$

With these values vectors $\mathbf{v}_F$ and $\mathbf{v}_{FE}$ can be completely defined:

$$\mathbf{v}_F = -32.91\hat{\mathbf{j}} = 32.91 \text{ cm/s} \angle 270°$$
$$\mathbf{v}_{FE} = 10.37\hat{\mathbf{i}} - 7.83\hat{\mathbf{j}} = 12.995 \text{ cm/s} \angle 322.9°$$

Figure 2.70 shows the velocity polygon, which was constructed by drawing the velocity vectors to scale. All the absolute velocities were drawn starting from the same point, o, called the pole of velocities.

After solving the position and velocity problems, we can calculate the accelerations of the links of the mechanism. We will start by analyzing the acceleration of points A and B.

The acceleration of point A (Eq. 2.283) is:

$$\mathbf{a}_A^n = \boldsymbol{\omega}_2 \wedge \mathbf{v}_A = \begin{vmatrix} \hat{\mathbf{i}} & \hat{\mathbf{j}} & \hat{\mathbf{k}} \\ 0 & 0 & 10 \\ 12.24 & -6.79 & 0 \end{vmatrix} = 67.9\hat{\mathbf{i}} + 122.4\hat{\mathbf{j}} \qquad (2.283)$$

Fig. 2.70 Velocity polygon of the mechanism shown in Fig. 2.66

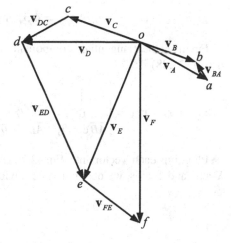

$$\mathbf{a}_A^t = 0$$

$$\mathbf{a}_A = \mathbf{a}_A^n = 67.9\hat{\mathbf{i}} + 122.4\hat{\mathbf{j}} = 140 \,\mathrm{cm/s^2} \,\angle 61°$$

The relationship between the accelerations of points A and B (Eq. 2.284) is:

$$\mathbf{a}_B^n + \mathbf{a}_B^t = \mathbf{a}_A^n + \mathbf{a}_A^t + \mathbf{a}_{BA}^n + \mathbf{a}_{BA}^t \tag{2.284}$$

We continue by analyzing the rest of the vectors in Eq. (2.284). The normal component of the acceleration of point B (Eq. 2.285) is:

$$\mathbf{a}_B^n = \omega_4 \wedge \mathbf{v}_B = \begin{vmatrix} \hat{\mathbf{i}} & \hat{\mathbf{j}} & \hat{\mathbf{k}} \\ 0 & 0 & \omega_4 \\ 10.17 & -3.55 & 0 \end{vmatrix} = 23.63\hat{\mathbf{i}} + 68.05\hat{\mathbf{j}} \tag{2.285}$$

$$\mathbf{a}_B^n = 72.04 \,\mathrm{cm/s^2} \,\angle 70.85°$$

The tangential component of the acceleration of point B (Eq. 2.286) is:

$$\mathbf{a}_B^t = \alpha_4 \wedge \mathbf{r}_{BO_4} = \begin{vmatrix} \hat{\mathbf{i}} & \hat{\mathbf{j}} & \hat{\mathbf{k}} \\ 0 & 0 & \alpha_4 \\ \overline{BO_4}\cos\theta_4 & \overline{BO_4}\sin\theta_4 & 0 \end{vmatrix} = 1.511\alpha_4\hat{\mathbf{i}} - 0.5249\alpha_4\hat{\mathbf{j}}$$

$$\tag{2.286}$$

The normal component of the acceleration of point B relative to point A (Eq. 2.287) is:

$$\mathbf{a}_{BA}^n = \omega_3 \wedge \mathbf{v}_{BA} = \begin{vmatrix} \hat{\mathbf{i}} & \hat{\mathbf{j}} & \hat{\mathbf{k}} \\ 0 & 0 & \omega_3 \\ -2.1 & 3.27 & 0 \end{vmatrix} = 3.63\hat{\mathbf{i}} + 2.326\hat{\mathbf{j}} \tag{2.287}$$

$$\mathbf{a}_{BA}^n = 4.31 \,\mathrm{cm/s^2} \,\angle 32.64°$$

Finally, the tangential component of the acceleration of B relative to A (Eq. 2.288) is:

$$\mathbf{a}_{BA}^t = \alpha_3 \wedge \mathbf{r}_{BA} = \begin{vmatrix} \hat{\mathbf{i}} & \hat{\mathbf{j}} & \hat{\mathbf{k}} \\ 0 & 0 & \alpha_3 \\ \overline{AB}\cos\theta_3 & \overline{AB}\sin\theta_3 & 0 \end{vmatrix} = 1.888\alpha_3\hat{\mathbf{i}} - 2.947\alpha_3\hat{\mathbf{j}} \tag{2.288}$$

Plugging each vector into Eq. (2.284) and projecting its components onto the X-axis and Y-axis, we obtain a system with two unknowns, α_3 and α_4, Eq. (2.289).

$$(23.63\hat{\mathbf{i}} + 68.05\hat{\mathbf{j}}) + (1.511\alpha_4\hat{\mathbf{i}} - 0.5249\alpha_4\hat{\mathbf{j}})$$
$$= (67.9\hat{\mathbf{i}} + 122.4\hat{\mathbf{j}}) + (3.63\hat{\mathbf{i}} + 2.326\hat{\mathbf{j}})$$
$$+ (1.888\alpha_3\hat{\mathbf{i}} - 2.947\alpha_3\hat{\mathbf{j}})$$

$$\left. \begin{aligned} 23.63 + 1.511\alpha_4 &= 67.9 + 3.63 + 1.888\alpha_3 \\ 68.05 - 0.5249\alpha_4 &= 122.4 + 2.326 - 2.947\alpha_3 \end{aligned} \right\} \qquad (2.289)$$

Solving this system, we obtain Eq. (2.290):

$$\left. \begin{aligned} \alpha_3 &= 31.9 \text{ rad/s}^2 \\ \alpha_4 &= 71.48 \text{ rad/s}^2 \end{aligned} \right\} \qquad (2.290)$$

With these values we can calculate the absolute acceleration vectors of points B and C and the acceleration vector of point B with respect to point A (Eq. 2.291):

$$\mathbf{a}_B = 131.67\hat{\mathbf{i}} + 30.53\hat{\mathbf{j}} = 135.16 \text{ cm/s}^2 \; \angle 13.1°$$

$$\mathbf{a}_{BA} = 63.85\hat{\mathbf{i}} - 91.69\hat{\mathbf{j}} = 111.73 \text{ cm/s}^2 \; \angle 304.85°$$

$$\mathbf{a}_C^n = \omega_4 \wedge \mathbf{v}_C = \begin{vmatrix} \hat{\mathbf{i}} & \hat{\mathbf{j}} & \hat{\mathbf{k}} \\ 0 & 0 & 6.71 \\ -13.31 & 4.62 & 0 \end{vmatrix} = -31\hat{\mathbf{i}} - 89.31\hat{\mathbf{j}}$$

$$\mathbf{a}_C^t = \alpha_4 \wedge \mathbf{r}_{CO_4} = \begin{vmatrix} \hat{\mathbf{i}} & \hat{\mathbf{j}} & \hat{\mathbf{k}} \\ 0 & 0 & 71.48 \\ 2.3\cos 70.85° & 2.3\sin 70.85° & 0 \end{vmatrix} = -15.53\hat{\mathbf{i}} + 53.61\hat{\mathbf{j}}$$

$$\mathbf{a}_C = \mathbf{a}_C^n + \mathbf{a}_C^t = -186.3\hat{\mathbf{i}} - 35.7\hat{\mathbf{j}} = 189.68 \text{ cm/s}^2 \; \angle 183.33° \qquad (2.291)$$

We will continue by studying the accelerations of points D and E. The relationship between the accelerations of points D and C is given by Eq. (2.292):

$$\mathbf{a}_D = \mathbf{a}_D^n + \mathbf{a}_D^t = \mathbf{a}_C^n + \mathbf{a}_C^t + \mathbf{a}_{DC}^n + \mathbf{a}_{DC}^t \qquad (2.292)$$

The remaining vectors (Eqs. 2.293–2.295) in Eq. (2.292) are:

$$\mathbf{a}_D^n = \omega_6 \wedge \mathbf{v}_D = \begin{vmatrix} \hat{\mathbf{i}} & \hat{\mathbf{j}} & \hat{\mathbf{k}} \\ 0 & 0 & -18.09 \\ -21.7 & 0 & 0 \end{vmatrix} = 392.7\hat{\mathbf{j}}$$

$$\mathbf{a}_D^t = \alpha_6 \wedge \mathbf{r}_{DO_6} = \begin{vmatrix} \hat{\mathbf{i}} & \hat{\mathbf{j}} & \hat{\mathbf{k}} \\ 0 & 0 & \alpha_6 \\ 1.2\cos 270° & 1.2\sin 270° & 0 \end{vmatrix} = 1.2\alpha_6\hat{\mathbf{i}}$$

$$\mathbf{a}_D = 1.2\alpha_6\hat{\mathbf{i}} + 392.7\hat{\mathbf{j}} \qquad (2.293)$$

$$\mathbf{a}_{DC}^t = \alpha_5 \wedge \mathbf{r}_{DC} = \begin{vmatrix} \hat{\mathbf{i}} & \hat{\mathbf{j}} & \hat{\mathbf{k}} \\ 0 & 0 & \alpha_5 \\ 5\cos 298.8° & 5\sin 298.8° & 0 \end{vmatrix} = 4.38\alpha_5\hat{\mathbf{i}} + 2.41\alpha_5\hat{\mathbf{j}}$$

$$\qquad (2.294)$$

$$\mathbf{a}_{DC}^n = \omega_5 \wedge \mathbf{v}_{DC} = \begin{vmatrix} \hat{\mathbf{i}} & \hat{\mathbf{j}} & \hat{\mathbf{k}} \\ 0 & 0 & -1.92 \\ -8.41 & -4.625 & 0 \end{vmatrix} = -8.88\hat{\mathbf{i}} + 16.15\hat{\mathbf{j}} \qquad (2.295)$$

Substituting these values in Eq. (2.292) and separating the resulting vectors into their components, we obtain Eq. (2.296):

$$(1.2\alpha_6\hat{\mathbf{i}} + 392.7\hat{\mathbf{j}}) = (-172.82\hat{\mathbf{i}} - 40.08\hat{\mathbf{j}})$$
$$+ (4.38\alpha_5\hat{\mathbf{i}} + 2.41\alpha_5\hat{\mathbf{j}}) + (-8.88\hat{\mathbf{i}} + 16.15\hat{\mathbf{j}})$$

$$\left. \begin{array}{l} 1.2\alpha_6 = -172.82 + 4.38\alpha_5 - 8.88 \\ 392.7 = -40.08 + 2.41\alpha_5 + 16.15 \end{array} \right\} \qquad (2.296)$$

Solving the system, we obtain Eq. (2.297):

$$\left. \begin{array}{l} \alpha_5 = 172.9 \text{ rad/s}^2 \\ \alpha_6 = 479.6 \text{ rad/s}^2 \end{array} \right\} \qquad (2.297)$$

The acceleration vectors of points D and E with respect to the frame and the acceleration vector of point D relative to point C (Eq. 2.298) are:

$$\mathbf{a}_D = 575.52\hat{\mathbf{i}} + 392.7\hat{\mathbf{j}} = 696.73 \text{ cm/s}^2 \, \angle 34.3°$$

$$\mathbf{a}_{DC} = 748.42\hat{\mathbf{i}} + 432.84\hat{\mathbf{j}} = 864.57 \text{ cm/s}^2 \, \angle 30.04°$$

$$\mathbf{a}_E^n = \omega_6 \wedge \mathbf{v}_E = \begin{vmatrix} \hat{\mathbf{i}} & \hat{\mathbf{j}} & \hat{\mathbf{k}} \\ 0 & 0 & -18.09 \\ -10.38 & -25.07 & 0 \end{vmatrix} = -453.52\hat{\mathbf{i}} + 187.77\hat{\mathbf{j}}$$

$$\mathbf{a}_E^t = \alpha_6 \wedge \mathbf{r}_{EO_6} = \begin{vmatrix} \hat{\mathbf{i}} & \hat{\mathbf{j}} & \hat{\mathbf{k}} \\ 0 & 0 & 479.6 \\ 1.5\cos 337.5° & 1.5\sin 337.5° & 0 \end{vmatrix} = 275.29\hat{\mathbf{i}} + 664.73\hat{\mathbf{j}}$$

$$\mathbf{a}_E = \mathbf{a}_E^n + \mathbf{a}_E^t = -178.2\hat{\mathbf{i}} + 852.5\hat{\mathbf{j}} \qquad (2.298)$$

$$\mathbf{a}_E = 870.93 \text{ cm/s}^2 \ \angle 101.8° \tag{2.299}$$

Finally, we need to find the acceleration in the crank-shaft mechanism formed by links 6, 7 and 8. We define acceleration vectors for point E (Eq. 2.300) and F of link 7:

$$\mathbf{a}_F = \mathbf{a}_F^n + \mathbf{a}_F^t = \mathbf{a}_E^n + \mathbf{a}_E^t + \mathbf{a}_{FE}^n + \mathbf{a}_{FE}^t \tag{2.300}$$

where:

$$\mathbf{a}_F = \mathbf{a}_F^t = a_F \hat{\mathbf{j}} \tag{2.301}$$

$$\mathbf{a}_{FE}^n = \boldsymbol{\omega}_7 \wedge \mathbf{v}_{FE} = \begin{vmatrix} \hat{\mathbf{i}} & \hat{\mathbf{j}} & \hat{\mathbf{k}} \\ 0 & 0 & \\ & & 0 \end{vmatrix} = 44.24\hat{\mathbf{i}} + 58.6\hat{\mathbf{j}} \tag{2.302}$$

$$\mathbf{a}_{FE}^t = \boldsymbol{\alpha}_7 \wedge \mathbf{r}_{FE} = \begin{vmatrix} \hat{\mathbf{i}} & \hat{\mathbf{j}} & \hat{\mathbf{k}} \\ 0 & 0 & \alpha_7 \\ & & 0 \end{vmatrix} = 1.836\alpha_7\hat{\mathbf{i}} - 1.386\alpha_7\hat{\mathbf{j}} \tag{2.303}$$

Substituting the values of Eqs. (2.301)–(2.303) in the relative acceleration (Eq. 2.300) and breaking the resulting vector into its components, we obtain the equation system of two equations with two unknowns, α_7 and a_F, Eq. (2.304):

$$(a_F\hat{\mathbf{j}}) = (-178.2\hat{\mathbf{i}} + 852.5\hat{\mathbf{j}}) + (44.24\hat{\mathbf{i}} + 58.6\hat{\mathbf{j}}) + (1.836\alpha_7\hat{\mathbf{i}} - 1.386\alpha_7\hat{\mathbf{j}})$$

$$\left. \begin{array}{l} 0 = -178.2 + 44.24 + 1.836\alpha_7 \\ a_F = 852.5 + 58.6 - 1.386\alpha_7 \end{array} \right\} \tag{2.304}$$

Solving the system we obtain:

$$\left. \begin{array}{l} \alpha_7 = 72.96 \text{ rad/s}^2 \\ a_F = 810 \text{ cm/s}^2 \end{array} \right\}$$

The relative acceleration vector of point F with respect to point E is:

$$\mathbf{a}_{FE} = 178.19\hat{\mathbf{i}} - 42.52\hat{\mathbf{j}} = 183.19 \text{ cm/s}^2 \ \angle 346.58°$$

Figure 2.71 represents the acceleration polygon built by drawing the absolute acceleration vectors, all of them starting at the pole of accelerations. Relative acceleration vectors are obtained by joining the extreme points of absolute acceleration vectors. Student are recommended to do this exercise in order to better understand vector directions in this problem.

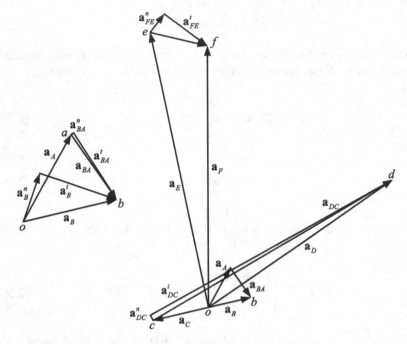

Fig. 2.71 Acceleration polygon of the mechanism shown in Fig. 2.66

Example 16 The mechanism in Fig. 2.72 is part of a calculating machine that carries out the "inverse" (1/y) arithmetic operation. Find the solution to the position, velocity and acceleration problems at the instant shown, knowing that the input is equal to y = 1.87 cm and that link 4 moves with a constant linear velocity of 0.5 cm/s in an ascending direction.

We start by solving the position problem using the trigonometric method (Eq. 2.305). The expressions are:

$$\left.\begin{array}{l}\overline{O_2A}\sin\theta_2 = y \\ \overline{O_2A}\cos\theta_2 = \sqrt{3}\end{array}\right\} \tag{2.305}$$

Therefore:

$$\tan\theta_2 = \frac{y}{\sqrt{3}} \Rightarrow \theta_2 = \arctan\frac{1.87}{\sqrt{3}} = 47.19°$$

$$\overline{O_2A} = \frac{y}{\sin\theta_2} = \frac{1.87}{\sin 47.19°} = 2.55\,\text{cm}$$

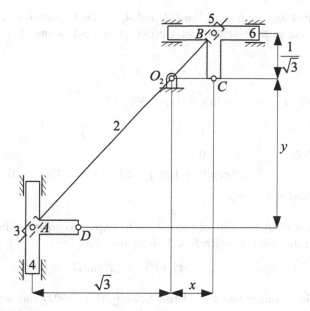

Fig. 2.72 Calculating machine that carries out the "inverse" $(x = 1/y)$ arithmetic operation

We analyze the position of point B (Eq. 2.306):

$$\left.\begin{array}{l} \overline{O_2B}\sin\theta_2 = \frac{1}{\sqrt{3}} \\ \overline{O_2B}\cos\theta_2 = x \end{array}\right\} \tag{2.306}$$

From where we obtain:

$$\overline{O_2B} = \frac{1/\sqrt{3}}{\sin\theta_2} = \frac{1/\sqrt{3}}{\sin 47.19°} = 0.79\,\text{cm}$$

So:

$$x = \overline{O_2B}\cos\theta_2 = 0.79\cos 47.19° = 0.535\,\text{cm}$$

It can be verified that the value of y is always the inverse value of x.

Once the position of the links in the mechanism have been defined, we can solve the velocity problem.

Link 4 makes a translational motion and follows a vertical trajectory at a constant velocity of 0.5 cm/s. Therefore, the velocity of point A of link 4 is:

$$\mathbf{v}_A = 0.5\hat{\mathbf{j}}$$

Since point 4 is common to links 3 and 4, it can be denominated A_3. The expression of the relative velocity of the two coincident points of links 2 and 3 (Eq. 2.307) is:

$$\mathbf{v}_{A_3} = \mathbf{v}_{A_2} + \mathbf{v}_{A_3 A_2} \tag{2.307}$$

The velocity of point A of link 2 (Eq. 2.308) is:

$$
\mathbf{v}_{A_2} = \mathbf{\omega}_2 \wedge \mathbf{r}_{AO_2} =
\begin{vmatrix}
\hat{\mathbf{i}} & \hat{\mathbf{j}} & \hat{\mathbf{k}} \\
0 & 0 & \omega_2 \\
2.55\cos(\theta_2 + 180°) & 2.55\sin(\theta_2 + 180°) & 0
\end{vmatrix}
\tag{2.308}
$$
$$
= 1.87\omega_2\hat{\mathbf{i}} - 1.73\omega_2\hat{\mathbf{j}}
$$

The direction of the velocity of point A_3 with respect to point A_2 (Eq. 2.309) is defined by the direction in which link 3 slides over link 2:

$$\mathbf{v}_{A_3 A_2} = v_{A_3 A_2} \cos 47.19°\hat{\mathbf{i}} + v_{A_3 A_2} \sin 47.19°\hat{\mathbf{j}} \tag{2.309}$$

Plugging these values into the relative velocity (Eq. 2.307) and separating the resulting vector into its components yields the system of two algebraic equations with two unknowns, ω_2 and $v_{A_3 A_2}$, Eq. (2.310):

$$
\left.
\begin{array}{l}
0 = 1.87\omega_2 + 0.679 v_{A_3 A_2} \\
0.5 = -1.73\omega_2 + 0.734 v_{A_3 A_2}
\end{array}
\right\}
\tag{2.310}
$$

Solving the system we obtain:

$$
\left.
\begin{array}{l}
\omega_2 = -0.13 \text{ rad/s} \\
v_{A_3 A_2} = 0.367 \text{ cm/s}
\end{array}
\right\}
$$

With these values we can calculate vectors $\mathbf{v}_{A_2}$ and $\mathbf{v}_{A_3 A_2}$:

$$\mathbf{v}_{A_2} = -0.2431\hat{\mathbf{i}} + 0.2249\hat{\mathbf{j}} = 0.331 \text{ cm/s} \angle 137.19°$$
$$\mathbf{v}_{A_3 A_2} = 0.249\hat{\mathbf{i}} + 0.269\hat{\mathbf{j}} = 0.36 \text{ cm/s} \angle 47.19°$$

To find the velocity of links 5 and 6, we have to relate the velocities of the two coincident points at B (Eq. 2.311) (one of link 2 and another of link 5):

$$\mathbf{v}_{B_5} = \mathbf{v}_{B_2} + \mathbf{v}_{B_5 B_2} \tag{2.311}$$

Since point B_5 also belongs to link 6 and all the points in this link share the same velocity with horizontal direction, we can assert that:

$$\mathbf{v}_{B_5} = v_{B_5}\hat{\mathbf{i}}$$

Therefore, the velocity of link 2 (Eq. 2.312) is:

$$\mathbf{v}_{B_2} = \mathbf{\omega}_2 \wedge \mathbf{r}_{BO_2} = \begin{vmatrix} \hat{\mathbf{i}} & \hat{\mathbf{j}} & \hat{\mathbf{k}} \\ 0 & 0 & -0.13 \\ \overline{O_2B}\cos\theta_2 & \overline{O_2B}\sin\theta_2 & 0 \end{vmatrix} = 0.075\hat{\mathbf{i}} - 0.069\hat{\mathbf{j}}$$

(2.312)

$$\mathbf{v}_{B_2} = 0.102\,\text{cm/s} \angle 317.19°$$

The direction of vector $\mathbf{v}_{B_5B_2}$ is defined by the direction in which link 5 slides over link 2 (Eq. 2.313):

$$\mathbf{v}_{B_5B_2} = v_{B_5B_2}\cos\theta_2\hat{\mathbf{i}} + v_{B_5B_2}\sin\theta_2\hat{\mathbf{j}} = 0.679v_{B_5B_2}\hat{\mathbf{i}} + 0.733v_{B_5B_2}\hat{\mathbf{j}} \quad (2.313)$$

Substituting the values obtained in Eq. (2.311) and separating the vectors into their components, we obtain Eq. (2.314):

$$\left.\begin{array}{l} v_{B_5} = 0.075 + 0.679v_{B_5B_2} \\ 0 = -0.069 + 0.733v_{B_5B_2} \end{array}\right\}$$

(2.314)

From where we can calculate the magnitudes of $\mathbf{v}_{B_5}$ and $\mathbf{v}_{B_5B_2}$:

$$\left.\begin{array}{l} v_{B_5} = 0.139\,\text{cm/s} \\ v_{B_5B_2} - 0.094\,\text{cm/s} \end{array}\right\}$$

With these values, we can define the velocity vectors:

$$\mathbf{v}_{B_5} = 0.139\hat{\mathbf{i}} = 0.139\,\text{cm/s} \angle 0°$$

$$\mathbf{v}_{B_5B_2} = 0.0639\hat{\mathbf{i}} + 0.0688\hat{\mathbf{j}} = 0.094\,\text{cm/s} \angle 47.19°$$

Finally, we solve the acceleration problem. We have to take into account that the acceleration of link 4 is null. The relative acceleration (Eqs. 2.315 and 2.316) of the two coincident points, A_2 and A_3, is:

$$\mathbf{a}_{A_3} = \mathbf{a}_{A_2} + \mathbf{a}_{A_3A_2}$$

(2.315)

$$\mathbf{a}_{A_3}^n + \mathbf{a}_{A_3}^t = \mathbf{a}_{A_2}^n + \mathbf{a}_{A_2}^t + \mathbf{a}_{A_3A_2}^n + \mathbf{a}_{A_3A_2}^t + \mathbf{a}_{A_3A_2}^c$$

(2.316)

Since point A_3 also belongs to link 4:

$$\mathbf{a}_{A_3} = 0$$

The rest of the vectors in Eq. (2.316) are (Eqs. 2.317–2.321):

$$\mathbf{a}_{A_2}^n = \boldsymbol{\omega}_2 \wedge \mathbf{v}_{A_2} = \begin{vmatrix} \hat{\mathbf{i}} & \hat{\mathbf{j}} & \hat{\mathbf{k}} \\ 0 & 0 & -0.13 \\ -0.2431 & 0.2249 & 0 \end{vmatrix} = 0.029\hat{\mathbf{i}} + 0.031\hat{\mathbf{j}} \quad (2.317)$$

$$\mathbf{a}_{A_2} = 0.043 \, \text{cm/s}^2 \, \angle 47.19°$$

$$\mathbf{a}_{A_2}^t = \boldsymbol{\alpha}_2 \wedge \mathbf{r}_{AO_2} = \begin{vmatrix} \hat{\mathbf{i}} & \hat{\mathbf{j}} & \hat{\mathbf{k}} \\ 0 & 0 & \alpha_2 \\ 2.55\cos(\theta_2 + 180°) & 2.55\sin(\theta_2 + 180°) & 0 \end{vmatrix}$$

$$= 1.87\alpha_2\hat{\mathbf{i}} - 1.73\alpha_2\hat{\mathbf{j}} \quad (2.318)$$

The relative motion between points A_2 and A_3 follows a straight-line trajectory along link 2. Therefore, the normal component of relative acceleration $\mathbf{a}_{A_3A_2}^n$ is zero and the direction of the tangential component is defined by link 2:

$$\mathbf{a}_{A_3A_2}^n = 0 \quad (2.319)$$

$$\mathbf{a}_{A_3A_2}^t = a_{A_3A_2}\cos\theta_2\hat{\mathbf{i}} + a_{A_3A_2}\sin\theta_2\hat{\mathbf{j}} \quad (2.320)$$

The Coriolis component of the acceleration is given by:

$$\mathbf{a}_{A_3A_2}^c = 2\boldsymbol{\omega}_2 \wedge \mathbf{v}_{A_3A_2} = 2\begin{vmatrix} \hat{\mathbf{i}} & \hat{\mathbf{j}} & \hat{\mathbf{k}} \\ 0 & 0 & -0.13 \\ 0.249 & 0.269 & 0 \end{vmatrix} = 0.07\hat{\mathbf{i}} - 0.0647\hat{\mathbf{j}} \quad (2.321)$$

Substituting the values obtained in Eq. (2.316) and separating the resulting vectors into their components yields (Eq. 2.322):

$$\left. \begin{array}{l} 0 = 0.029 + 1.87\alpha_2 + 0.07 + 0.6796a_{A_3A_2} \\ 0 = 0.031 - 1.73\alpha_2 - 0.0647 + 0.7336a_{A_3A_2} \end{array} \right\} \quad (2.322)$$

$$\left. \begin{array}{l} \alpha_2 = -0.0375 \, \text{rad/s}^2 \\ a_{A_3A_2} = -0.0425 \, \text{cm/s}^2 \end{array} \right\}$$

In order to calculate the absolute acceleration of point B of link 5 (B_5), we use Eqs. (2.323) and (2.324):

$$\mathbf{a}_{B_5} = \mathbf{a}_{B_2} + \mathbf{a}_{B_5B_2} \quad (2.323)$$

$$\mathbf{a}_{B_5}^n + \mathbf{a}_{B_5}^t = \mathbf{a}_{B_2}^n + \mathbf{a}_{B_2}^t + \mathbf{a}_{B_5B_2}^n + \mathbf{a}_{B_5B_2}^t + \mathbf{a}_{B_5B_2}^c \quad (2.324)$$

We analyze the acceleration components in Eq. (2.324) starting with point B of link 5 (Eq. 2.325). Since point B_5 also belongs to link 6 and all the points in this link follow a horizontal trajectory:

$$\mathbf{a}^n_{B_5} = 0 \tag{2.325}$$

$$\mathbf{a}^t_{B_5} = a_B\hat{\mathbf{i}} \tag{2.326}$$

The acceleration components of point B of link 2 will be:

$$\mathbf{a}^n_{B_2} = \omega_2 \wedge \mathbf{v}_{B_2} = \begin{vmatrix} \hat{\mathbf{i}} & \hat{\mathbf{j}} & \hat{\mathbf{k}} \\ 0 & 0 & -0.13 \\ 0.075 & -0.069 & 0 \end{vmatrix} = -0.0089\hat{\mathbf{i}} - 0.0097\hat{\mathbf{j}} \tag{2.327}$$

$$\mathbf{a}^n_{B_2} = 0.013 \,\text{cm/s}^2 \,\angle 227.19°$$

$$\mathbf{a}^t_{B_2} = \alpha_2 \wedge \mathbf{r}_{BO_2} = \begin{vmatrix} \hat{\mathbf{i}} & \hat{\mathbf{j}} & \hat{\mathbf{k}} \\ 0 & 0 & -0.0375 \\ 0.79\cos 47.19° & 0.79\sin 47.19° & 0 \end{vmatrix}$$
$$= 0.0217\hat{\mathbf{i}} - 0.02\hat{\mathbf{j}} \tag{2.328}$$

$$\mathbf{a}^t_{B_2} = 0.0296 \,\text{cm/s}^2 \,\angle 317.19°$$

The relative motion of point B_5 with respect to link 2 follows a straight trajectory defined by link 2. Therefore, the normal component of the acceleration of point B_5 relative to point B_2 is zero and the direction of the tangential component is defined by link 2.

$$\mathbf{a}^n_{B_5B_2} = 0 \tag{2.329}$$

$$\mathbf{a}^t_{B_5B_2} = a^t_{B_5B_2}\cos\theta_2\hat{\mathbf{i}} + a^t_{B_5B_2}\sin\theta_2\hat{\mathbf{j}} \tag{2.330}$$

The Coriolis component of the acceleration of point B_5 with respect to point B_2 can be calculated with the following expression:

$$\mathbf{a}^c_{B_5B_2} = 2\omega_2 \wedge \mathbf{v}_{B_5B_2} = 2\begin{vmatrix} \hat{\mathbf{i}} & \hat{\mathbf{j}} & \hat{\mathbf{k}} \\ 0 & 0 & -0.13 \\ 0.0639 & 0.0688 & 0 \end{vmatrix} = 0.0178\hat{\mathbf{i}} - 0.0166\hat{\mathbf{j}} \tag{2.331}$$

$$\mathbf{a}^c_{B_5B_2} = 0.0243 \,\text{cm/s}^2 \,\angle 317.19°$$

Plugging the expression of these acceleration components Eqs. (2.325)–(2.331) into Eq. (2.324) and separating each vector into its x and y components, we obtain Eq. (2.332):

$$\left.\begin{array}{l} a_{B_5} = -0.0089 + 0.0217 + 0.0178 + 0.6796 a^t_{B_5 B_2} \\ 0 = -0.0097 - 0.02 - 0.0166 + 0.7336 a^t_{B_5 B_2} \end{array}\right\} \tag{2.332}$$

Solving the system, we find the unknowns:

$$\left.\begin{array}{l} a_{B_5} = 0.0734 \, \text{cm/s}^2 \\ a^t_{B_5 B_2} = 0.0631 \, \text{cm/s}^2 \end{array}\right\}$$

Hence, the acceleration of link 6 is given by:

$$\mathbf{a}_6 = \mathbf{a}_{B_5} = 0.0734 \hat{\mathbf{i}} = 0.0734 \, \text{cm/s}^2 \, \angle 0°$$

Chapter 3
Analytical Methods for the Kinematic Analysis of Planar Linkages. Raven's Method

Abstract Graphical methods have played an essential role along the history of Machine and Mechanism Theory. However, mathematical methods of mechanism analysis have gained major importance mainly due to the appearance of programmable calculators and personal computers. Above all, graphical methods have a great educational value and are interesting when it comes to solving mechanisms in one position, since, as well as being simple, they give us a clear view of their operation. Finding the mathematical solution to a mechanism is more complicated, as it requires a higher investment of time and errors are easier to make and more difficult to detect. However, the ability of the computer to save and re-use all operations implies that mathematical analysis can save a lot of time when we have to study a mechanism by varying parameters such as the length of the bars or the position, velocity and acceleration of the input link. It is also possible to easily obtain variable diagrams along one revolution. Mathematical methods basically try to obtain an analytical expression of the variables that we want to determine, such as position, velocity and the acceleration of any link in terms of the dimensions of the mechanism $(r_1, r_2, r_3, \ldots)$ and the position, velocity and acceleration of the motor link (θ_2, ω_2 and α_2). Along this chapter, several mechanisms will be solved using Raven's method in order to foment comprehension.

3.1 Analytical Methods

Among the various mathematical approaches to mechanism analysis, we can distinguish trigonometric analysis, analysis by complex numbers and vector analysis. All of them determine the position of the links in a mechanism in terms of any of the parameters mentioned above $(\theta_2, r_1, r_2, r_3, \ldots)$. By double-differentiating the obtained equations, we first calculate the speed, then the acceleration of each link.

© Springer International Publishing Switzerland 2016
A. Simón Mata et al., *Fundamentals of Machine Theory and Mechanisms*,
Mechanisms and Machine Science 40, DOI 10.1007/978-3-319-31970-4_3

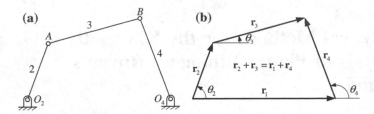

Fig. 3.1 a Four-bar linkage. **b** Closed vector polygon with the unknowns, θ_3 and θ_4

The main advantage of the analysis by complex numbers and vectors is the simplicity of their approach to problem solving. Both methods start by defining vectors with the length and angle of the links in the mechanism, so that they form a closed polygon or loop. The mechanism position problem (Fig. 3.1a) can be solved by defining the vector loop equation and finding the unknowns (Fig. 3.1b). The time derivative of these equations yields the velocity equation. Next, the time derivative of the velocity equation yields the acceleration equation.

In order to better understand the behavior of mechanisms, we recommend the reader to download the free demo version of WinMecC, a simulation program able to perform the analysis of planar articulated mechanisms with one degree of freedom. WinMecC allows the user to obtain the position, velocity, and acceleration of any point or link of a complex mechanism (with more than four links) either at a given position or along a full kinematic cycle. The program also displays an animated model of the mechanism being analyzed as well as the kinematic diagrams of any point or link.

3.1.1   Trigonometric Method

This method is suitable to calculate the position of links in mechanisms. However, it is not used to calculate velocities and accelerations since the equations obtained are too complex and they are difficult to derive.

Appendix A comprises the resolution of a four-bar, slider-crank and crank-shaft mechanism as an illustrative example of this method.

3.1.2   Raven's Method

Analysis by complex numbers starts from the vector loop equation and replaces vectors by complex numbers expressed in exponential form and then converted into trigonometric form. By separating the real and imaginary parts and defining a linear equation system, we can clear the unknowns. This method is called Raven's method. In many cases the main problem is the difficulty to solve the obtained equation systems, especially in mechanisms with more than four links. The equation solving problem is solved to some extent when the analysis of the mechanism is carried out with a computer, since, if necessary, numerical resolution methods can be used. Along this chapter, several mechanisms will be solved using Raven's method in order to foment comprehension.

3.1.2.1 Analysis of a Four-Bar Linkage. Position, Velocity and Angular Acceleration of Its Links

We start by writing the generic equations that will help us to calculate the position, velocity and acceleration of the links in a four-bar mechanism using Raven's method. We know the dimensions of the links (r_1, r_2, r_3 and r_4) as well as the position, velocity and acceleration of link 2 (θ_2, ω_2 and α_2), which is the input link. We aim to calculate the position of links 3 and 4 (θ_3 and θ_4), their speed (ω_3 and ω_4) and acceleration (α_3 and α_4). As explained before, we start by considering the loop equation of the mechanism. This equation can be written in different ways depending on the direction of the vectors. Figure 3.2 shows two possible vector loop equations of the four-bar linkage in Fig. 3.1a.

- Angular position of links:

  Consider the vector equation in Fig. 3.2b. Vectors can be expressed as complex numbers in complex exponential form. This way the vector loop equation needed to solve the kinematic problem by using Raven's method (Eq. 3.1) is:

$$r_1 e^{i\theta_1} = r_2 e^{i\theta_2} + r_3 e^{i\theta_3} + r_4 e^{i\theta_4} \qquad (3.1)$$

Each complex number in Eq. (3.1) can be expressed in trigonometric form (Eq. 3.2) allowing us to separate the real and imaginary parts.

$$r_1(\cos\theta_1 + i\sin\theta_1) = r_2(\cos\theta_2 + i\sin\theta_2) + r_3(\cos\theta_3 + i\sin\theta_3)$$
$$+ r_4(\cos\theta_4 + i\sin\theta_4) \qquad (3.2)$$

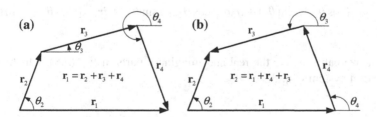

Fig. 3.2 a One of the vector loop equations that can be formulated for a four-bar linkage. **b** A different vector loop equation for the same four-bar linkage

This yields the system (Eq. 3.3) consisting of two algebraic equations and two unknowns:

$$\left.\begin{array}{l} r_1 \cos \theta_1 = r_2 \cos \theta_2 + r_3 \cos \theta_3 + r_4 \cos \theta_4 \\ r_1 \sin \theta_1 = r_2 \sin \theta_2 + r_3 \sin \theta_3 + r_4 \sin \theta_4 \end{array}\right\} \tag{3.3}$$

From (Eq. 3.3), angular position values (θ_3 and θ_4) can be determined in terms of r_1, r_2, r_3 and r_4. Freudenstein's method can be used to solve this system (see Appendix B for an explanation of the method). Mathematical software programs could be used to find the solution as well.

This problem can also be solved by means of the trigonometric method explained in Appendix A of this book.

• Angular velocity of links:

In order to obtain the equation for the velocity analysis, we time differentiate the vector loop (Eq. 3.4) in its complex exponential form:

$$r_1 e^{i\theta_1} = r_2 e^{i\theta_2} + r_3 e^{i\theta_3} + r_4 e^{i\theta_4} \tag{3.4}$$

Since $\theta_1 = 0$, the equation remains (Eq. 3.5):

$$r_1 = r_2 e^{i\theta_2} + r_3 e^{i\theta_3} + r_4 e^{i\theta_4} \tag{3.5}$$

After time differentiating (Eq. 3.5), we obtain (Eq. 3.6):

$$0 = ir_2 \frac{d\theta_2}{dt} e^{i\theta_2} + ir_3 \frac{d\theta_3}{dt} e^{i\theta_3} + ir_4 \frac{d\theta_4}{dt} e^{i\theta_4} \tag{3.6}$$

Where the time-derivative of the position angle of every link is its angular velocity and we obtain (Eq. 3.7):

$$0 = ir_2 \omega_2 e^{i\theta_2} + ir_3 \omega_3 e^{i\theta_3} + ir_4 \omega_4 e^{i\theta_4} \tag{3.7}$$

The complex numbers in Eq. (3.7) can be written in trigonometric form (Eq. 3.8):

$$0 = ir_2 \omega_2 (\cos \theta_2 + i \sin \theta_2) + ir_3 \omega_3 (\cos \theta_3 + i \sin \theta_3) + ir_4 \omega_4 (\cos \theta_4 + i \sin \theta_4) \tag{3.8}$$

Again, we can separate the real and imaginary parts in Eq. (3.8), which yields the equation system (Eq. 3.9):

$$0 = -r_2\omega_2 \sin\theta_2 - r_3\omega_3 \sin\theta_3 - r_4\omega_4 \sin\theta_4 \left.\vphantom{\begin{matrix}a\\b\end{matrix}}\right\}$$
$$0 = r_2\omega_2 \cos\theta_2 + r_3\omega_3 \cos\theta_3 + r_4\omega_4 \cos\theta_4$$

$$(3.9)$$

Hence, the angular velocity of links 3 and 4 (Eqs. 3.10 and 3.11) can be calculated from the angular velocity of the input link, ω_2, the angular position of the links, θ_2, θ_3 and θ_4, and the length of the links, r_1, r_2, r_3 and r_4.

$$\omega_3 = \frac{-r_2 \sin(\theta_2 - \theta_4)}{r_3 \sin(\theta_3 - \theta_4)} \omega_2 \qquad (3.10)$$

$$\omega_4 = \frac{r_2 \sin(\theta_2 - \theta_3)}{r_4 \sin(\theta_3 - \theta_4)} \omega_2 \qquad (3.11)$$

- Angular acceleration of links:

Finally, starting from the second time derivative of the vector loop equation and operating in a similar way as in the previous two sections, a third system of two algebraic equations is defined allowing us to find the values of α_3 and α_4 in terms of α_2, ω_2, ω_3, ω_4, θ_2, θ_3, θ_4, r_1, r_2, r_3 and r_4.

Differentiating Eq. (3.6) yields (Eq. 3.12):

$$\begin{aligned}
0 = &-r_2 \frac{d\theta_2}{dt}^2 e^{i\theta_2} + ir_2 \frac{d^2\theta_2}{dt^2} e^{i\theta_2} \\
&- r_3 \frac{d\theta_3}{dt}^2 e^{i\theta_3} + ir_3 \frac{d^2\theta_3}{dt^2} e^{i\theta_3} \\
&- r_4 \frac{d\theta_4}{dt}^2 e^{i\theta_4} + ir_4 \frac{d^2\theta_4}{dt^2} e^{i\theta_4}
\end{aligned} \qquad (3.12)$$

The time differential of the angular velocity of link i is the angular acceleration of the link, α_i, yielding (Eq. 3.13).

$$0 = -r_2\omega_2^2 e^{i\theta_2} + ir_2\alpha_2 e^{i\theta_2} - r_3\omega_3^2 e^{i\theta_3} + ir_3\alpha_3 e^{i\theta_3} - r_4\omega_4^2 e^{i\theta_4} + ir_4\alpha_4 e^{i\theta_4} \qquad (3.13)$$

The complex numbers in Eq. (3.13) can be written in trigonometric form (Eq. 3.14):

$$\begin{aligned}
0 = &(-r_2\omega_2^2 + ir_2\alpha_2)(\cos\theta_2 + i\sin\theta_2) \\
&+ (-r_3\omega_3^2 + ir_3\alpha_3)(\cos\theta_3 + i\sin\theta_3) \\
&+ (-r_4\omega_4^2 + ir_4\alpha_4)(\cos\theta_4 + i\sin\theta_4)
\end{aligned} \qquad (3.14)$$

By separating the real and imaginary parts, we obtain (Eq. 3.15):

$$0 = -r_2\omega_2^2 \cos\theta_2 - r_2\alpha_2 \sin\theta_2 - r_3\omega_3^2 \cos\theta - r_3\alpha_3 \sin\theta_3 - r_4\omega_4^2 \cos\theta_4 - r_4\alpha_4 \sin\theta_4 \left.\right\}$$
$$0 = -r_2\omega_2^2 \sin\theta_2 + r_2\alpha_2 \cos\theta - r_3\omega_3^2 \sin\theta_3 + r_3\alpha_3 \cos\theta - r_4\omega_4^2 \sin\theta_4 + r_4\alpha_4 \cos\theta_4 \right.$$

$$(3.15)$$

Thus, angular acceleration of links 3 and 4 (Eqs. 3.16 and 3.17) can be found by clearing α_3 and α_4 and plugging the expressions already obtained for ω_3 and ω_4 into their equations.

$$\alpha_3 = \frac{\omega_3}{\omega_2}\alpha_2 - \frac{r_2\omega_2^2 \cos(\theta_2 - \theta_4) + r_3\omega_3^2 \cos(\theta_3 - \theta_4) + r_4\omega_4^2}{r_3 \sin(\theta_3 - \theta_4)} \qquad (3.16)$$

$$\alpha_4 = \frac{\omega_4}{\omega_2}\alpha_2 - \frac{r_2\omega_2^2 \cos(\theta_2 - \theta_3) + r_4\omega_4^2 \cos(\theta_3 - \theta_4) + r_3\omega_3^2}{r_4 \sin(\theta_3 - \theta_4)} \qquad (3.17)$$

3.1.2.2 Analysis of a Four-Bar Linkage: Position, Velocity and Acceleration of Any of Its Points

To calculate the position, velocity and acceleration of a point in a mechanism, we previously calculate positions, velocities and accelerations of its links. Therefore, to calculate the kinematics of a point in a four-bar linkage, we need to know the dimensions of the links (r_1, r_2, r_3 and r_4), the positions of the bars ($\theta_1, \theta_2, \theta_3$ and θ_4), and their velocities (ω_2, ω_3 and ω_4) and acceleration (α_2, α_3 and α_4).

• Position of a point:

In order to define the position of the points, we define a Cartesian coordinate system. Figure 3.3a shows the chosen system, with its origin in O_2 and the X-axis defined by line O_2O_4.

The position of any point P_i in any link i is defined by a vector, $\mathbf{g}_i$, pointing from the origin of vector $\mathbf{r}_i$ (Fig. 3.3b). This vector has a constant magnitude, g_i, and an angle, θ_{g_i}, which varies along time. We can write this angle (Eq. 3.18) in terms of angle θ_i of the link i and fixed angle φ_i between vectors $\mathbf{g}_i$ and $\mathbf{r}_i$:

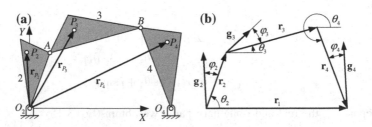

Fig. 3.3 a Four-bar linkage with points P_2, P_3 and P_4 on links 2, 3 and 4 respectively. **b** Closed vector polygon with φ_i angles and $\mathbf{g}_i$ vectors

$$\theta_g = \theta_i + \varphi_i \tag{3.18}$$

For point P_2 on link 2, its position is given by vector $\mathbf{r}_{P_2}$, which coincides with $\mathbf{g}_2$. This vector can be expressed in complex exponential form as (Eq. 3.19):

$$\mathbf{r}_{P_2} = \mathbf{g}_2 = g_2 e^{i(\theta_2 + \varphi_2)} \tag{3.19}$$

Separating the real and imaginary parts, we find the Cartesian coordinates:

$$\left. \begin{array}{l} x_{P_2} = g_2 \cos(\theta_2 + \varphi_2) \\ y_{P_2} = g_2 \sin(\theta_2 + \varphi_2) \end{array} \right\} \tag{3.20}$$

Position vector of point P_3 in link 3 (Eq. 3.21) is:

$$\mathbf{r}_{P_3} = \mathbf{r}_2 + \mathbf{g}_3 = r_2 e^{i\theta_2} + g_3 e^{i(\theta_3 + \varphi_3)} \tag{3.21}$$

And the Cartesian coordinates will be (Eq. 3.22):

$$\left. \begin{array}{l} x_{P_3} = r_2 \cos\theta_2 + g_3 \cos(\theta_3 + \varphi_3) \\ y_{P_3} = r_2 \sin\theta_2 + g_3 \sin(\theta_3 + \varphi_3) \end{array} \right\} \tag{3.22}$$

Finally, position vector of point P_4 of link 4 (Eq. 3.23) is:

$$\mathbf{r}_{P_4} = \mathbf{r}_1 + \mathbf{g}_4 = r_1 e^{i\theta_1} + g_4 e^{i(\theta_4 + 180° - \varphi_4)} = r_1 + g_4 e^{i(\theta_4 + 180° - \varphi_4)} \tag{3.23}$$

The Cartesian components of position vector $\mathbf{r}_{P_4}$ (Eq. 3.24) are:

$$\left. \begin{array}{l} x_{P_4} = r_1 + g_4 \cos(\theta_4 + 180° - \varphi_4) \\ y_{P_4} = g_4 \sin(\theta_4 + 180° - \varphi_4) \end{array} \right\} \tag{3.24}$$

- Linear velocity of a point:

The position vector expression of every point in every link can be differentiated in its complex exponential form in order to obtain the components of the linear velocity.

For point P_2 on link 2 its velocity vector will be (Eq. 3.25):

$$\mathbf{v}_{P_2} = \frac{d\mathbf{r}_{P_2}}{dt} = \frac{d}{dt}\left(g_2 e^{i(\theta_2 + \varphi_2)}\right) = i g_2 \frac{d(\theta_2 + \varphi_2)}{dt} e^{i(\theta_2 + \varphi_2)} = i g_2 \omega_2 e^{i(\theta_2 + \varphi_2)} \tag{3.25}$$

And the Cartesian components of the velocity (Eq. 3.26) can be obtained as:

$$\left. \begin{array}{l} v_{P_2 x} = -g_2 \omega_2 \sin(\theta_2 + \varphi_2) \\ v_{P_2 y} = g_2 \omega_2 \cos(\theta_2 + \varphi_2) \end{array} \right\} \tag{3.26}$$

Linear velocity of point P_3 on link 3 (Eq. 3.27) can be calculated in a similar way:

$$\mathbf{v}_{P_3} = \frac{d\mathbf{r}_{P_3}}{dt} = \frac{d}{dt}\left(r_2 e^{i\theta_2} + g_3 e^{i(\theta_3 + \varphi_3)}\right) = ir_2\omega_2 e^{i\theta_2} + ig_3\omega_3 e^{i(\theta_3 + \varphi_3)} \qquad (3.27)$$

The Cartesian components (Eq. 3.28) are:

$$\left.\begin{array}{l} v_{P_3x} = -r_2\omega_2 \sin\theta_2 - g_3\omega_3 \sin(\theta_3 + \varphi_3) \\ v_{P_3y} = r_2\omega_2 \cos\theta_2 + g_3\omega_3 \cos(\theta_3 + \varphi_3) \end{array}\right\} \qquad (3.28)$$

Finally, the linear velocity of point P_4 on link 4 (Eq. 3.29) is:

$$\mathbf{v}_{P_4} = \frac{d\mathbf{r}_{P_4}}{dt} = \frac{d}{dt}\left(r_1 + g_4 e^{i(\theta_4 + 180° - \varphi_4)}\right) = ig_4\omega_4 e^{i(\theta_4 + 180° - \varphi_4)} \qquad (3.29)$$

With the Cartesian components (Eq. 3.30):

$$\left.\begin{array}{l} v_{P_4x} = -g_4\omega_4 \sin(\theta_4 + 180° - \varphi_4) \\ v_{P_4y} = g_4\omega_4 \cos(\theta_4 + 180° - \varphi_4) \end{array}\right\} \qquad (3.30)$$

The magnitude and direction of the velocity vector of every point P_i is defined by Eqs. (3.31) and (3.32):

$$v_{P_i} = \sqrt{v_{P_ix}^2 + v_{P_iy}^2} \qquad (3.31)$$

$$\theta_{vP_i} = \arctan\frac{v_{P_iy}}{v_{P_ix}} \qquad (3.32)$$

As mentioned before, to find the linear velocity of any point on any link, it is necessary to know the angular velocity of the link. It is easy to verify that the magnitudes of the velocities of points P_2 and P_4 respectively have a value of $g_2\omega_2$ and $g_4\omega_4$ and their directions are $\theta_{vP_2} = \theta_2 + \varphi_2 \pm 90°$ and $\theta_{vP_4} = \theta_4 + 180° - \varphi_4 \pm 90°$. It can also be verified that Eq. (3.27) is the same as $\mathbf{v}_{P_3} = \mathbf{v}_A + \mathbf{v}_{P_3A}$, where $v_A = r_2\omega_2$ and $v_{P_3A} = g_3\omega_3$, and the direction of vectors $\mathbf{v}_A$ and $\mathbf{v}_{P_3A}$ are perpendicular to link 2 and $\overline{P_3A}$ respectively.

• Linear acceleration of a point:

Linear acceleration of a point is obtained by time differentiating its velocity vector. For point P_2 on link 2 the acceleration vector (Eq. 3.33) can be obtained as:

$$\mathbf{a}_{P_2} = \frac{d\mathbf{v}_{P_2}}{dt} = \frac{d}{dt}\left(ig_2\omega_2 e^{i(\theta_2+\varphi_2)}\right)$$

$$= i^2 g_2\omega_2 \frac{d(\theta_2+\varphi_2)}{dt} e^{i(\theta_2+\varphi_2)} + ig_2 \frac{d\omega_2}{dt} e^{i(\theta_2+\varphi_2)} \qquad (3.33)$$

$$= -g_2\omega_2^2 e^{i(\theta_2+\varphi_2)} + ig_2\alpha_2 e^{i(\theta_2+\varphi_2)}$$

The Cartesian components (Eq. 3.34) are:

$$\left. \begin{array}{l} a_{P_2x} = -g_2\omega_2^2\cos(\theta_2+\varphi_2) - g_2\alpha_2\sin(\theta_2+\varphi_2) \\ a_{P_2y} = -g_2\omega_2^2\sin(\theta_2+\varphi_2) + g_2\alpha_2\cos(\theta_2+\varphi_2) \end{array} \right\} \qquad (3.34)$$

The acceleration vector of point P_3 on link 3 (Eq. 3.35) is:

$$\mathbf{a}_{P_3} = \frac{d\mathbf{v}_{P_3}}{dt} = \frac{d}{dt}\left(ir_2\omega_2 e^{i\theta_2} + ig_3\omega_3 e^{i(\theta_3+\varphi_3)}\right)$$

$$= i^2 r_2\omega_2^2 e^{i\theta_2} + ir_2 \frac{d\omega_2}{dt} e^{i\theta_2} + i^2 g_3\omega_3^2 e^{i(\theta_3+\varphi_3)} + ig_3 \frac{d\omega_2}{dt} e^{i(\theta_3+\varphi_3)} \qquad (3.35)$$

$$= -r_2\omega_2^2 e^{i\theta_2} + ir_2\alpha_2 e^{i\theta_2} - g_3\omega_3^2 e^{i(\theta_3+\varphi_3)} + ig_3\alpha_3 e^{i(\theta_3+\varphi_3)}$$

And the components of this vector (Eq. 3.36) are:

$$\left. \begin{array}{l} a_{P_3x} = -r_2\omega_2^2\cos\theta_2 - r_2\alpha_2\sin\theta_2 - g_3\omega_3^2\cos(\theta_3+\varphi_3) - g_3\alpha_3\sin(\theta_3+\varphi_3) \\ a_{P_3y} = -r_2\omega_2^2\sin\theta_2 + r_2\alpha_2\cos\theta_2 - g_3\omega_3^2\sin(\theta_3+\varphi_3) + g_3\alpha_3\cos(\theta_3+\varphi_3) \end{array} \right\}$$

$$(3.36)$$

Finally the acceleration vector of point P_4 on link 4 (Eq. 3.37) is:

$$\mathbf{a}_{P_4} = \frac{d\mathbf{v}_{P_4}}{dt} = \frac{d}{dt}\left(ig_4\omega_4 e^{i(\theta_4+180°-\varphi_4)}\right)$$

$$= i^2 g_4\omega_4 \frac{d(\theta_4'+\varphi_4)}{dt} e^{i(\theta_4+180°-\varphi_4)} + ig_4 \frac{d\omega_4}{dt} e^{i(\theta_4+180°-\varphi_4)} \qquad (3.37)$$

$$= -g_4\omega_4^2 e^{i(\theta_4+180°-\varphi_4)} + ig_4\alpha_4 e^{i(\theta_4+180°-\varphi_4)}$$

And the Cartesian components of this vector (Eq. 3.38) are:

$$\left. \begin{array}{l} a_{P_4x} = -g_4\omega_4^2\cos(\theta_4+180°-\varphi_4) - g_4\alpha_4\sin(\theta_4+180°-\varphi_4) \\ a_{P_4y} = -g_4\omega_4^2\sin(\theta_4+180°-\varphi_4) + g_4\alpha_4\cos(\theta_4+180°-\varphi_4) \end{array} \right\} \qquad (3.38)$$

The magnitude and direction of any point P_i on any link i (Eqs. 3.39 and 3.40) can be obtained as:

$$a_{P_i} = \sqrt{a_{P_ix}^2 + a_{P_iy}^2} \tag{3.39}$$

$$\theta_{a_{P_i}} = \arctan \frac{a_{P_iy}}{a_{P_ix}} \tag{3.40}$$

3.1.2.3 Analysis of a Crank-Shaft Mechanism. Position, Velocity and Acceleration of Its Links

Consider the crank-shaft mechanism with offset in Fig. 3.4a in which r_2 and r_3 are the lengths of links 2 and 3 respectively. The measure of the offset is given by vector $\mathbf{r}_1$ with magnitude $r_1 = y$ and angle $\theta_1 = 270°$ (Fig. 3.4b). The angle of the path is $\theta_4 = 0°$ and magnitude r_4 gives us the position of link 4 along this path. The angles θ_2 and θ_3 define the angular position of links 2 and 3 considering the positive direction counterclockwise.

Values r_1, r_2, r_3, θ_1 and θ_4 are known as they are constant. Angle θ_2 is variable but as it is the input, it is also known. Values r_4 and θ_3 are variables and unknowns.

- Position of the links:

The vector loop equation of the mechanism (Eq. 3.41) in Fig. 3.4a is:

$$\mathbf{r}_1 + \mathbf{r}_4 = \mathbf{r}_2 + \mathbf{r}_3 \tag{3.41}$$

Expressing the vectors in their complex exponential form (Eq. 3.42):

$$r_1 e^{i\theta_1} + r_4 e^{i\theta_4} = r_2 e^{i\theta_2} + r_3 e^{i\theta_3} \tag{3.42}$$

And using the trigonometric form (Eq. 3.43):

$$r_1(\cos\theta_1 + i\sin\theta_1) + r_4(\cos\theta_4 + i\sin\theta_4)$$
$$= r_2(\cos\theta_2 + i\sin\theta_2) + + r_3(\cos\theta_3 + i\sin\theta_3) \tag{3.43}$$

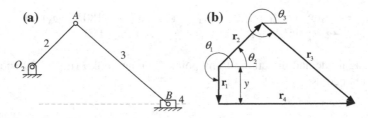

Fig. 3.4 a Kinematic skeleton of a crank-shaft mechanism with offset y. **b** Closed vector polygon with the position unknowns, θ_3 and r_4 (magnitude of $\mathbf{r}_4$)

The real and imaginary parts can be separated (Eq. 3.44) and taking into account that $\theta_1 = 270°$ and $\theta_4 = 0°$, we obtain:

$$\left.\begin{array}{l} r_4 = r_2 \cos \theta_2 + r_3 \cos \theta_3 \\ -r_1 = r_2 \sin \theta_2 + r_3 \sin \theta_3 \end{array}\right\} \tag{3.44}$$

The values of r_4 and θ_3 (Eqs. 3.45 and 3.46) in terms of θ_2 (angular position of the input link) and r_1, r_2 and r_3 are obtained by solving equation system (Eq. 3.44):

$$r_4 = r_2 \cos \theta_2 + r_3 \cos \theta_3 \tag{3.45}$$

$$\theta_3 = \arcsin \frac{-(r_1 + r_2 \sin \theta_2)}{r_3} \tag{3.46}$$

- Link velocity:

By time differentiating Eq. (3.42), we find (Eq. 3.47):

$$\frac{dr_4}{dt} e^{i\theta_4} = ir_2 \frac{d\theta_2}{dt} e^{i\theta_2} + ir_3 \frac{d\theta_3}{dt} e^{i\theta_3} \tag{3.47}$$

The time differential of r_4 is the linear velocity of link 4, v_4, yielding (Eq. 3.48):

$$v_4 e^{i\theta_4} = ir_2 \omega_2 e^{i\theta_2} + ir_3 \omega_3 e^{i\theta_3} \tag{3.48}$$

Using the trigonometric form (Eq. 3.49):

$$v_4(\cos \theta_4 + i \sin \theta_4) = ir_2 \omega_2 (\cos \theta_2 + i \sin \theta_2) + ir_3 \omega_3 (\cos \theta_3 + i \sin \theta_3) \tag{3.49}$$

Separating real and imaginary parts (Eq. 3.50) and taking into account that $\theta_4 = 0°$, we obtain:

$$\left.\begin{array}{l} v_4 = -r_2 \omega_2 \sin \theta_2 - r_3 \omega_3 \sin \theta_3 \\ 0 = r_2 \omega_2 \cos \theta_2 + r_3 \omega_3 \cos \theta_3 \end{array}\right\} \tag{3.50}$$

Thus, we can find angular velocity ω_3 of link 3 (Eq. 3.51), and velocity v_4 of link 4 (Eq. 3.52) from the angular velocity of the input link and the angular position and length of links 2 and 3 (ω_2, θ_2, θ_3, r_2 and r_3).

$$v_4 = -r_2 \omega_2 \sin \theta_2 - r_3 \omega_3 \sin \theta_3 \tag{3.51}$$

$$\omega_3 = \frac{-r_2 \omega_2 \cos \theta_2}{r_3 \cos \theta_3} \tag{3.52}$$

- Link acceleration:

Finally, we differentiate (Eq. 3.47) and find (Eq. 3.53). Using the trigonometric form (Eq. 3.55) and separating the real and imaginary parts (Eq. 3.56) in a similar way as in the previous sections, we find a third system of two equations and two unknowns, α_3 and a_4, in terms of α_2, ω_2, ω_3, θ_2, θ_3, r_2 and r_3:

$$a_4 e^{i\theta_4} = i^2 r_2 \omega_2^2 e^{i\theta_2} + i r_2 \frac{d\omega_2}{dt} e^{i\theta_2} + i^2 r_3 \omega_3^2 e^{i\theta_3} + i r_3 \frac{d\omega_3}{dt} e^{i\theta_3} \qquad (3.53)$$

Knowing that $i^2 = -1$, we can write (Eq. 3.54):

$$a_4 e^{i\theta_4} = -r_2 \omega_2^2 e^{i\theta_2} + i r_2 \alpha_2 e^{i\theta_2} - r_3 \omega_3^2 e^{i\theta_3} + i r_3 \alpha_3 e^{i\theta_3} \qquad (3.54)$$

$$\begin{aligned} a_4(\cos\theta_4 + i\sin\theta_4) &= -r_2\omega_2^2(\cos\theta_2 + i\sin\theta_2) + ir_2\alpha_2(\cos\theta_2 + i\sin\theta_2) \\ &\quad - r_3\omega_3^2(\cos\theta_3 + i\sin\theta_3) + ir_3\alpha_3(\cos\theta_3 + i\sin\theta_3) \end{aligned} \qquad (3.55)$$

$$\left. \begin{aligned} a_4 &= -r_2\omega_2^2\cos\theta_2 - r_2\alpha_2\sin\theta_2 - r_3\omega_3^2\cos\theta_3 - r_3\alpha_3\sin\theta_3 \\ 0 &= -r_2\omega_2^2\sin\theta_2 + r_2\alpha_2\cos\theta_2 - r_3\omega_3^2\cos\theta_3 - r_3\alpha_3\sin\theta_3 \end{aligned} \right\} \qquad (3.56)$$

By solving this system we find α_3 (Eq. 3.57) and a_4 (Eq. 3.58):

$$\alpha_3 = \frac{r_2\omega_2^2\sin\theta_2 - r_2\alpha_2\cos\theta_2 + r_3\omega_3^2\cos\theta_3}{r_3\sin\theta_3} \qquad (3.57)$$

$$a_4 = -r_2\omega_2^2\cos\theta_2 - r_2\alpha_2\sin\theta_2 - r_3\omega_3^2\cos\theta_3 - r_3\alpha_3\sin\theta_3 \qquad (3.58)$$

3.1.2.4 Kinematic Analysis of a Crank-Shaft Linkage. Position, Velocity and Acceleration of Any Point on a Link

In this case, we will determine the position, velocity and linear acceleration of any point on links 2 and 3 in Fig. 3.5a, since link 4 moves with pure translational motion and, thus, all its points have the same velocity and acceleration.

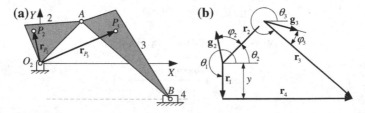

Fig. 3.5 **a** Crank-shaft linkage with points P_2 and P_3 on links 2 and 3 respectively. **b** Closed vector polygon with φ_i angles and $\mathbf{g}_i$ vectors

- Linear position of a point:

We define the position of any point P_i on any link i of the linkage in terms of its constant distance g_i and angle φ_i between vector and link i.

For point P_2 on link 2, position vector coincides with vector $\mathbf{g}_2$ (Eq. 3.59). We can write $\mathbf{r}_{P_2}$ in its complex exponential form:

$$\mathbf{r}_{P_2} = \mathbf{g}_2 = g_2 e^{i(\theta_2 + \varphi_2)} \tag{3.59}$$

Relative to the Cartesian system shown in Fig. 3.5a, the components of the position vector (Eq. 3.60) are:

$$\left.\begin{array}{l} x_{P_2} = g_2 \cos(\theta_2 + \varphi_2) \\ y_{P_2} = g_2 \sin(\theta_2 + \varphi_2) \end{array}\right\} \tag{3.60}$$

For a point P_3 on link 3, position vector is $\mathbf{h}_3$ (Eq. 3.61):

$$\mathbf{r}_{P_2} = \mathbf{r}_2 + \mathbf{g}_3 = r_2 e^{i\theta_2} + g_3 e^{i(\theta_3 + \varphi_3)} \tag{3.61}$$

The Cartesian components of position vector (Eq. 3.62) are:

$$\left.\begin{array}{l} x_{P_3} = r_2 \cos\theta_2 + g_3 \cos(\theta_3 + \varphi_3) \\ y_{P_3} = r_2 \sin\theta_2 + g_3 \sin(\theta_3 + \varphi_3) \end{array}\right\} \tag{3.62}$$

- Linear velocity of a point:

The position vector of any link can be differentiated in its complex exponential form, yielding the linear velocity vector components of any point.

Linear velocity of point P_2 on link 2 (Eq. 3.63) is:

$$\mathbf{v}_{P_2} = \frac{d\mathbf{r}_{P_2}}{dt} = ig_2 \frac{d(\theta_2 + \varphi_2)}{dt} e^{i(\theta_2 + \varphi_2)} - ig_2 \omega_2 e^{i(\theta_2 + \varphi_2)} \tag{3.63}$$

The Cartesian components of this vector (Eq. 3.64) are:

$$\left.\begin{array}{l} v_{P_2 x} = -g_2 \omega_2 \sin(\theta_2 + \varphi_2) \\ v_{P_2 y} = g_2 \omega_2 \cos(\theta_2 + \varphi_2) \end{array}\right\} \tag{3.64}$$

The velocity of a point on link 3 (Eq. 3.65) is given by:

$$\mathbf{v}_{P_3} = \frac{d\mathbf{r}_{P_3}}{dt} = \frac{d}{dt}\left(r_2 e^{i\theta_2} + g_3 e^{i(\theta_3 + \varphi_3)}\right) = ir_2 \omega_2 e^{i\theta_2} + ig_3 \omega_3 e^{i(\theta_3 + \varphi_3)} \tag{3.65}$$

And its Cartesian components (Eq. 3.66) are:

$$\left.\begin{array}{l} v_{P_3x} = -r_2\omega_2 \sin\theta_2 - g_3\omega_3 \sin(\theta_3 + \varphi_3) \\ v_{P_3y} = r_2\omega_2 \cos\theta_2 + g_3\omega_3 \cos(\theta_3 + \varphi_3) \end{array}\right\} \qquad (3.66)$$

- Acceleration of a point:

By time-differentiating the velocity vector of any point on any link, we obtain the linear acceleration of any point.

The acceleration of point P_2 on link 2 (Eq. 3.67) is:

$$\begin{aligned} \mathbf{a}_{P_2} &= \frac{d\mathbf{v}_{P_2}}{dt} = \frac{d}{dt}\left(ig_2\omega_2 e^{i(\theta_2 + \varphi_2)}\right) \\ &= i^2 g_2\omega_2 \frac{d(\theta_2 + \varphi_2)}{dt} e^{i(\theta_2 + \varphi_2)} + ig_2 \frac{d\omega_2}{dt} e^{i(\theta_2 + \varphi_2)} \qquad (3.67) \\ &= -g_2\omega_2^2 e^{i(\theta_2 + \varphi_2)} + ig_2\alpha_2 e^{i(\theta_2 + \varphi_2)} \end{aligned}$$

With the Cartesian components (Eq. 3.68):

$$\left.\begin{array}{l} a_{P_2x} = -g_2\omega_2^2 \cos(\theta_2 + \varphi_2) - g_2\alpha_2 \sin(\theta_2 + \varphi_2) \\ a_{P_2y} = -g_2\omega_2^2 \sin(\theta_2 + \varphi_2) + g_2\alpha_2 \cos(\theta_2 + \varphi_2) \end{array}\right\} \qquad (3.68)$$

A similar reasoning can be used in order to find the acceleration of a point on link 3 (Eq. 3.69). By differentiating Eq. (3.65) we obtain:

$$\begin{aligned} \mathbf{a}_{P_3} &= \frac{d\mathbf{v}_{P_3}}{dt} = \frac{d}{dt}\left(ir_2\omega_2 e^{i\theta_2} + ig_3\omega_3 e^{i(\theta_3 + \varphi_3)}\right) \\ &= i^2 r_2\omega_2^2 e^{i\theta_2} + ir_2 \frac{d\omega_2}{dt} e^{i\theta_2} + i^2 g_3\omega_3^2 e^{i(\theta_3 + \varphi_3)} + ig_3 \frac{d\omega_2}{dt} e^{i(\theta_3 + \varphi_3)} \qquad (3.69) \\ &= -r_2\omega_2^2 e^{i\theta_2} + ir_2\alpha_2 e^{i\theta_2} - g_3\omega_3^2 e^{i(\theta_3 + \varphi_3)} + ig_3\alpha_3 e^{i(\theta_3 + \varphi_3)} \end{aligned}$$

Its Cartesian components (Eq. 3.70) are:

$$\left.\begin{array}{l} a_{P_3x} = -r_2\omega_2^2 \cos\theta_2 - r_2\alpha_2 \sin\theta_2 - g_3\omega_3^2 \cos(\theta_3 + \varphi_3) - g_3\alpha_3 \sin(\theta_3 + \varphi_3) \\ a_{P_3y} = -r_2\omega_2^2 \sin\theta_2 + r_2\alpha_2 \cos\theta_2 - g_3\omega_3^2 \sin(\theta_3 + \varphi_3) + g_3\alpha_3 \cos(\theta_3 + \varphi_3) \end{array}\right\}$$
$$(3.70)$$

The magnitude and direction of the acceleration vector of any point P_i on any link i will be given by the expressions (Eqs. 3.71 and 3.72):

$$a_{P_i} = \sqrt{a_{P_ix}^2 + a_{P_iy}^2} \qquad (3.71)$$

$$\theta_{aP_i} = \arctan\frac{a_{P_iy}}{a_{P_ix}} \qquad (3.72)$$

3.1.2.5 Kinematic Analysis of a Slider Linkage. Position, Velocity and Acceleration of Its Links

We will carry out the kinematic analysis of the slider linkage in Fig. 3.6a, where the motion of link 4 is an eccentric rotation about O_4 with offset c. In this case, r_1 and r_2 are constant, θ_2 is the known angular position of the input link and we are asked to find r_4 and θ_4.

- Angular position of the links

   The vector loop equation (Eq. 3.73) is defined as (Fig. 3.6b):

$$\mathbf{r}_1 = \mathbf{r}_2 + \mathbf{r}_4 + \mathbf{r}_c \qquad (3.73)$$

   Vectors can be expressed in their exponential form (Eq. 3.74):

$$r_1 e^{i\theta_1} = r_2 e^{i\theta_2} + r_4 e^{i\theta_4} + r_c e^{i\theta_c} \qquad (3.74)$$

   Substituting $\theta_1 = 0°$ and $\theta_c = \theta_4 + 270°$ in Eq. (3.71), we obtain (Eq. 3.75):

$$r_1 = r_2 e^{i\theta_2} + r_4 e^{i\theta_4} + r_c e^{i(\theta_4 + 270°)} \qquad (3.75)$$

   Using the trigonometric form (Eq. 3.76):

$$r_1 = r_2(\cos\theta_2 + i\sin\theta_2) + r_4(\cos\theta_4 + i\sin\theta_4) + r_c(\sin\theta_4 - i\cos\theta_4) \qquad (3.76)$$

   Real and imaginary parts can be separated giving way to a system of two equations and two unknowns (θ_4 and r_4) (Eq. 3.77). By solving this system, we obtain the values of the unknowns in terms of r_1, r_2, r_c and θ_2:

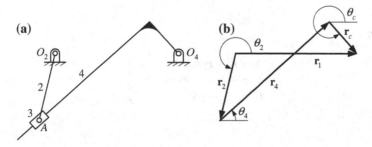

Fig. 3.6 a Slider linkage with offset r_c and eccentric rotation about O_4. **b** Closed vector polygon with the position unknowns, θ_4 and r_4 (magnitude of $\mathbf{r}_4$)

$$\left. \begin{array}{l} r_1 = r_2\cos\theta_2 + r_4\cos\theta_4 + r_c sin\theta_4 \\ 0 = r_2 \sin\theta_2 + r_4 \sin\theta_4 - r_c \cos\theta_4 \end{array} \right\} \qquad (3.77)$$

Unknowns θ_4 and r_4 can be obtained as follows:

- Clear r_4 in terms of θ_4 (Eq. 3.78) in the imaginary term equation:

$$r_4 = \frac{r_c \cos\theta_4 - r_2 \sin\theta_2}{\sin\theta_4} \qquad (3.78)$$

- Plug r_4 into the real term equation (Eq. 3.79):

$$r_1 = r_2 \cos\theta_2 + \frac{r_c \cos\theta_4 - r_2 \sin\theta_2}{\sin\theta_4}\cos\theta_4 + r_c \sin\theta_4 \qquad (3.79)$$

- Compute (Eq. 3.80):

$$\begin{aligned} (r_1 - r_2 \cos\theta_2)\sin\theta_4 &= r_c \cos^2\theta_4 - r_2 \sin\theta_2 \cos\theta_4 + r_c \sin^2\theta_4 \\ &= r_c - r_2 \sin\theta_2 \cos\theta_4 \end{aligned} \qquad (3.80)$$

If we substitute $\sin\theta_4$ and $\cos\theta_4$ for their corresponding half angle identities, (Eq. 3.81) is obtained:

$$2(r_1 - r_2 \cos\theta_2)\frac{\tan\theta_4/2}{1+\tan^2\theta_4/2} = r_c - r_2 \sin\theta_2 \frac{1-\tan^2\theta_4/2}{1+\tan^2\theta_4/2} \qquad (3.81)$$

- Next, rearrange, define parameters A, B and C (Eq. 3.83) and we find (Eq. 3.82):

$$0 = A \tan^2\frac{\theta_4}{2} + B \tan\frac{\theta_4}{2} + C \qquad (3.82)$$

where:

$$\left. \begin{array}{l} A = r_c + r_2 \sin\theta_2 \\ B = -2(r_1 - r_2 \cos\theta_2) \\ C = r_c - r_2 \sin\theta_2 \end{array} \right\} \qquad (3.83)$$

Thus, we obtain unknown θ_4 (Eq. 3.84):

$$\theta_4^{I,II} = 2 \arctan \frac{-B \pm \sqrt{B^2 - 4AC}}{2A} \tag{3.84}$$

Every value of the motor link position angle, θ_2, gives way to two values of r_4 and θ_4, two possible configurations of the linkage with offset r_c above or below O_4.

- Angular velocity of the links:

The vector loop (Eq. 3.74) can be time-differentiated in its complex exponential form (Eq. 3.85):

$$0 = ir_2\omega_2 e^{i\theta_2} + ir_4\omega_4 e^{i\theta_4} + v_{P_2P_4} e^{i\theta_4} + ir_c\omega_4 e^{i(\theta_4 + 270°)} \tag{3.85}$$

It is worth emphasizing that, in this case, r_4 is not constant over time. Its derivative, dr_4/dt, yields the velocity of the slider over link 4, in other words, $v_{P_2P_4}$. By using trigonometric form (Eq. 3.86):

$$\begin{aligned} 0 = {} & ir_2\omega_2(\cos\theta_2 + i\sin\theta_2) + v_{P_2P_4}(\cos\theta_4 + i\sin\theta_4) \\ & + ir_4\omega_4(\cos\theta_4 + i\sin\theta_4) + ir_c\omega_4(\sin\theta_4 - i\cos\theta_4) \end{aligned} \tag{3.86}$$

Separating the real and imaginary parts of Eq. (3.86), we obtain a system of two equations with two unknowns, $v_{P_2P_4}$ and ω_4 (Eq. 3.87).

$$\left. \begin{aligned} 0 &= -r_2\omega_2 \sin\theta_2 + v_{P_2P_4} \cos\theta_4 - r_4\omega_4 \sin\theta_4 + r_c\omega_4 \cos\theta_4 \\ 0 &= r_2\omega_2 \cos\theta_2 + v_{P_2P_4} \sin\theta_4 + r_4\omega_4 \cos\theta_4 + r_c\omega_4 \sin\theta_4 \end{aligned} \right\} \tag{3.87}$$

The value of ω_4 (Eq. 3.88) is obtained by clearing $v_{P_2P_4}$ from the second equation and plugging the result into the first one.

$$\omega_4 = \frac{-r_2}{r_4}\omega_2 \cos(\theta_2 - \theta_4) \tag{3.88}$$

Thus:

$$v_{P_2P_4} = r_2\omega_2 \sin(\theta_2 - \theta_4) - r_c\omega_4 \tag{3.89}$$

It is necessary to point out that if offset r_c is placed above the line connecting centers O_2 and O_4, then $\theta_c = \theta_4 + 90°$ and Eq. (3.89) becomes (Eq. 3.90):

$$v_{P_2P_4} = r_2\omega_2 \sin(\theta_2 - \theta_4) + r_c\omega_4 \tag{3.90}$$

• Angular acceleration of the links

Proceeding in a similar way as in the previous section, we time-differentiate the vector-loop (Eq. 3.85) and find (Eq. 3.91).

$$
\begin{aligned}
0 = {}& -r_2\omega_2^2 e^{i\theta_2} + ir_2\alpha_2 e^{i\theta_2} + a_{P_2P_4}e^{i\theta_4} - r_4\omega_4^2 e^{i\theta_4} + ir_4\alpha_4 e^{i\theta_4} + i2\omega_4 v_{P_2P_4}e^{i\theta_4} \\
& - r_c\omega_4^2 e^{i(\theta_4 + 270^\circ)} + ir_c\alpha_4 e^{i(\theta_4 + 270^\circ)}
\end{aligned}
\tag{3.91}
$$

Using trigonometric form, we find (Eq. 3.92):

$$
\begin{aligned}
0 = {}& -r_2\omega_2^2(\cos\theta_2 + i\sin\theta_2) + ir_2\alpha_2(\cos\theta_2 + i\sin\theta_2) \\
& + a_{P_2P_4}(\cos\theta_4 + i\sin\theta_4) - r_4\omega_4^2(\cos\theta_4 + i\sin\theta_4) \\
& + ir_4\alpha_4(\cos\theta_4 + i\sin\theta_4) + i2\omega_4 v_{P_2P_4}(\cos\theta_4 + i\sin\theta_4) \\
& - r_c\omega_4^2(\sin\theta_4 - i\cos\theta_4) + ir_c\alpha_4(\sin\theta_4 - i\cos\theta_4)
\end{aligned}
\tag{3.92}
$$

Real and imaginary parts can be separated. Thus, obtain a system of two equations and two unknowns, $a_{P_2P_4}$ and α_4 (Eq. 3.93):

$$
\left.
\begin{aligned}
0 = {}& -r_2\omega_2^2\cos\theta_2 - r_2\alpha_2\sin\theta_2 + a_{P_2P_4}\cos\theta_4 - r_4\omega_4^2\cos\theta_4 \\
& -r_4\alpha_4\sin\theta_4 - 2\omega_4 v_{P_2P_4}\sin\theta_4 - r_c\omega_4^2\sin\theta_4 + r_c\alpha_4\cos\theta_4 \\
0 = {}& -r_2\omega_2^2\sin\theta_2 + r_2\alpha_2\cos\theta_2 + a_{P_2P_4}\sin\theta_4 - r_4\omega_4^2\sin\theta_4 \\
& + r_4\alpha_4\cos\theta_4 + 2\omega_4 v_{P_2P_4}\cos\theta_4 + r_c\omega_4^2\cos\theta_4 + r_c\alpha_4\sin\theta_4
\end{aligned}
\right\}
\tag{3.93}
$$

Clearing $a_{P_2P_4}$ from one of the equations (Eq. 3.94) and plugging its expression into the other one, we find the value of α_4 (Eq. 3.95).

$$
a_{P_2P_4} = r_2\omega_2^2\cos(\theta_2 - \theta_4) + r_2\alpha_2\sin(\theta_2 - \theta_4) + r_4\omega_4^2 - r_c\alpha_4
\tag{3.94}
$$

$$
\alpha_4 = \frac{r_2\omega_2^2\sin(\theta_2 - \theta_4) - r_2\alpha_2\sin(\theta_2 - \theta_4) - 2\omega_4 v_{P_2P_4} - r_c\omega_4^2}{r_4}
\tag{3.95}
$$

3.1.2.6 Kinematic Analysis of a Slider Linkage. Position, Velocity and Acceleration of Any Point

In this section we will develop the kinematic analysis of the mechanism in Fig. 3.7a using the Raven method.

• Linear position of a point:

The position of point P_i on a link is given by constant distance g_i and angle φ_i (Fig. 3.7b). The position vector of point P_2 on link 2 (Eq. 3.96), $\mathbf{r}_{P_2}$, coincides with $\mathbf{g}_2$.

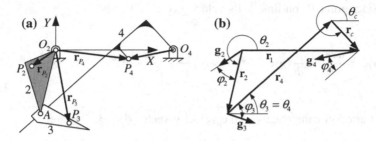

Fig. 3.7 **a** Slider linkage with points P_2, P_3 and P_4 on links 2, 3 and 4 respectively. **b** Closed vector polygon with φ_i angles and $\mathbf{g}_i$ vectors

$$\mathbf{r}_{P_2} = \mathbf{g}_2 = g_2 e^{i(\theta_2 - \varphi_2)} \tag{3.96}$$

The components of this vector (Eq. 3.97) are:

$$\left.\begin{array}{l} x_{P_2} = g_2 \cos(\theta_2 - \varphi_2) \\ y_{P_2} = g_2 \sin(\theta_2 - \varphi_2) \end{array}\right\} \tag{3.97}$$

Position vector $\mathbf{r}_{P_3}$ of point P_3 on link 3 in terms of $\mathbf{r}_2$ and $\mathbf{g}_3$ (Eq. 3.98) is:

$$\mathbf{r}_{P_3} = \mathbf{r}_2 + \mathbf{g}_3 = r_2 e^{i\theta_2} + g_3 e^{i(\theta_3 - \varphi_3)} \tag{3.98}$$

The angular position of link 3 coincides with link 4, hence $\theta_3 = \theta_4$.
The Cartesian components of position $\mathbf{r}_{P_3}$ vector (Eq. 3.99) are:

$$\left.\begin{array}{l} x_{P_3} = r_2 \cos \theta_2 + g_3 \cos(\theta_3 - \varphi_3) \\ y_{P_3} = r_2 \sin \theta_2 + g_3 \sin(\theta_3 - \varphi_3) \end{array}\right\} \tag{3.99}$$

Position vector of point P_4 on link 4 (Eq. 3.100) is:

$$\mathbf{r}_{P_4} = \mathbf{r}_1 + \mathbf{g}_4 = r_1 e^{i\theta_1} + g_4 e^{i(\theta_4 + 180° - \varphi_4)} = r_1 + g_4 e^{i(\theta_4 + 180° - \varphi_4)} \tag{3.100}$$

The Cartesian components of the vector (Eq. 3.101) are:

$$\left.\begin{array}{l} x_{P_4} = r_1 + g_4 \cos(\theta_4 + 180° - \varphi_4) \\ y_{P_4} = g_4 \sin(\theta_4 + 180° - \varphi_4) \end{array}\right\} \tag{3.101}$$

- Linear velocity of a point:

The position vector expression of a point can be differentiated in its exponential form in order to find the linear velocity vector components of such a point.

- Consider point P_2 on link 2. Its velocity (Eq. 3.102) is:

$$\mathbf{v}_{P_2} = \frac{d\mathbf{r}_{P_2}}{dt} = \frac{d}{dt}\left(g_2 e^{i(\theta_2 - \varphi_2)}\right) = ig_2 \frac{d(\theta_2 - \varphi_2)}{dt} e^{i(\theta_2 - \varphi_2)} = ig_2\omega_2 e^{i(\theta_2 - \varphi_2)}$$

$$(3.102)$$

The Cartesian components of the velocity vector (Eq. 3.103) are:

$$\left.\begin{aligned} v_{P_2 x} &= -g_2\omega_2 \sin(\theta_2 - \varphi_2) \\ v_{P_2 y} &= g_2\omega_2 \cos(\theta_2 - \varphi_2) \end{aligned}\right\}$$

$$(3.103)$$

- For point P_3 on ink 3 the velocity vector (Eq. 3.104) is:

$$\mathbf{v}_{P_3} = \frac{d\mathbf{r}_{P_3}}{dt} = \frac{d}{dt}\left(r_2 e^{i\theta_2} + g_3 e^{i(\theta_3 - \varphi_3)}\right) = ir_2\omega_2 e^{i\theta_2} + ig_3\omega_3 e^{i(\theta_3 - \varphi_3)} \quad (3.104)$$

With the Cartesian components (Eq. 3.105):

$$\left.\begin{aligned} v_{P_3 x} &= -r_2\omega_2 \sin\theta_2 - g_3\omega_3 \sin(\theta_3 - \varphi_3) \\ v_{P_3 y} &= r_2\omega_2 \cos\theta_2 + g_3\omega_3 \cos(\theta_3 - \varphi_3) \end{aligned}\right\}$$

$$(3.105)$$

Again, note that the angular velocity of link 3, ω_3, coincides with the one of link 4, ω_4, already calculated.
- Finally, the velocity of P_4 on link 4 (Eq. 3.106) is:

$$\mathbf{v}_{P_4} = ig_4\omega_4 e^{i(\theta_4 + 180° - \varphi_4)}$$

$$(3.106)$$

And its Cartesian components (Eq. 3.107) are:

$$\left.\begin{aligned} v_{P_4 x} &= -g_4\omega_4 \sin(\theta_4 + 180° - \varphi_4) \\ v_{P_4 y} &= g_4\omega_4 \cos(\theta_4 + 180° - \varphi_4) \end{aligned}\right\}$$

$$(3.107)$$

- Linear acceleration of a point:
  Starting from the velocity vector expression and computing it the same way as above, we can find the linear acceleration of any point on any link.

  - Consider point P_2 on link 2. Its linear acceleration vector (Eq. 3.108) is:

$$\mathbf{a}_{P_2} = \frac{d\mathbf{v}_{P_2}}{dt} = \frac{d}{dt}\left(ig_2\omega_2 e^{i(\theta_2-\varphi_2)}\right)$$

$$= i^2 g_2\omega_2 \frac{d(\theta_2+\varphi_2)}{dt}e^{i(\theta_2-\varphi_2)} + ig_2\frac{d\omega_2}{dt}e^{i(\theta_2-\varphi_2)} \qquad (3.108)$$

$$= -g_2\omega_2^2 e^{i(\theta_2-\varphi_2)} + ig_2\alpha_2 e^{i(\theta_2-\varphi_2)}$$

Its Cartesian components (Eq. 3.109) are:

$$\left.\begin{aligned} a_{P_2x} &= -g_2\omega_2^2\cos(\theta_2-\varphi_2) - g_2\alpha_2\sin(\theta_2-\varphi_2)\\ a_{P_2y} &= -g_2\omega_2^2\sin(\theta_2-\varphi_2) + g_2\alpha_2\cos(\theta_2-\varphi_2) \end{aligned}\right\} \qquad (3.109)$$

– Acceleration $\mathbf{a}_{P_3}$ of a point on link 3 (Eq. 3.110) is:

$$\mathbf{a}_{P_3} = \frac{d\mathbf{v}_3}{dt} = \frac{d}{dt}\left(ir_2\omega_2 e^{i\theta_2} + ig_3\omega_3 e^{i(\theta_3-\varphi_3)}\right)$$

$$= i^2 r_2\omega_2^2 e^{i\theta_2} + ir_2\frac{d\omega_2}{dt}e^{i\theta_2} + i^2 g_3\omega_3^2 e^{i(\theta_3-\varphi_3)} + ig_3\frac{d\omega_2}{dt}e^{i(\theta_3-\varphi_3)} \qquad (3.110)$$

$$= -r_2\omega_2^2 e^{i\theta_2} + ir_2\alpha_2 e^{i\theta_2} - g_3\omega_3^2 e^{i(\theta_3-\varphi_3)} + ig_3\alpha_3 e^{i(\theta_3-\varphi_3)}$$

Its Cartesian components (Eq. 3.111) are:

$$\left.\begin{aligned} a_{P_3x} &= -r_2\omega_2^2\cos\theta_2 - r_2\alpha_2\sin\theta_2 - g_3\omega_3^2\cos(\theta_3-\varphi_3) - g_3\alpha_3\sin(\theta_3-\varphi_3)\\ a_{P_3y} &= -r_2\omega_2^2\sin\theta_2 + r_2\alpha_2\cos\theta_2 - g_3\omega_3^2\sin(\theta_3-\varphi_3) + g_3\alpha_3\cos(\theta_3-\varphi_3) \end{aligned}\right\}$$
$$(3.111)$$

Angular velocity of link 3 coincides with the one of link 4: $\alpha_3 = \alpha_4$.
– Finally, the acceleration vector of point P_4 on link 4 (Eq. 3.112) is:

$$\mathbf{a}_{P_4} = \frac{d\mathbf{v}_{P_4}}{dt} = \frac{d}{dt}\left(ig_4\omega_4 e^{i(\theta_4+180°-\varphi_4)}\right)$$

$$= i^2 g_4\omega_4 \frac{d(\theta_4'+\varphi_4)}{dt}e^{i(\theta_4+180°-\varphi_4)} + ig_4\frac{d\omega_4}{dt}e^{i(\theta_4+180°-\varphi_4)} \qquad (3.112)$$

$$= -g_4\omega_4^2 e^{i(\theta_4+180°-\varphi_4)} + ig_4\alpha_4 e^{i(\theta_4+180°-\varphi_4)}$$

With the Cartesian components (Eq. 3.113):

$$\left.\begin{aligned} a_{P_4x} &= -g_4\omega_4^2\cos(\theta_4+180°-\varphi_4) - g_4\alpha_4\sin(\theta_4+180°-\varphi_4)\\ a_{P_4y} &= -g_4\omega_4^2\sin(\theta_4+180°-\varphi_4) + g_4\alpha_4\cos(\theta_4+180°-\varphi_4) \end{aligned}\right\}$$
$$(3.113)$$

3.1.3   Complex Mechanism Analysis

When a mechanism has more than four links, we have to define more than one vector-loop equation. For mechanisms with one degree of freedom, the number of equations needed *NVLE* (Eq. 3.114) is:

$$NVLE = \frac{N-2}{2} \tag{3.114}$$

where N is the number of links in the mechanism. We take out the frame and the input links and divide the expression by 2 since every equation is used to find two unknowns.

3.1.3.1   Angular Position

Figure 3.8a shows a mechanism with six links. The velocity and acceleration of the links can be determined by using the same analytical method as with the four-link mechanisms studied before. In Fig. 3.8b several vectors have been defined with known magnitudes and angles. They also include unknown angles θ_3, θ_4, θ_5 and θ_6 that define the position of links 3, 4, 5 and 6 respectively. These vectors give rise to two loop equations (Eq. 3.115):

$$\left.\begin{array}{r}\mathbf{r}_2 + \mathbf{r}_3 + \mathbf{r}_{4'} + \mathbf{r}_5 = \mathbf{r}_1 \\ \mathbf{r}_6 + \mathbf{r}_4 + \mathbf{r}_5 = \mathbf{r}_{1'}\end{array}\right\} \tag{3.115}$$

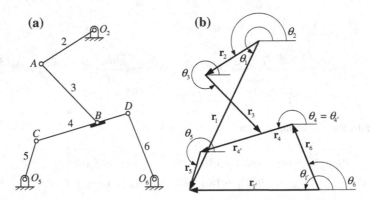

Fig. 3.8 a Six-bar linkage skeleton. **b** Closed vector polygons that include the length and angle of all the links. The unknowns are θ_3, θ_4, θ_5 and θ_6 while the rest of angles and all magnitudes are known

Vectors can be expressed in their complex exponential form (Eq. 3.116):

$$\left.\begin{array}{l} r_2 e^{i\theta_2} + r_3 e^{i\theta_3} + r_{4'} e^{i\theta_4} + r_5 e^{i\theta_5} = r_1 e^{i\theta_1} \\ r_4 e^{i\theta_4} + r_5 e^{i\theta_5} + r_6 e^{i\theta_6} = r_{1'} e^{i\theta_{1'}} \end{array}\right\} \quad (3.116)$$

Separating the real and imaginary parts, we obtain the system of four equations (Eq. 3.117) that allows us to find unknowns θ_3, θ_4, θ_5 and θ_6 in terms of link lengths, constant angles θ_1 and $\theta_{1'}$ and input angle θ_2.

$$\left.\begin{array}{l} r_2 \cos\theta_2 + r_3 \cos\theta_3 + r_{4'} \cos\theta_4 + r_5 \cos\theta_5 = r_1 \cos\theta_1 \\ r_2 \sin\theta_2 + r_3 \sin\theta_3 + r_{4'} \sin\theta_4 + r_5 \sin\theta_5 = r_1 \sin\theta_1 \\ r_4 \cos\theta_4 + r_5 \cos\theta_5 + r_6 \cos\theta_6 = r_{1'} \cos\theta_1 \\ r_4 \sin\theta_4 + r_5 \sin\theta_5 + r_6 \sin\theta_6 = r_{1'} \sin\theta_{1'} \end{array}\right\} \quad (3.117)$$

Despite the fact that the mechanism in this example only has six links, the resolution process of the position problem is quite tedious. If the mechanism had eight links, the number of loop equations would be 3, so we would obtain a system of equations similar to the one in Eq. (3.117) but with 6 equations and 6 unknowns, complicating the resolution substantially.

3.1.3.2 Angular Velocity

Differentiating Eq. (3.116), we obtain Eq. (3.118) from which to determine the velocities of the links:

$$\left.\begin{array}{l} ir_2\omega_2 e^{i\theta_2} + ir_3\omega_3 e^{i\theta_3} + ir_{4'}\omega_4 e^{i\theta_4} + ir_5\omega_5 e^{i\theta_5} = 0 \\ ir_4\omega_4 e^{i\theta_4} + ir_5\omega_5 e^{i\theta_5} + ir_6\omega_6 e^{i\theta_6} = 0 \end{array}\right\} \quad (3.118)$$

The sum of the real and imaginary parts of the equations leads us to setting up the system of four equations with four unknowns ($\omega_3, \omega_4, \omega_5$ and ω_6) (Eq. 3.119):

$$\left.\begin{array}{l} -r_2\omega_2 \sin\theta_2 - r_3\omega_3 \sin\theta_3 - r_{4'}\omega_4 \sin\theta_4 - r_5\omega_5 \sin\theta_5 = 0 \\ r_2\omega_2 \cos\theta_2 + r_3\omega_3 \cos\theta_3 + r_{4'}\omega_4 \cos\theta_4 + r_5\omega_5 \cos\theta_5 = 0 \\ -r_4\omega_4 \sin\theta_4 - r_5\omega_5 \sin\theta_5 - r_6\omega_6 \sin\theta_6 = 0 \\ r_4\omega_4 \cos\theta_4 + r_5\omega_5 \cos\theta_5 + r_6\omega_6 \cos\theta_6 = 0 \end{array}\right\} \quad (3.119)$$

The solution of this system gives the angular velocity of links 3, 4, 5 and 6.

3.1.3.3 Angular Acceleration

The velocity equations can be differentiated to obtain (Eq. 3.120):

$$
\left.
\begin{aligned}
0 &= -r_2\omega_2^2 e^{i\theta_2} + ir_2\alpha_2 e^{i\theta_2} - r_3\omega_3^2 e^{i\theta_3} + ir_3\alpha_3 e^{i\theta_3} \\
&\quad - r_{4'}\omega_4^2 e^{i\theta_4} + ir_{4'}\alpha_4 e^{i\theta_4} - r_5\omega_5^2 e^{i\theta_5} + ir_5\alpha_5 e^{i\theta_5} \\
0 &= -r_4\omega_4^2 e^{i\theta_4} + ir_4\alpha_4 e^{i\theta_4} - r_5\omega_5^2 e^{i\theta_5} + ir_5\alpha_5 e^{i\theta_5} - r_6\omega_6^2 e^{i\theta_6} + ir_6\alpha_6 e^{i\theta_6}
\end{aligned}
\right\}
\quad (3.120)
$$

Again, separating the real and imaginary parts we obtain the system with four unknowns, α_3, α_4, α_5 and α_6 (Eq. 3.121).

$$
\left.
\begin{aligned}
0 &= -r_2\omega_2^2\cos\theta_2 - r_2\alpha_2\sin\theta_2 - r_3\omega_3^2\cos\theta_3 - r_3\alpha_3\sin\theta_3 \\
&\quad - r_{4'}\omega_4^2\cos\theta_4 - r_{4'}\alpha_4\sin\theta_4 - r_5\omega_5^2\cos\theta_5 - r_5\alpha_5\sin\theta_5 \\
0 &= -r_2\omega_2^2\sin\theta_2 + r_2\alpha_2\cos\theta_2 - r_3\omega_3^2\sin\theta_3 + r_3\alpha_3\cos\theta_3 \\
&\quad - r_{4'}\omega_4^2\sin\theta_4 + r_{4'}\alpha_4\cos\theta_4 - r_5\omega_5^2\sin\theta_5 + r_5\alpha_5\cos\theta_5 \\
0 &= -r_4\omega_4^2\cos\theta_4 - r_4\alpha_4\sin\theta_4 - r_5\omega_5^2\cos\theta_5 - r_5\alpha_5\sin\theta_5 \\
&\quad - r_6\omega_6^2\cos\theta_6 - r_6\alpha_6\sin\theta_6 \\
0 &= -r_4\omega_4^2\sin\theta_4 + r_4\alpha_4\cos\theta_4 - r_5\omega_5^2\sin\theta_5 + r_5\alpha_5\cos\theta_5 \\
&\quad - r_6\omega_6^2\sin\theta_6 + r_6\alpha_6\cos\theta_6
\end{aligned}
\right\}
\quad (3.121)
$$

The solution of the system yields the values of the accelerations of links 3, 4, 5 and 6. It is not necessary to emphasize the difficulty in solving this equation system manually. Raven's method for mechanism analysis may require numerical methods for equation solving like those implemented in some computer programs or even some pocket calculators.

3.2 Examples with Their Solutions

In this section we will carry out the kinematic analysis of different mechanisms by applying the methods developed in this chapter up to now.

Example 1 In the four-bar linkage in Fig. 3.9a, use Raven's method to calculate angles θ_3 and θ_4, angular velocity and acceleration of links 3 and 4 and the velocity

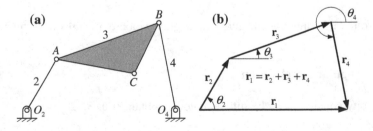

Fig. 3.9 a Four-bar linkage skeleton. **b** Vector loop equation for Raven's analysis

and acceleration of point C, when the dimension of the linkages are: $r_1 = \overline{O_2O_4} = 15\,\text{cm}$, $r_2 = \overline{O_2A} = 6\,\text{cm}$, $r_3 = \overline{AB} = 11\,\text{cm}$, $r_4 = \overline{O_4B} = 9\,\text{cm}$, $\overline{AC} = 8\,\text{cm}$ and $\angle BAC = 30°$. The position angle of the input link is $\theta_2 = 60°$, it moves with velocity $\omega_2 = -20\,\text{rad/s}$ (clockwise) and its angular acceleration is $\alpha_2 = 150\,\text{rad/s}^2$ (counterclockwise).

We will use the vector loop equation in Fig. 3.9b. Equation (3.122) determine the position of the links in the four-bar linkage:

$$\left.\begin{aligned} r_1 \cos\theta_1 &= r_2 \cos\theta_2 + r_3 \cos\theta_3 + r_4 \cos\theta_4 \\ r_1 \sin\theta_1 &= r_2 \sin\theta_2 + r_3 \sin\theta_3 + r_4 \sin\theta_4 \end{aligned}\right\} \qquad (3.122)$$

Taking into account the dimensions of the links and the value of input angle θ_2:

$$\left.\begin{aligned} 15 &= 6\cos 60° + 11\cos\theta_3 + 9\cos\theta_4 \\ 0 &= 6\sin 60° + 11\sin\theta_3 + 9\sin\theta_4 \end{aligned}\right\}$$

The unknowns of the system can be cleared by an iterative numerical method like the Newton-Raphson's one. However, we can also make use of Freudenstein's method developed in Appendix B of this book.

The latter yields the following values for the unknowns:

$$\theta_3 = 19.4°$$
$$\theta_4 = 280.4°$$

After solving the position problem, we can calculate the values of angular velocity, using Eqs. (3.10) and (3.11):

$$\omega_3 = \frac{-r_2 \sin(\theta_2 - \theta_4)}{r_3 \sin(\theta_3 - \theta_4)}\omega_2$$

$$\omega_4 = \frac{r_2 \sin(\theta_2 - \theta_3)}{r_4 \sin(\theta_3 - \theta_4)}\omega_2$$

If we substitute the known values in the previous equations, we obtain:

$$\omega_3 = \frac{-6\sin(60° - 280.4°)}{11\sin(19.4° - 280.4°)} \cdot -20 = 7.16\,\text{rad/s}$$

$$\omega_4 = \frac{-6\sin(60° - 19.4°)}{9\sin(19.4° - 280.4°)} \cdot -20 = 8.78\,\text{rad/s}$$

Finally, we can find the angular acceleration values by using Eqs. (3.16) and (3.17):

$$\alpha_3 = \frac{\omega_3}{\omega_2}\alpha_2 - \frac{r_2\omega_2^2\cos(\theta_2 - \theta_4) + r_3\omega_3^2\cos(\theta_3 - \theta_4) + r_4\omega_4^2}{r_3\sin(\theta_3 - \theta_4)}$$

$$\alpha_4 = \frac{\omega_4}{\omega_2}\alpha_2 - \frac{r_2\omega_2^2\cos(\theta_2 - \theta_3) + r_4\omega_4^2\cos(\theta_3 - \theta_4) + r_3\omega_3^2}{r_4\sin(\theta_3 - \theta_4)}$$

Operating with the known values:

$$\alpha_3 = \frac{7.16}{-20}150 - \frac{6(-20)^2\cos(60° - 280.4°) + 11(7.16)^2\cos(19.4° - 280.4°) + 9(-8.78)^2}{11\sin(19.4° - 280.4°)}$$

$$= 58.8\,\text{rad/s}^2$$

$$\alpha_4 = \frac{-8.78}{-20}150 - \frac{6(-20)^2\cos(60° - 19.4°) + 9(-8.78)^2\cos(19.4° - 280.4°) + 11(7.16)^2}{9\sin(19.4° - 280.4°)}$$

$$= 322.2\,\text{rad/s}^2$$

The velocity of point C can be calculated by means of equation (Eq. 3.28), where $g_3 = 8\,\text{cm}$ and $\varphi_3 = 330°$. By computing with the known values we obtain:

$$\left.\begin{array}{l} v_{C_x} = -6\cdot -20\sin 60° - 8\cdot 7.16\sin(19.4° + 330°) = 114.4\,\text{cm/s} \\ v_{C_y} = -6\cdot -20\cos 60° - 8\cdot 7.16\cos(19.4° + 330°) = -3.7\,\text{cm/s} \end{array}\right\}$$

The magnitude and angle of the point C velocity vector will be:

$$v_C = 114.46\,\text{cm/s}\,\angle 358.1°$$

The acceleration of point C is given by equation (Eq. 3.36). By operating with the known values:

$$\left.\begin{array}{l} a_{Cx} = -6\cdot -20^2\cos 60° - 6\cdot 150\sin 60° \\ \qquad - 8\cdot 7.16^2\cos(19.4° - 30°) - 8\cdot 58.8\sin(19.4° - 30°) \\ \qquad = -2296\,\text{cm/s}^2 \\ a_{Cy} = -6\cdot -20^2\sin 60° + 6\cdot 150\cos 60° \\ \qquad - 8\cdot 7.16^2\sin(19.4° - 30°) + 8\cdot 58.8\cos(19.4° - 30°) \\ \qquad = -1090.16\,\text{cm/s}^2 \end{array}\right\}$$

We can calculate the magnitude and angle of the point C acceleration as:

$$a_C = 2541.6\,\text{cm/s}^2\,\angle 205.4°$$

Example 2 Calculate the acceleration of points G_3 and G_4 on the crank-shaft linkage in Fig. 3.10a when $r_2 = \overline{O_2A} = 3\,\text{cm}$, $r_3 = \overline{AB} = 7\,\text{cm}$, $\overline{AG_3} = 2\,\text{cm}$, $\theta_2 = 60°$, $\omega_2 = -20\,\text{rad/s}$ and $\alpha_2 = -100\,\text{rad/s}$.

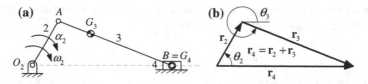

Fig. 3.10 a Crank-shaft linkage. **b** Vector loop equation for the kinematic analysis using Raven's method

To calculate the acceleration of points G_3 and G_4 of the linkage, we have to find the acceleration of the links first. We start by finding the solution to the position problem. The loop equation defined in Fig. 3.10b yields the system (Eq. 3.123):

$$\left.\begin{array}{l} r_4 = r_2 \cos \theta_2 + r_3 \cos \theta_3 \\ -r_1 = r_2 \sin \theta_2 + r_3 \sin \theta_3 \end{array}\right\} \tag{3.123}$$

In this example $r_1 = 0$, $r_2 = 3 \, \text{cm}$, $r_3 = 7 \, \text{cm}$ and $\theta_2 = 60°$. The unknowns are θ_3 and r_4.

$$\left.\begin{array}{l} r_4 = 3 \cos 60° + 7 \cos \theta_3 \\ 0 = 3 \sin 60° + 7 \sin \theta_3 \end{array}\right\}$$

This system is extremely easy to solve.

$$\theta_3 = 338.2°$$
$$r_4 = 8 \, \text{cm}$$

To calculate the velocity of point C, we will make use of Eqs. (3.51) and (3.52):

$$\omega_3 = \frac{-3 \cdot -20 \cos 60°}{7 \cos 338.2°} = 4.62 \, \text{rad/s}$$
$$v_4 = -3 \cdot -20 \sin 60° - 7 \cdot 4.62 \sin 338.2° = 63.97 \, \text{cm/s}$$

For the acceleration problem, we make use of Eqs. (3.57) and (3.58):

$$\alpha_3 = \frac{3(-20)^2 \sin 60° - 3 \cdot -100 \cos 60° + 7 \cdot 4.62^2 \sin 338.2°}{7 \cos 338.2°}$$
$$= 174.44 \, \text{rad/s}^2$$
$$a_4 = -3 \cdot (-20)^2 \cos 60° - 3 \cdot -100 \sin 60° - 7 \cdot 4.62^2 \cos 338.2° - 7 \cdot 174.4 \sin 338.2°$$
$$= -25.45 \, \text{cm/s}^2$$

Since all the points on link 4 have the same acceleration, the magnitude and angle of the acceleration vector of point G_4 will be:

$$\mathbf{a}_{G_4} = 25.45\,\text{cm/s}^2 \,\angle 180°$$

Finally, to calculate the acceleration of point G_3, we have to make use of Eq. (3.70). If we plug in the known values:

$$\left.\begin{aligned}
a_{G_3x} &= -3 \cdot (-20)^2 \cos 60° - 3 \cdot -100 \sin 60° - 2 \cdot 4.62^2 \cos 338.2° - 2 \cdot 174.4 \sin 338.2° \\
&= -250.26\,\text{cm/s}^2 \\
a_{G_3y} &= -3 \cdot (-20)^2 \sin 60° - 3 \cdot -100 \cos 60° - 2 \cdot 4.62^2 \sin 338.2° - 2 \cdot 174.4 \cos 338.2° \\
&= -849.45\,\text{cm/s}^2
\end{aligned}\right\}$$

The magnitude and angle of the acceleration vector of point G_3 will be:

$$\mathbf{a}_{G_3} = 885.55\,\text{cm/s}^2 \,\angle 253.58°$$

Example 3 Figure 3.11a shows the mixing machine that was studied in Example 4 of Chap. 2. Find the solution to the position problem through Raven's method when angle $\theta_2 = 0°$. The dimensions are in centimeters.

The position unknowns are θ_3 and distance $\overline{O_4A}$. We set up the vector loop equation shown in Fig. 3.11c where the unknowns are the magnitude and angle of vector $\mathbf{r}_3$:

$$\mathbf{r}_2 + \mathbf{r}_3 = \mathbf{r}_1 \tag{3.124}$$

We write the vector loop (Eq. 3.124) in its complex exponential form (Eq. 3.125):

$$r_2 e^{i\theta_2} + r_3 e^{i\theta_3} = r_1 e^{i\theta_1} \tag{3.125}$$

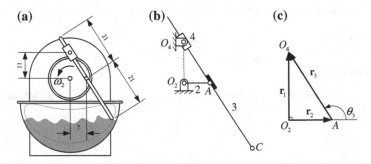

Fig. 3.11 a Mixing machine. **b** Kinematic skeleton. **c** Vector loop equation

Taking into account that $\theta_1 = 90°$, $r_1 = 11$ cm and $r_2 = 7$ cm, we can write the real and imaginary parts:

$$\left.\begin{array}{c} r_3 \sin \theta_3 = 11 \\ 7 + r_3 \cos \theta_3 = 0 \end{array}\right\}$$

The values of the unknowns are:

$$\theta_3 = \arctan \frac{11}{-7} = 122.4°$$

$$r_3 = \frac{11}{\sin 122.4°} = 13.04 \, \text{cm}$$

Example 4 In the mechanism in Fig. 3.12a, use Raven's method to solve the position knowing that $\overline{O_2A} = 1$ cm, $\overline{O_2B} = 0.5$ cm and $x = 0.25$ cm (input).

In this example, the number of links in the linkage is six. Therefore, we will need 2 loop equations to solve 4 position unknowns. Figure 3.12b shows 2 loop equations (Eq. 3.126) that can be used to solve the problem:

$$\left.\begin{array}{c} \mathbf{r}_2 = \mathbf{r}_1 + \mathbf{y} \\ \mathbf{r}_{2'} = \mathbf{r}_{1'} + \mathbf{x} \end{array}\right\} \tag{3.126}$$

Their complex exponential form (Eq. 3.127) is:

$$\left.\begin{array}{c} r_2 e^{i\theta_2} = r_1 e^{i0°} + y e^{i90°} \\ r_{2'} e^{i(\theta_2 + 270°)} = r_{1'} e^{i270°} + x e^{i0°} \end{array}\right\} \tag{3.127}$$

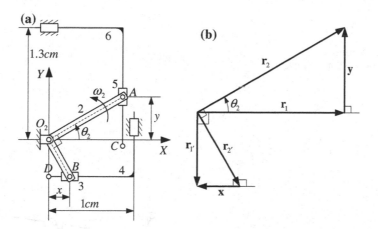

Fig. 3.12 **a** Calculating machine. **b** Vector loops for the mechanism analysis

Breaking these equations into real and imaginary parts and taking into account that $x = 0.25$ cm, $r_2 = 1$ cm and $r_{2'} = 0.5$ cm and that the unknowns of the problem are y, r_1, $r_{1'}$ and θ_2, we obtain the following equations:

$$\left.\begin{aligned}
1\cos\theta_2 &= r_1 \\
1\sin\theta_2 &= y \\
0.5\cos(\theta_2 + 270°) &= 0.25 \\
0.5\sin(\theta_2 + 270°) &= -r_{1'}
\end{aligned}\right\}$$

Finally, we calculate the values of the unknowns:

$$\theta_2 = \arccos\frac{0.25}{0.5} - 270° = 30°$$
$$r_{1'} = -0.5\sin(30° + 270°) = 0.433 \text{ cm}$$
$$r_1 = 1\cos 30° = 0.866 \text{ cm}$$
$$y = 1\sin 30° = 0.5 \text{ cm}$$

Example 5 In the elliptical trainer in Fig. 3.13a, use Raven's method to solve the position problem when $\overline{O_4B} = 70$ cm, $\overline{O_2A} = 20$ cm, $\overline{AB} = 140$ cm, $\overline{DF} = 70$ cm, $\overline{O_2D} = 40$ cm, $\overline{O_2O_4} = 152.1$ cm, $\theta_1 = 208.25°$ (angle of a vector from O_4 to O_2) and input angle $\theta_2 = 300°$.

As in the previous example, the number of links is six, hence, the number of loop equations needed to solve the problem is two. Figure 3.14b represents two loop equations (Eq. 3.128) defined to solve this problem:

$$\left.\begin{aligned}
\mathbf{r}_1 + \mathbf{r}_2 &= \mathbf{r}_4 + \mathbf{r}_3 \\
\mathbf{r}_1 + \mathbf{r}_{2'} &= \mathbf{r}_4 + \mathbf{r}_{3'} + \mathbf{r}_5
\end{aligned}\right\} \qquad (3.128)$$

Fig. 3.13 Kinematic skeleton of an elliptical trainer

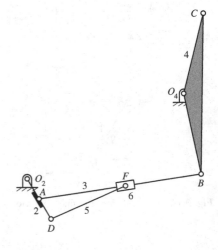

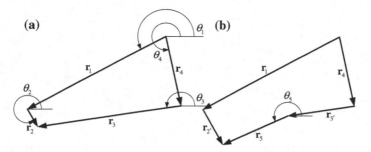

Fig. 3.14 a Vector loop for the elliptical trainer mechanism with unknowns θ_3 and θ_4. **b** Vector loop with unknowns $r_{3'}$ and θ_5

Using the exponential form we find (Eq. 3.129). All the unknowns of the problem appear in the equations (θ_3, θ_4, $r_{3'}$ and θ_5):

$$\left. \begin{array}{l} r_1 e^{i\theta_1} + r_2 e^{i\theta_2} = r_4 e^{i\theta_4} + r_3 e^{i\theta_3} \\ r_1 e^{i\theta_1} + r_{2'} e^{i\theta_{2'}} = r_4 e^{i\theta_4} + r_{3'} e^{i\theta_3} + r_5 e^{i\theta_5} \end{array} \right\} \qquad (3.129)$$

Breaking both equations into their real and imaginary parts and taking into account that $r_1 = \overline{O_2 O_4}$, $r_2 = \overline{O_2 A}$, $r_3 = \overline{AB}$, $r_4 = \overline{O_4 B}$, $r_5 = \overline{DF}$ and $r_{2'} = \overline{O_2 D}$, the following equations are obtained:

$$\left. \begin{array}{l} 152.1 \cos 208.25° + 20 \cos 300° = 70 \cos \theta_4 + 140 \cos \theta_3 \\ 152.1 \sin 208.25° + 20 \sin 300° = 70 \sin \theta_4 + 140 \sin \theta_3 \end{array} \right\} \qquad (3.130)$$

$$\left. \begin{array}{l} 152.1 \cos 208.25° + 40 \cos 300° = 70 \cos \theta_4 + r_{3'} \cos \theta_3 + 70 \cos \theta_5 \\ 152.1 \sin 208.25° + 40 \sin 300° = 70 \sin \theta_4 + r_{3'} \sin \theta_3 + 70 \sin \theta_5 \end{array} \right\} \qquad (3.131)$$

Freudenstein's method (see Appendix B) can be applied to Eq. (3.130) in order to obtain the values of θ_3 and θ_4:

$$\theta_3 = 188.6°$$
$$\theta_4 = 281.9°$$

With these values we can clear the rest of the unknowns, $r_{3'}$ and θ_5, from (3.131):

$$r_{3'} = 65.1\,\text{cm}$$
$$\theta_5 = 204°$$

Chapter 4
Graphical and Analytical Methods for Dynamic Analysis of Planar Linkages

Abstract We can basically distinguish two types of problems in the study of forces acting on a mechanism. On one side, we can determine the type of motion produced by a known system of forces (direct problem). On the other side, and provided that we know the variables defining the motion of the links, we can find the forces producing it (inverse problem). In this chapter we will focus on the second problem. Forces acting on a mechanism can be due to several reasons, such as the weight of the links (gravity forces), external forces, friction forces, acceleration of the links (inertial forces) and so on. However, the weight of the links is usually a negligible force compared to the other forces mentioned. Furthermore, if the mechanism is well lubricated, it is possible to neglect friction forces and obtain sufficiently accurate results, simplifying the problem significantly. For this reason we will only consider external and inertial forces in this book. We will first study the action of external forces applied to a mechanism in static equilibrium. We will begin by developing a method for the static analysis of linkages which, in addition to facilitating the comprehension of force transmission in mechanisms, will be useful to explain methods that will subsequently be applied when considering inertial forces of the links, that is, when performing a dynamic analysis. We will only study the action of forces in articulated mechanisms with planar motion and assume that all forces occur on the same plane, not on parallel ones. When the neglected moments are considerable, it will be necessary to make a second analysis on a plane perpendicular to the one on which forces are acting.

4.1 Machine Statics

In this chapter, we will develop a method for the analysis of forces acting on a mechanism in static equilibrium. First of all, it is interesting to study the way in which forces are transmitted between linkages in a mechanism.

© Springer International Publishing Switzerland 2016
A. Simón Mata et al., *Fundamentals of Machine Theory and Mechanisms*,
Mechanisms and Machine Science 40, DOI 10.1007/978-3-319-31970-4_4

4.1.1 Force Transmission in a Mechanism

Force transmission in a mechanism takes place through the joints of kinematic pairs. Reaction forces between the different linkage members appear in these joints. The direction of these forces will depend on the type of kinematic pair. In articulated planar mechanisms, only hinge and sliding pairs will be found. We will study the reaction force that appears in these pairs.

- In the case of a prismatic pair, the direction of the reaction force is known because it is independent of the forces acting on each link. Forces are always transmitted perpendicularly to the slider path and equilibrium equations have to be solved in order to know the magnitude (Fig. 4.1a, b)
- In the case of a rotation pair, the direction of the reaction depends on the forces acting on each link. We cannot know the direction and magnitude of the reaction force until we solve the equilibrium equations (Fig. 4.1c).

From now on, we will use the F_{jk} nomenclature to refer to the force exerted by link j on link k.

4.1.2 Static Equilibrium Conditions

For a rigid body to be in static equilibrium:

- The vector sum of all the forces acting on it must be zero (Eq. 4.1):

$$\sum_i \mathbf{F}_i = 0 \qquad (4.1)$$

- The sum of the torques about any axis of all forces acting on the body must also be zero (Eq. 4.2):

$$\sum_i \mathbf{M}_i = 0 \qquad (4.2)$$

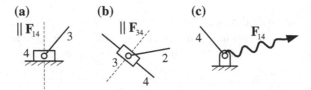

Fig. 4.1 The direction of the force that appears in a joint connecting two links is known in sliding pairs (**a** and **b**), and unknown in rotation pairs (**c**)

Fig. 4.2 The magnitude of
the moment of force about
point P can be calculated as
$M^P = Fh$

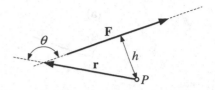

On a plane, it is possible to calculate the moment of a force $\mathbf{F}$ about any point P by simply multiplying its magnitude by the distance, h, from the line of action of the force to the point (Eq. 4.3), measured on a direction perpendicular to it (Fig. 4.2), since:

$$\mathbf{M}^P = \mathbf{r} \wedge \mathbf{F} \rightarrow M^P = Fr\sin\theta = Fh \qquad (4.3)$$

The line of action of the moment vector is perpendicular to the plane formed by vectors $\mathbf{F}$ and $\mathbf{r}$ and its direction depends on the right hand rule.

To sum up, a rigid body is in static equilibrium if the following algebraic sums are zero (Eq. 4.4), provided that all the forces acting on it are on the same plane:

$$\left.\begin{array}{c} \sum_i F_{i_x} = 0 \\ \sum_i F_{i_y} = 0 \\ \sum_k M_{k_z} = 0 \end{array}\right\} \qquad (4.4)$$

where P is a point on the plane and F_{i_x} and F_{i_y} are the Cartesian coordinates of force $\mathbf{F}_i$.

4.1.2.1   Equilibrium of a Mechanism Considering the Equilibrium of Every Link that Forms It

In order for a mechanism to be in equilibrium, it is necessary and sufficient that all its links are in equilibrium.

A system of three equations can be defined for every link allowing us to find the same amount of unknowns. This way, we can calculate the reactions in the different links.

Let us see how to graphically define the equilibrium equations on a link where no outer torque is applied.

- Link with two forces:
  Consider 3 to be a link on which forces $\mathbf{F}_{23}$ and $\mathbf{F}_{43}$, exerted by links 2 and 4 respectively, are applied (Fig. 4.3a). In this case, in order for Eq. (4.1) to happen, both forces must have the same line of action, the same magnitude and opposite directions (Fig. 4.3b). As both forces act through the joining elements, in order for the moment sum on link 3 to be null (Eq. 4.2), it is necessary that both forces

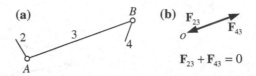

Fig. 4.3 a Link 3 with two joints A and B. **b** Force polygon with forces $\mathbf{F}_{23}$ and $\mathbf{F}_{43}$

go in the direction defined by the line that joins points A and B. In any other case, a pair of forces will appear creating an unbalanced torque on link 3.

- A link with three forces:
  Consider a link on which three forces are acting, link 3 in (Fig. 4.4a). We know the course line, magnitude and direction of one of the forces, $\mathbf{F}_{23}$ for example, as well as the direction of another one, $\mathbf{F}_{43}$ for example.

As we analyze the problem, it is noted that we need to determine the magnitude and direction of force $\mathbf{F}_{13}$ as well as the magnitude of force $\mathbf{F}_{43}$. Hence, we have a problem with three unknowns that we can solve by applying the equilibrium equations to link 3.

In order to achieve (Eq. 4.2), these three forces have to converge at a point. Otherwise, the resultant of two of the forces will generate a torque with the third one. This concept can be applied to our problem to find the direction of $\mathbf{F}_{13}$, which has to go through the joint of links 1 and 3 as well as through point N, the intersection of the lines of action of forces $\mathbf{F}_{23}$ and $\mathbf{F}_{43}$ (Fig. 4.4a).

To meet the force sum condition, (Eq. 4.1), the force polygon has to close. Knowing the direction of all three forces and the magnitude of $\mathbf{F}_{23}$, we can draw the polygon and determine the magnitude and direction of $\mathbf{F}_{13}$ and $\mathbf{F}_{43}$ as shown in (Fig. 4.4b).

Example 1 In the mechanism in (Fig. 4.5) we know how force $\mathbf{P}_4$ acts on link 4. We want to find the reactions at the four joints as well as the torque applied to link 2 so that the mechanism remains stable in the given instant.

We will start by studying link 4. Besides external force $\mathbf{P}_4$, there are two reaction forces acting on this link: $\mathbf{F}_{14}$ and $\mathbf{F}_{34}$ (Fig. 4.6a).

Fig. 4.4 a Link 3 with joints A, B and O_3. **b** Force polygon for forces $\mathbf{F}_{23}$, $\mathbf{F}_{43}$ and $\mathbf{F}_{13}$ acting on link 3

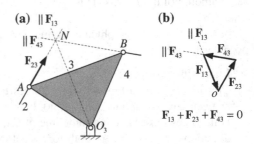

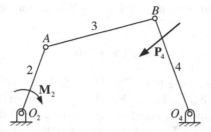

Fig. 4.5 Four-bar mechanism with known force P_4 acting on link 4 and unknown torque M_2 acting on link 2 to keep the mechanism in static equilibrium

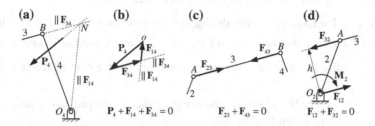

Fig. 4.6 **a** The lines of action of the three forces acting on link 4 converge at point N. **b** Force polygon of the forces acting on link 4. **c** Forces acting on link 3. **d** Forces and moment M_2 acting on link 2

To maintain the system in equilibrium, the vector sum of all the forces acting on it (Eq. 4.5) must be zero:

$$P_4 + F_{14} + F_{34} = 0 \tag{4.5}$$

Also, all three forces have to intersect at a point. Initially, however, we only know the magnitude and direction of force P_4. Hence, we have four unknowns, the directions and magnitudes of forces F_{14} and F_{34}. Link 3 will give us the direction of force F_{34} which reduces the number of unknowns to three.

Forces F_{43} and F_{23} act upon link 3 (Eq. 4.6). As already seen in this chapter, the direction of these two forces has to follow straight line AB so that (Eq. 4.2). This way we already know the direction of F_{43}.

Force F_{34} has the same magnitude and opposite direction as F_{43}. Thus, back on link 4, we can determine the direction of force F_{14} since it has to go through the intersection point of the lines of action of forces P_4 and F_{34} (Fig. 4.6a). We can obtain the magnitude F_{14} and F_{34} direction of forces and from the force polygon (Fig. 4.6b).

Next, and taking into account that $\mathbf{F}_{43} = -\mathbf{F}_{34}$ we can define the force equilibrium of link 3 as:

$$\mathbf{F}_{23} + \mathbf{F}_{43} = 0 \tag{4.6}$$

This allows us to determine force $\mathbf{F}_{23}$ (Fig. 4.6c).

Last, we will study link 2. Forces $\mathbf{F}_{32}$, $\mathbf{F}_{12}$ and moment $\mathbf{M}_2$, act on this link. Force $\mathbf{F}_{32} = -\mathbf{F}_{23}$ is fully determined, as for magnitude and direction. As the sum of forces must be zero, $\mathbf{F}_{32} + \mathbf{F}_{12} = 0$, we can obtain $\mathbf{F}_{12}$, which will have the same magnitude as $\mathbf{F}_{32}$ but with an opposite direction (Fig. 4.6d).

The magnitude of moment $\mathbf{M}_2$ can be found knowing that the sum of the moments about point O_2 has to be zero: $M_2 = F_{32}h$ in clockwise direction (Fig. 4.6d). This way the problem remains fully determined.

Example 2 Solve Example 1 considering that the external force is acting on link 3.

Figure 4.7 shows the solution with the same approach used in Example 1.

Example 3 Solve the previous example considering that the external force is acting on link 2.

Following a procedure similar to the one described in the previous examples, we reach the solution shown in (Fig. 4.8).

Example 4 Force $\mathbf{P}_4$ is acting on link 4 of the slider-crank mechanism in (Fig. 4.9a). Calculate the reactions in the different links as well as moment $\mathbf{M}_2$ which keeps the mechanism in static equilibrium.

Fig. 4.7 a Four-bar mechanism with the reactions that act on each link when known force $\mathbf{P}_3$ acts on link 3 and unknown torque $\mathbf{M}_2$ acts on link 2 to keep the mechanism in static equilibrium. **b** Polygon of forces acting on link 3

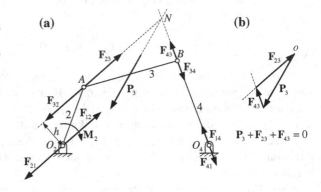

Fig. 4.8 Four-bar mechanism with external force $\mathbf{P}_2$ and torque $\mathbf{M}_2$ that keeps the mechanism in static equilibrium. As both act on link 2, there are no reactions on links 3 and 4

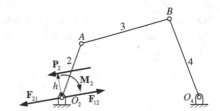

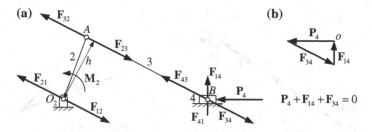

Fig. 4.9 **a** Reactions on each link of a crank-shaft mechanism when force $\mathbf{P}_4$ acts on link 4 and torque $\mathbf{M}_2$ acts on link 2 to keep the mechanism in static equilibrium. **b** Polygon of forces acting on link 4

1. We define the equilibrium of link 3. Since only two forces, $\mathbf{F}_{43}$ and $\mathbf{F}_{23}$, are applied, they have to be in the direction of the link.
2. Once we have obtained the direction of $\mathbf{F}_{34} = -\mathbf{F}_{43}$, we can define the equilibrium of link 4 and find the magnitudes of $\mathbf{F}_{34}$ and $\mathbf{F}_{14}$ (Fig. 4.9b).
3. Back to link 3, we can determine $\mathbf{F}_{23} = -\mathbf{F}_{43}$.

Finally, we study the equilibrium of link 2 to find $\mathbf{F}_{12} = -\mathbf{F}_{32}$ and $M_2 = F_{32}h$ (clockwise) (Fig. 4.9a).

Example 5 Consider Witworth's quick-return mechanism, shown in (Fig. 4.10a), where known force $\mathbf{P}_4$ is applied to link 4. Find the reactions in the links and the value of force $\mathbf{P}_6$ that acts on link 6 to keep the mechanism in static equilibrium.

Example 6 Force $\mathbf{P}_6$ is applied to link 6 of the mechanism in (Fig. 4.11a). Find the reactions in the links as well as the torque needed for the mechanism to be in static equilibrium.

Figure 4.11b–f show the force polygons for each one of the links. The torque needed for the mechanism to be in static equilibrium can be calculated from Eq. (4.7):

$$-M_2 + F_{32}h = 0 \rightarrow M_2 = F_{32}h \tag{4.7}$$

4.1.2.2 Equilibrium of a Mechanism as a Single Free Body

In (Fig. 4.12) three external forces, $\mathbf{P}_2$, $\mathbf{P}_3$ and $\mathbf{P}_4$, act on a four-bar mechanism. To apply the equilibrium equations to the mechanism (Eqs. 4.8 and 4.9), we have to consider the external forces as well as the reactions at the supports, $\mathbf{F}_{12}$ and $\mathbf{F}_{14}$, and the reaction torque, $\mathbf{M}_2$ (Fig. 4.12):

$$\sum_i \mathbf{F}_i = 0 \rightarrow \mathbf{F}_{14} + \mathbf{F}_{12} + \mathbf{P}_2 + \mathbf{P}_3 + \mathbf{P}_4 = 0 \tag{4.8}$$

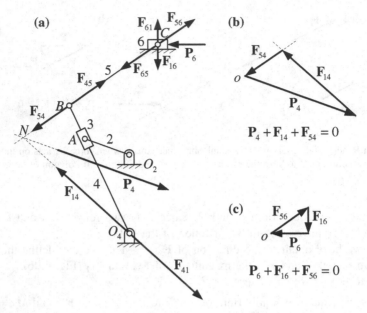

Fig. 4.10 a Reactions on each link of Witworth's mechanism when force $\mathbf{P}_4$ acts on link 4 and force $\mathbf{P}_6$ acts on link 6 to keep the mechanism in static equilibrium. **b** Polygon of forces acting on link 4. **c** Polygon of forces acting on link 6

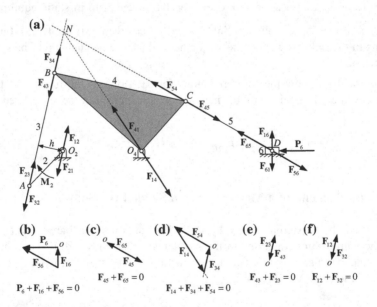

Fig. 4.11 a Reactions on each link of a six-bar mechanism when force $\mathbf{P}_4$ acts on link 4 and torque $\mathbf{M}_2$ acts on link 2 to keep the mechanism in static equilibrium. **b–f** Equilibrium equations and polygon of force vectors for links 6, 5, 4, 3 and 2

Fig. 4.12 Four-bar
mechanism with three
external forces, two reaction
forces at the supports and a
reaction torque on link 2

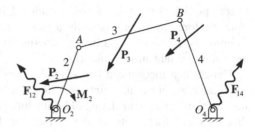

$$\sum_j \mathbf{M}_j^{O_2} = 0 \xrightarrow{z} F_{14}h_{14} - P_4h_4 - P_3h_3 + P_2h_2 + M_2 = 0 \qquad (4.9)$$

To define the moment equation about point O_2, force $\mathbf{F}_{14}$ is considered to be known (Fig. 4.13). Normally, $\mathbf{F}_{14}$ is an unknown vector and distance h_{14} between this force and support O_2 is also undetermined. Therefore, in (Eqs. 4.8 and 4.9) there are too many unknowns ($\mathbf{F}_{12}$, $\mathbf{F}_{14}$, h_{14} and M_2) and reactions cannot be calculated by using this method.

Usually, we use these equilibrium equations of the complete mechanism to verify the results obtained by any other method. However, they can also be used to find $\mathbf{F}_{12}$ and M_2, provided we have already calculated $\mathbf{F}_{14}$.

4.1.3 Superposition Principle

In the problems solved so far, we only considered one external force acting on the mechanism. When there are several forces acting on different links, the problem becomes more complicated as it is more difficult to know the direction of the reactions in the links. A simple way to solve this kind of problem is to apply the Principle of Superposition.

The Principle of Superposition applied to force analysis states that the effect produced in a mechanism by several forces acting on it is the same as the sum of the

Fig. 4.13 External forces
acting on a four-bar
mechanism and a support
reaction at point O_4 with the
distances from their lines of
action to point O_2

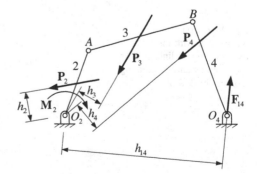

effects produced by every force considered individually. This means that if two different force systems separately applied to a mechanism keep it in static equilibrium, the system formed by the superposition of these two systems will equally keep the mechanism in equilibrium.

Hence, if a mechanism has several forces acting on its links and we want to calculate reactions at the joints and the equilibrium torque, we can do it by finding the reaction forces and the equilibrium torque caused by each force, considering that no other force acts on the rest of the links. Adding up the reactions and equilibrating moments found for each force, we reach the final solution to the problem.

When adding up reactions due to different forces, special caution has to be taken in handling sub-indexes. Force $\mathbf{F}_{jk}$ due to external load $\mathbf{P}$ has to be added up with $\mathbf{F}'_{jk}$ due to $\mathbf{P}'$ and never with $\mathbf{F}'_{kj}$.

In the case of several forces acting on the same link, the problem can be simplified by considering just the resultant.

Example 7 Forces acting on the links of the four-bar mechanism in (Fig. 4.14) are known. Determine the reactions at the four joints as well as the torque needed to apply to link 2, so that the mechanism is in static equilibrium.

This problem has already been partially solved in Example 1 (force $\mathbf{P}_4$), Example 2 (force $\mathbf{P}_3$) and Example 3 (force $\mathbf{P}_2$). Therefore, we know the reactions and the equilibrating torque of each force treated separately. The problem is reduced to a vector sum of the different reactions at each joint and an algebraic sum of the different equilibrating torques. We will use notation $\mathbf{F}_{jk}^{(2)}$ to refer to the reaction forces due to $\mathbf{P}_2$, $\mathbf{F}_{jk}^{(3)}$ to those due to $\mathbf{P}_3$ and $\mathbf{F}_{jk}^{(4)}$ to those caused by $\mathbf{P}_4$. The reactions at the joints will be expressed in Eq.(4.10) (Fig. 4.15):

$$\left.\begin{aligned}
\mathbf{F}_{14} &= \mathbf{F}_{14}^{(2)} + \mathbf{F}_{14}^{(3)} + \mathbf{F}_{14}^{(4)} = \mathbf{F}_{14}^{(3)} + \mathbf{F}_{14}^{(4)} \\
\mathbf{F}_{32} &= \mathbf{F}_{32}^{(2)} + \mathbf{F}_{32}^{(3)} + \mathbf{F}_{32}^{(4)} = \mathbf{F}_{32}^{(3)} + \mathbf{F}_{32}^{(4)} \\
\mathbf{F}_{34} &= \mathbf{F}_{34}^{(2)} + \mathbf{F}_{34}^{(3)} + \mathbf{F}_{34}^{(4)} = \mathbf{F}_{34}^{(3)} + \mathbf{F}_{34}^{(4)} \\
\mathbf{F}_{12} &= \mathbf{F}_{12}^{(2)} + \mathbf{F}_{12}^{(3)} + \mathbf{F}_{12}^{(4)}
\end{aligned}\right\}
\qquad (4.10)$$

Fig. 4.14 Four-bar mechanism with three known forces acting on links 2, 3 and 4 as well as an equilibrating moment on link 2

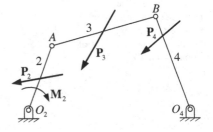

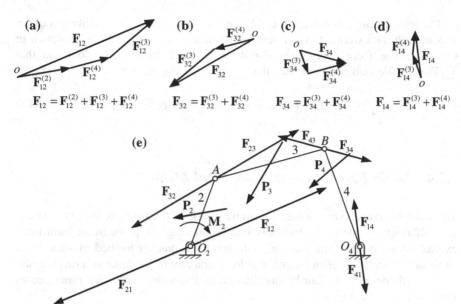

Fig. 4.15 **a–d** Polygon of forces of each link. **e** Forces at the joints due to the action of forces P_2, P_3 and P_4 on the mechanism

The same nomenclature is used for the equilibrating torques: $M_2^{(2)}$, $M_2^{(3)}$ and $M_2^{(4)}$. The magnitude of the equilibrating torque is shown in Fig. 4.16a and is calculated in Eq. (4.11):

$$M_2 = M_2^{(2)} + M_2^{(3)} + M_2^{(4)} \xrightarrow{z} M_2 = P_2 h_2^{(2)} + F_{32}^{(3)} h_{32}^{(3)} + F_{32}^{(4)} h_{32}^{(4)} \qquad (4.11)$$

where $h_2^{(2)}$, $h_{32}^{(3)}$ and $h_{32}^{(4)}$ are the distances from the lines of action of forces P_2, $F_{32}^{(3)}$ and $F_{32}^{(4)}$ to point O_2.

The torque that balances the system can also be calculated as (Fig. 4.16b); (Eq. 4.12):

$$M_2 = F_{32} h_{32} + P_2 h_2 \qquad (4.12)$$

Fig. **4.16 a** Moments of forces P_2, $F_{32}^{(3)}$ and $F_{32}^{(4)}$ about O_2. **b** Moments of forces P_2 and F_{32} about O_2

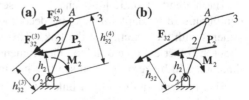

The solution of the mechanism can be verified by defining the equilibrium of the system with the external forces acting on it and the support reactions as shown in Fig. 4.13. The force polygon in Fig. 4.19 has to form a closed loop, so that $\sum_i \mathbf{F}_i = 0$. We also have to check that $\sum_i \mathbf{M}_i = 0$ (Eq. 4.13):

$$\sum_i \mathbf{M}_i^{O_2} = 0 \xrightarrow{z} F_{14}h_{14} - P_4h_4 - P_3h_3 + P_2h_2 - M_2 = 0 \qquad (4.13)$$

4.1.4 Static Force Analysis. Graphical Method

The static force analysis of a mechanism by means of the superposition principle is easy although too slow as it basically implies solving one problem for each force exerted on the mechanism. Here we will develop a quicker method of static force analysis that works with any number of forces on any of the links in a mechanism.

We will consider the four-bar mechanism in Example 7 on which three known forces are exerted (Fig. 4.14).

As seen before, if we analyze the mechanism considering it a single rigid body, the number of unknowns to be determined increases to five: the magnitude and direction of $\mathbf{F}_{12}$ and $\mathbf{F}_{14}$ and the magnitude of torque M_2. Therefore, to solve the problem, we have to solve the equilibrium of each link.

We will start with links 3 and 4. The force equations (Eqs. 4.14 and 4.15) have to be met:

$$\mathbf{P}_3 + \mathbf{F}_{43} + \mathbf{F}_{23} = 0 \qquad (4.14)$$

$$\mathbf{P}_4 + \mathbf{F}_{14} + \mathbf{F}_{34} = 0 \qquad (4.15)$$

Note that there are eight algebraic unknowns: magnitude and direction of forces $\mathbf{F}_{23}$, $\mathbf{F}_{34}$, $\mathbf{F}_{43}$ and $\mathbf{F}_{14}$. However, due to $\mathbf{F}_{34} = -\mathbf{F}_{43}$, the number of unknowns is reduced to six. The problem can, thus, be solved by formulating the equilibrium of forces (Eqs. 4.14 and 4.15) and the equilibrium of moments on both links simultaneously.

In general, we can carry out the static analysis of a planar articulated mechanism with kinematic pairs of one degree of freedom by solving the equilibrium equations of each couple of links. We will start with the one placed farthest from the input link and finish with the equilibrium of the input link itself.

The method of static force analysis presented here consists of formulating the equilibrium of a couple of links by decomposing their joining force in two components with the direction of the links. This way, when we calculate the moments about the other extreme of a link, the moment of the force component with the link direction will be zero. We can see this in Fig. 4.17.

The moment equilibrium of link 4 is considered through equation $\sum_i \mathbf{M}_i^{O_4} = 0$. As we don't know the line of action of force $\mathbf{F}_{34}$, we don't know its distance to

Fig. 4.17 a Link 3: the moment of force component $\mathbf{F}_{43}^{N3}$ about point A is zero. **b** Link 4: the moment of force component $\mathbf{F}_{34}^{N3}$ about O_4 is zero

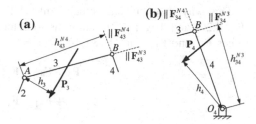

point O_4 either. So, instead of considering the moment of $\mathbf{F}_{34}$, we will consider the moment of two components of the force, one in the direction of link 3, $\mathbf{F}_{34}^{N3}$, and another one in the direction of link 4, $\mathbf{F}_{34}^{N4}$. This way only the first one generates a torque about O_4 with a known distance between its line of action and point O_4, as seen in (Fig. 4.17b).

In order to formulate the moment equilibrium equation, we need to know the direction of unknown force $\mathbf{F}_{34}^{N3}$. In this case, since there are only two forces, its direction will be the one producing an opposite torque to the one created by $\mathbf{P}_4$. However, if there are other forces or moments, we will have to assume an arbitrary direction on line AB. If once calculated, the magnitude of the force turns out to be positive, the direction is validated; if, on the contrary, the direction is negative, we have to change it.

We write moment equation $\sum_i \mathbf{M}_i^{O_4} = 0$ (Eq. 4.16) considering that the positive direction is counterclockwise:

$$-F_{34}^{N3} h_{34}^{N3} + P_4 h_4 = 0 \qquad (4.16)$$

Thus, we clear F_{34}^{N3} (Eq. 4.17):

$$F_{34}^{N3} = P_4 \frac{h_4}{h_{34}^{N3}} \qquad (4.17)$$

Next we will consider the equilibrium of link 3. The moment equation (Eq. 4.18) is $\sum_i \mathbf{M}_i^A = 0$ (Fig. 4.17a). Knowing that force $\mathbf{F}_{43}^{N3}$ is the same as $\mathbf{F}_{34}^{N3}$ but with an opposite direction, we can determine $\mathbf{F}_{43}^{N4}$, since $\mathbf{F}_{23}$ produces no torque about point A. Following the same criterion for the moment sign as before and assuming that the direction of $\mathbf{F}_{43}^{N4}$ is upwards (Fig. 4.17a), we obtain:

$$F_{43}^{N4} h_{43}^{N4} - P_3 h_3 = 0 \qquad (4.18)$$

Then, F_{43}^{N4} is (Eq. 4.19):

$$F_{43}^{N4} = P_3 \frac{h_3}{h_{43}^{N4}} \qquad (4.19)$$

This way, $\mathbf{F}_{43}$ and $\mathbf{F}_{34}$ remain fully defined. The force polygons of links 3 (Eq. 4.20) and 4 (Eq. 4.21) allow us to calculate the magnitude and direction of forces $\mathbf{F}_{23}$ and $\mathbf{F}_{14}$ (Fig. 4.18a, b).

$$\mathbf{P}_3 + \mathbf{F}_{43} + \mathbf{F}_{23} = 0 \qquad (4.20)$$

$$\mathbf{P}_4 + \mathbf{F}_{34} + \mathbf{F}_{14} = 0 \qquad (4.21)$$

Back to the original problem, we can see that the number of unknowns that we need to find to determine the equilibrium of the whole mechanism has been reduced to three. We can find $\mathbf{F}_{12}$ by constructing the force polygon formed by $\mathbf{P}_2$, $\mathbf{P}_3$, $\mathbf{P}_4$, $\mathbf{F}_{12}$ and $\mathbf{F}_{14}$ and which has to form a closed loop (Eq. 4.22) so that the mechanism is in equilibrium as shown in Fig. 4.19:

$$\mathbf{P}_2 + \mathbf{P}_3 + \mathbf{P}_4 + \mathbf{F}_{12} + \mathbf{F}_{14} = 0 \qquad (4.22)$$

The value of the torque, M_2, that balances the system, can be determined by formulating the moment equilibrium equation about O_2 (Eq. 4.23), considering all forces acting on the mechanism (Fig. 4.13).

$$\sum_i \mathbf{M}_i^{O_2} = 0 \xrightarrow{z} M_2 + F_{14}h_{14} + P_2h_2 - P_3h_3 - P_4h_4 = 0 \qquad (4.23)$$

Fig. 4.18 a Force polygon of forces on link 3. **b** Force polygon of forces on link 4

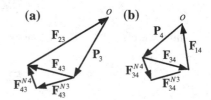

Fig. 4.19 Force polygon of forces on mechanism in Example 7

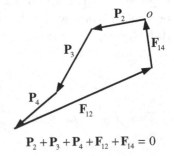

$$\mathbf{P}_2 + \mathbf{P}_3 + \mathbf{P}_4 + \mathbf{F}_{12} + \mathbf{F}_{14} = 0$$

Fig. 4.20 **a** Forces acting on link 2 and their distances to O_2. **b** Force polygon of link 2

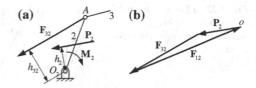

In order to find force $\mathbf{F}_{12}$ and torque M_2, we can also write the equilibrium equations of link 2 (Eqs. 4.24 and 4.25) (Fig. 4.20a). Assuming that M_2 has a clockwise direction:

$$\mathbf{P}_2 + \mathbf{F}_{12} + \mathbf{F}_{32} = 0 \tag{4.24}$$

$$\sum_i \mathbf{M}_i^{O_2} = 0 \xrightarrow{z} -M_2 + F_{32}h_{32} + P_2h_2 = 0 \tag{4.25}$$

By drawing the force polygon (Fig. 4.20b), we can calculate force $\mathbf{F}_{12}$ in (Eq. 4.24) and the value of M_2 can be cleared from (Eq. 4.25).

Example 8 Known forces $\mathbf{P}_3$, $\mathbf{P}_4$ and $\mathbf{P}_6$ are exerted on the mechanism in Fig. 4.21. Find the support reaction forces and the torque that we need to apply to link 2 to keep the mechanism is static equilibrium.

- We start by calculating the equilibrium of links 5 and 6 to obtain force $\mathbf{F}_{54}$ (Figs. 4.21 and 4.22c). The steps to be followed for this task are the same as in Example 4.
- Next, we analyze the static forces on links 3 and 4.
- $\sum_i \mathbf{M}_i^{O_4} = 0$ in link 4, therefore:

$$F_{34}^{N3} = \frac{P_4h_4 + F_{54}h_{54}}{h_{34}^{N3}}$$

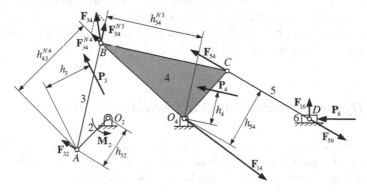

Fig. 4.21 Forces $\mathbf{P}_3$, $\mathbf{P}_4$ and $\mathbf{P}_6$ acting on a six-link mechanism and the reactions at the joints with the necessary distances to formulate the moment equilibrium equations of the links

Fig. 4.22 **a** Force polygon of
link 3. **b** Force polygon of the
mechanism. **c** Force polygon
of link 6

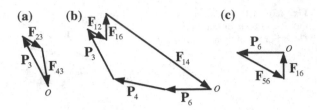

- $\sum_i \mathbf{M}_i^A = 0$ in link 3, hence:

$$F_{43}^{N4} = P_3 \frac{h_3}{h_{43}^{N4}}$$

- Adding up vectors $\mathbf{F}_{34}^{N3}$ and $\mathbf{F}_{34}^{N4}$, we determine $\mathbf{F}_{34}$ (Fig. 4.21).
- As we know that $\mathbf{F}_{34} = -\mathbf{F}_{43}$, we can draw the force polygon of link 2 to find $\mathbf{F}_{32}$ (Fig. 4.22a).
- On link 2, we know that $\mathbf{F}_{12} = -\mathbf{F}_{32}$.
- We can determine $\mathbf{F}_{14}$ by constructing the force polygon of link 4. However, it can also be determined through a force polygon with all the forces acting on the mechanism (Fig. 4.22b).
- The magnitude of the equilibrating torque $M_2 = F_{32}h_{32}$ is with the direction shown in Fig. 4.21.

4.2 Dynamic Analysis

So far, we have made a static analysis of forces in a mechanism. This study is interesting in itself and has allowed us to develop a series of interesting resolution techniques.

However, our real concern is the dynamic analysis of the mechanism. We will formulate force and moment equilibrium, considering not only the external forces and torques acting on it but also those actions generated by the motion of the links, in other words, taking into account inertial forces.

4.2.1 Dynamic Equilibrium of a Particle with Mass

Newton's Second Law allows us to write Eq. (4.26) to establish the dynamic equilibrium of a particle or a solid body in which we suppose all its mass to be concentrated at its center of mass. This law can be written as:

$$\sum_i \mathbf{F}_i = m\mathbf{a} \qquad (4.26)$$

D'Alembert proposed a modification to this approach by applying a new force called force of inertia $\mathbf{F}^{In}$, whose value is $-m\mathbf{a}$ and went on to consider it an additional external force. This way the equilibrium of the system is written as (Eqs. 4.27–4.28):

$$\sum_i \mathbf{F}_i + \mathbf{F}^{In} = 0 \qquad (4.27)$$

$$\sum_i \mathbf{F}_i - m\mathbf{a} = 0 \qquad (4.28)$$

The inertia force is, consequently, a vector with magnitude ma and an opposite direction to the acceleration.

4.2.2 Inertia Components of a Link with Planar Motion

When a link has planar motion, we will consider the forces acting on it to be on the same plane.

We can consider that a link is formed by a set of particles, each one of them having an inertial force as we have defined before. Consider we know the mass, m, of the rigid body in (Fig. 4.23a). We also know its center of mass that coincides with the center of gravity, G, as well as its angular velocity and acceleration, ω and α, the acceleration of point G and the relative acceleration of any point P with respect to G. Consider a mass, dm, associated to a particle at point P.

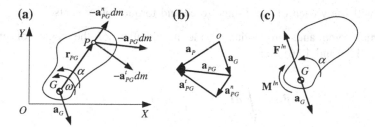

Fig. 4.23 **a** Rigid body with its center of mass located at point G. **b** Acceleration of point P can be calculated in terms of the acceleration of point G and the acceleration of P with respect to G. **c** The effects of inertia on a solid body are a force $\mathbf{F}^{In}$ and a torque $\mathbf{M}^{In}$

The acceleration of point P in terms of the acceleration of G (Eq. 4.29) is:

$$\mathbf{a}_P = \mathbf{a}_G + \mathbf{a}_{PG} \tag{4.29}$$

We can write the same equation breaking into its normal and tangential components (Eq. 4.30) (Fig. 4.23b):

$$\mathbf{a}_P = \mathbf{a}_G + \mathbf{a}_{PG}^n + \mathbf{a}_{PG}^t \tag{4.30}$$

where vector $\mathbf{a}_{PG}^n = \boldsymbol{\omega} \wedge (\boldsymbol{\omega} \wedge \mathbf{r}_{PG})$ has magnitude $a_{PG}^n = \omega^2 \overline{PG}$ and its direction is defined from P to G. The tangential component $\mathbf{a}_{PG}^t = \boldsymbol{\alpha} \wedge \mathbf{r}_{PG}$ has magnitude $a_{PG}^t = \alpha \overline{PG}$ and its direction is perpendicular to $\overline{PG}$ following the direction of α.

We can formulate the dynamic equilibrium of the particle (Eqs. 4.31 and 4.32) by multiplying the terms in (Eq. 4.30) by dm and changing its sign:

$$-\mathbf{a}_P dm = -\mathbf{a}_G dm - \mathbf{a}_{PG}^n dm - \mathbf{a}_{PG}^t dm \tag{4.31}$$

$$d\mathbf{F}_P = d\mathbf{F}_G + d\mathbf{F}_{PG}^n + d\mathbf{F}_{PG}^t \tag{4.32}$$

We can study the effect of inertia on the whole body (Eq. 4.33) integrating (Eq. 4.32) along its volume.

$$\mathbf{F}^{In} = \mathbf{F}_P = \int_V d\mathbf{F}_G + \int_V d\mathbf{F}_{PG}^n + \int_V d\mathbf{F}_{PG}^t \tag{4.33}$$

We have to obtain not only the resultant of each force component but also the resultant of their torques about G (Eq. 4.34):

$$\mathbf{M}^{In} = \mathbf{r}_{PG} \wedge \mathbf{F}_P = \int_V \mathbf{r} \wedge d\mathbf{F}_G + \int_V \mathbf{r} \wedge d\mathbf{F}_{PG}^n + \int_V \mathbf{r} \wedge d\mathbf{F}_{PF}^t \tag{4.34}$$

We will study each of these three force and torque components:

- Inertial force and torque due to the acceleration of the center of gravity (Eqs. 4.35 and 4.36) are:

$$\mathbf{F}_G = \int_V d\mathbf{F}_G = -\int_V \mathbf{a}_G dm = -m\mathbf{a}_G \tag{4.35}$$

$$\mathbf{M}_G = \int_V \mathbf{r} \wedge d\mathbf{F}_G = -\int_V \mathbf{r} \wedge \mathbf{a}_G dm = 0 \tag{4.36}$$

The resultant torque value is zero since **r** is the radius vector to center of gravity G and, therefore, $\int_V \mathbf{r}\,dm$ is zero.

- The force term due to normal acceleration (Eq. 4.37) is:

$$\mathbf{F}^n = \int_V d\mathbf{F}^n_{PG} = -\int_V \boldsymbol{\omega} \wedge (\boldsymbol{\omega} \wedge \mathbf{r})\,dm = 0 \qquad (4.37)$$

The torque value (Eq. 4.38) will be:

$$\mathbf{M}^n = \int_V \mathbf{r} \wedge d\mathbf{F}^n_{PG} = -\int_V \mathbf{r} \wedge (\boldsymbol{\omega} \wedge (\boldsymbol{\omega} \wedge \mathbf{r}))\,dm = 0 \qquad (4.38)$$

Both, force and torque generated by the normal component of acceleration are null, due to $\int_V \mathbf{r}\,dm = 0$.

- The inertial force component due to the tangential acceleration (Eq. 4.39) is:

$$\mathbf{F}^t = \int_V d\mathbf{F}^t_{PG} = -\int_V \boldsymbol{\alpha} \wedge \mathbf{r}\,dm = 0 \qquad (4.39)$$

This component is zero for the same reason as the previous ones.
The torque about point G (Eq. 4.40) is:

$$\mathbf{M}^t = \int_V \mathbf{r} \wedge d\mathbf{F}^t_{PG} = \int_V \mathbf{r} \wedge (\boldsymbol{\alpha} \wedge \mathbf{r})\,dm \qquad (4.40)$$

Since we know that both vectors form an angle of 90°, we can obtain its magnitude (Eq. 4.41) as:

$$M^t = \int_V \alpha r^2 dm = I_G \alpha \qquad (4.41)$$

where I_G is the polar moment of inertia about its center of gravity.
Hence, the inertia effect on the link in (Fig. 4.23c) is reduced to two vectors:

- An inertial force (Eq. 4.42) with opposite direction to the linear acceleration of the center of gravity defined by:

$$\mathbf{F}^{In} = -m\mathbf{a}_G \qquad (4.42)$$

- An inertial torque (Eq. 4.43) opposite to the angular acceleration of the body given by:

$$\mathbf{M}^{In} = -I_G\boldsymbol{\alpha} = -I_G\alpha\hat{\mathbf{k}} \qquad (4.43)$$

4.2.3 Inertial Components of a Plane Link

All forces due to motion that are exerted on a plane rigid link of a mechanism can be reduced to a single force. Its magnitude (Eq. 4.44) depends on the acceleration of its center of gravity, $\mathbf{a}_G$, and its mass, m, as follows:

$$F^{In} = -ma_G \qquad (4.44)$$

Furthermore, this force has the same line of action as vector but with an opposite direction as shown in (Fig. 4.24).

Also, on every link that moves with angular acceleration a torque (Eq. 4.45) acts with a magnitude of:

$$M^{In} = -I_G\alpha \qquad (4.45)$$

where I_G is the mass moment of inertia of the link about its center of gravity, G, and α is its angular acceleration.

This way, the effects of inertia on a link with planar motion are fully defined. When carrying out a dynamic analysis by means of the force analysis method of the links, we can compute the inertial force and torque considering them one more term in the force and moment equilibrium equations respectively.

Example 9 Consider the crank-shaft mechanism in (Fig. 4.25a) with link 2 moving with constant angular velocity ω_2. We already know the acceleration polygon as well as the mass and moment of inertia about its center of gravity of each link. Find the effect of inertia on every link.

Fig. 4.24 The effect of inertia on link 3 can be considered a torque that acts on the link and a force acting at its center of gravity

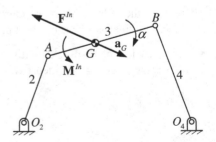

Fig. 4.25 a Crank-shaft
mechanism. b Acceleration
polygon

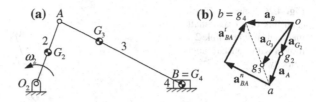

From the acceleration polygon we can obtain vectors $\mathbf{a}_{G_2}$, $\mathbf{a}_{G_3}$ and $\mathbf{a}_{G_4}$.
Knowing the mass of links, m_2, m_3 and m_4, we can calculate the inertial forces
(Eqs. 4.46–4.48):

$$\mathbf{F}_2^{In} = -m_2\mathbf{a}_{G_2} \qquad\qquad (4.46)$$

$$\mathbf{F}_3^{In} = -m_3\mathbf{a}_{G_3} \qquad\qquad (4.47)$$

$$\mathbf{F}_4^{In} = -m_4\mathbf{a}_{G_4} \qquad\qquad (4.48)$$

Moreover, to calculate the inertial torque value that acts on each link, we need to
calculate the angular acceleration value of each link first.

- The angular acceleration of link 2, α_2, is zero as it moves with constant angular
  velocity ω_2.
- The angular acceleration of link 4 is equally zero, since it moves along a straight
  path.
- The angular acceleration of link 3 (Eq. 4.49) can be calculated from the tangential acceleration of point B with respect to A, $\mathbf{a}_{BA}^t$:

$$\alpha_3 = \frac{a_{BA}^t}{BA} \qquad\qquad (4.49)$$

The direction of α_3 depends on the direction of vector $\mathbf{a}_{BA}^t$. In this case, it is
counterclockwise. Once we know the angular acceleration of each link, we can
calculate the magnitudes of the inertial torques (Eqs. 4.50–4.52) as follows:

$$M_2 = I_{G_2}\alpha_2 = 0 \qquad\qquad (4.50)$$

$$M_3 = I_{G_3}\alpha_3 > 0 \qquad\qquad (4.51)$$

$$M_4 = I_{G_4}\alpha_4 = 0 \qquad\qquad (4.52)$$

Figure 4.26 shows the inertial forces and torques on the mechanism. The inertial
forces are exerted on centers of gravity G_2, G_3 and G_4 and their lines of action
coincide with the ones of vectors $\mathbf{a}_{G_2}$, $\mathbf{a}_{G_3}$ and $\mathbf{a}_{G_4}$.

Fig. 4.26 Effect of inertia on each link of a crank-shaft mechanism when link 2 moves with constant angular velocity

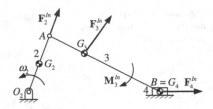

Note that despite the fact that link 2 moves with a constant angular velocity, it is under the action of inertial force $\mathbf{F}_2^{In}$, which is different to zero. This force is generated by normal acceleration and is referred to as centrifugal force.

Example 10 Find the effect of inertia acting on link 6 of the mechanism in (Fig. 4.27). Consider the acceleration polygon of the mechanism to be known (Fig. 4.28), as well as mass m_6 and moment of inertia I_{G_6} of the link. Its center of gravity G_6 coincides with steady point O_6.

We start by computing the value of inertia force $\mathbf{F}_6^{In}$. In order to do this, we have to find the acceleration of G_6. There is no need to use the acceleration polygon yet as we can see that the acceleration has to be zero due to the fact that G_6 coincides with steady point O_6. Therefore, the value of inertia force $\mathbf{F}_6^{In}$ is null.

However, there is an angular acceleration, α_6, which has to generate an inertial torque of magnitude $I_{G_6}\alpha_6$ as we saw in Sect. 4.2.3 in this chapter. To determine its

Fig. 4.27 Six-link mechanism

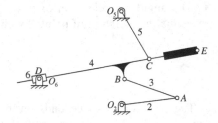

Fig. 4.28 Acceleration polygon of the six-link mechanism in Fig. 4.27

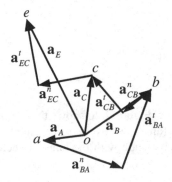

value we need to find α_6 first. Observe (Fig. 4.27) and note that the angular velocity of link 6 has to be equal to the angular velocity of link 4 as both links turn jointly.

Since the acceleration polygon has already been determined, we can find the value of the tangential acceleration of point E with respect to C, $\mathbf{a}^t_{EC}$. Next, we obtain the value of α_6 (Eq. 4.53):

$$\alpha_4 = \frac{a^t_{EC}}{EC} = \alpha_6 \tag{4.53}$$

Its direction will be counter-clockwise. With α_6 and moment of inertia I_{G_6} (stated at the problem wording), we can calculate the magnitude of the inertial torque on link 6 (Eq. 4.54). Its direction is opposite to α_6.

$$M^{In}_6 = I_{G_6}\alpha_6 \tag{4.54}$$

4.2.4 Inertia Force Analysis in a Mechanism

Once the inertial forces acting on the different links of a mechanism have been calculated, we are interested in learning about the effect of these forces on the supports, as well as the torque exerted on the input link that produces the velocity and acceleration of the mechanism. We can do this by considering the inertial force of each link as a static force and then solving the problem using the method developed in Sect. 4.1.4.

It is also interesting, once the inertial forces have been found, to obtain their resultant force. This force, which will be explained further ahead in this chapter, is referred to as trepidation and it has to be taken into account when calculating the support bench. Trepidation can cause such strong vibrations in machines that all this has to be thoroughly studied.

We will carry out a full analysis of the inertia forces acting on a four-bar mechanism as an illustrative example.

Example 11 Find the reaction forces on the links and the input torque on link 2 at the given position in the four-bar mechanism in (Fig. 4.29a). Consider the acceleration polygon to be known, as well as the masses, centers of gravity and moments of inertia of links 2, 3 and 4. Link 2 rotates with angular velocity ω_2 and angular acceleration α_2, both known.

First, we have to calculate the magnitude of the inertia force that acts on each link. To do so, we need to know the acceleration of their center of gravity. Figure 4.29b shows the acceleration polygon of the mechanism. In (Fig. 4.29c) the accelerations of the centers of gravity accelerations, $\mathbf{a}_{G_2}$, $\mathbf{a}_{G_3}$ and $\mathbf{a}_{G_4}$, have been determined using the acceleration image principle.

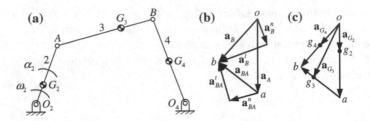

Fig. 4.29 a Four-bar mechanism with the center of gravity of each link. **b** Acceleration polygon with the accelerations of points A and B. **c** Acceleration polygon used to determine the acceleration of the centers of gravity

Therefore, the magnitudes of the inertia forces (Eqs. 4.55–4.57) are:

$$F_2^{In} = m_2 a_{G_2} \qquad (4.55)$$

$$F_3^{In} = m_3 a_{G_3} \qquad (4.56)$$

$$F_4^{In} = m_4 a_{G_4} \qquad (4.57)$$

The point of application of these forces is the center of gravity, G, of each link and their directions are opposite to vectors a_{G_2}, a_{G_3} and a_{G_4} respectively.

To determine the inertial torques in the system, we need to know the angular acceleration of links 2, 3 and 4:

- α_2 is given at the problem wording.
- $\alpha_4 = \frac{a_{BA}^t}{BA}$ (counterclockwise).
- $\alpha_4 = \frac{a_B^t}{O_4 B}$ (counterclockwise).

The magnitudes of the inertial torques (Eqs. 4.58–4.60), whose directions are always opposite to the direction of the angular accelerations, are:

$$M_2 = I_{G_2} \alpha_2 \qquad (4.58)$$

$$M_3 = I_{G_3} \alpha_3 \qquad (4.59)$$

$$M_4 = I_{G_4} \alpha_4 \qquad (4.60)$$

Figure 4.30 shows the mechanism and all its inertial forces and torques acting on the different links.

Next, we can calculate the reaction forces acting on the supports as well as the equilibrium torque on link 2 by using the static force analysis method.

In order to determine F_{34}^{N3}, we analyze link 4 considering the equilibrium of moments about point O_4. F_{43}^{N3} has the same magnitude as F_{34}^{N3} but with opposite direction (Fig. 4.31a).

Fig. 4.30 Inertial force and torques that act on each link of the four-bar mechanism in Fig. 4.29a

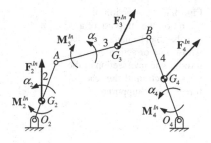

Fig. 4.31 **a** Link 4 with the reaction force in joint B and inertia force $\mathbf{F}_4^{In}$. **b** Link 3 with the reaction force in joint B and inertia force $\mathbf{F}_3^{In}$

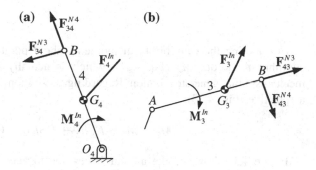

Formulating the equilibrium of link 3 and considering the equilibrium of moments about point A, we can determine $\mathbf{F}_{43}^{N4}$ (Fig. 4.31b). Then, drawing the polygon of forces acting on link 3, we can find $\mathbf{F}_{23}$ (Eq. 4.61) as the vector that closes the polygon (Fig. 4.32b):

$$\mathbf{F}_3^{In} + \mathbf{F}_{43}^{N3} + \mathbf{F}_{43}^{N4} + \mathbf{F}_{23} = 0 \tag{4.61}$$

Finally, we write the equilibrium equations of links 2 (Eq. 4.62) and 4 (Eq. 4.63) to obtain the reactions at the supports, $\mathbf{F}_{12}$ and $\mathbf{F}_{14}$ (Fig. 4.32a–c):

$$\mathbf{F}_2^{In} + \mathbf{F}_{12} + \mathbf{F}_{32} = 0 \tag{4.62}$$

$$\mathbf{F}_4^{In} + \mathbf{F}_{14} + \mathbf{F}_{34} = 0 \tag{4.63}$$

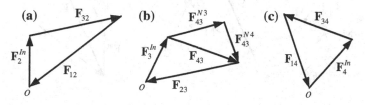

Fig. 4.32 **a** Polygon of forces acting on link 2. **b** Polygon of forces that act on link 3. **c** Polygon of forces acting on link 4

Fig. 4.33 The action of
inertia on the supports of a
four-bar mechanism is a force
whose line of action goes
through the intersection of the
lines of action of the reaction
forces at the supports

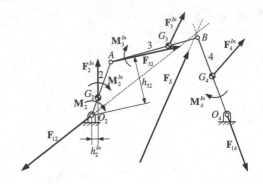

Considering the equilibrium of moments about point O_2 on link 2, we obtain equilibrating torque M_2 (Eq. 4.64), which is actually the torque acting on the mechanism that generates motion. Regarding those moments in clockwise direction as positive (Fig. 4.33):

$$M_2^{In} + M_2 - F_2^{In} h_2^{In} + F_{32} h_{32} = 0 \qquad (4.64)$$

Figure 4.33 shows all the inertial forces and torques, the reaction forces at the supports and the input torque. It is interesting to know the resultant force, $\mathbf{F}_S$, exerted by inertia at the supports. To calculate this force, which is actually the action of inertia on the bench (frame link), we just have to add up the reactions at the supports. This way, we obtain both its magnitude and direction, since $\mathbf{F}_S$ equals $\mathbf{F}_{12} + \mathbf{F}_{14}$ but with an opposite direction.

If we add up inertia forces $\mathbf{F}_2^{In}$, $\mathbf{F}_3^{In}$ and $\mathbf{F}_4^{In}$, we can also find the magnitude of $\mathbf{F}_S$ but not its line of action (Fig. 4.34b).

Nevertheless, we can determine the line of action of $\mathbf{F}_S$ (Eq. 4.66) without needing to calculate reactions at the supports first, but just calculating the moments

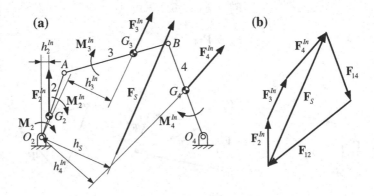

Fig. 4.34 a Distances to calculate the torque produced by $\mathbf{F}_S$ and by inertia forces. **b** Force polygon for inertia forces and reactions at the supports

about any point. The torque produced by $\mathbf{F}_S$ has to be equal to the sum of torques produced by inertial forces and moments. Figure 4.34a is a graphical example of this last statement. Calculating the moments about O_2 (Eq. 4.65) and considering the positive direction to be counterclockwise:

$$F_S h_S = F_2^{In} h_2^{In} + F_3^{In} h_3^{In} + F_4^{In} h_4^{In} - M_2^{In} - M_3^{In} - M_4^{In} \qquad (4.65)$$

Thus:

$$h_S = \frac{F_2^{In} h_2^{In} + F_3^{In} h_3^{In} + F_4^{In} h_4^{In} - M_2^{In} - M_3^{In} - M_4^{In}}{F_S} \qquad (4.66)$$

This way, $\mathbf{F}_S$ can be determined without carrying out a complete analysis of the inertial forces in the mechanism.

The value of inertial forces varies along the kinematic cycle of a mechanism. If $\mathbf{F}_S$ has to be considered when calculating the machine bench, we should carry out the inertial force analysis at the position of maximum acceleration of the output link as this is usually also the position of the highest magnitude of $\mathbf{F}_S$. When the mass of any of the links in the mechanism is considerably higher than the rest, it is possible that meets $\mathbf{F}_S$ its highest value at the position of maximum acceleration of such a link even though the remaining links have minimum acceleration. This implies that we should carry out some tests so that we can be sure that we are considering the worst case scenario of the effects of inertial forces on the machine bench.

Example 12 Link 2 of the quick return mechanism in Fig. 4.35a moves clockwise at angular speed ω_2 and angular acceleration α_2. At the instant shown in Fig. 4.35a, force $\mathbf{P}_6$ acts on link 6 opposing the motion of the slider. Find the forces on the links and the torque acting on link 2 that produce this motion.

Assume that we know moments of inertia I_{G_2}, I_{G_3}, I_{G_4}, I_{G_5} and I_{G_6} as well as masses m_2, m_3, m_4, m_5 and m_6 and centers of gravity G_2, G_3, G_4, G_5 and G_6. Figure 4.35b shows the acceleration polygon from which we can obtain the accelerations of the centers of gravity by means of the acceleration image principle. We can also calculate angular acceleration of the links from the tangential acceleration components.

Once we know all these accelerations, we can calculate the force and inertial torque acting on each link. Figure 4.36a shows all these forces and inertia moments acting on the mechanism.

We start the study by analyzing links 5 and 6 and calculating the force transmitted between both links.

Forces acting on link 6 are $\mathbf{P}_6$ and $\mathbf{F}_6^{In}$, completely known, $\mathbf{F}_{16}$ with known direction and unknown magnitude as well as $\mathbf{F}_{56}$ with unknown magnitude and direction (Fig. 4.36b). $\mathbf{F}_{56}$ can be broken into its components, $\mathbf{F}_{56}^{T6}$ and $\mathbf{F}_{56}^{N6}$. Since $\mathbf{F}_{16}$ has no horizontal component, we can determine the horizontal component of $\mathbf{F}_{56}$ as $\mathbf{F}_{56}^{T6} = \mathbf{F}_6^{In} + \mathbf{P}_6$. The direction of $\mathbf{F}_{56}^{T6}$ is opposite to $\mathbf{F}_6^{In}$ and $\mathbf{P}_6$.

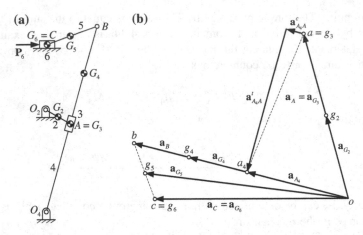

Fig. 4.35 **a** Quick-return mechanism with external force acting on link 6. **b** The accelerations of the centers of gravity have been determined on the acceleration polygon using the acceleration image method

Fig. 4.36 **a** Quick-return mechanism with inertia forces and torques acting on each link. **b** Forces acting on link 6. **c** Free-body diagram of link 5

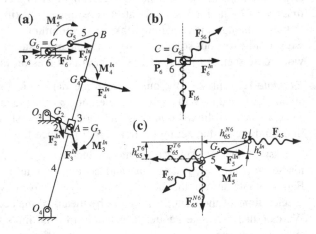

To determine the vertical component, $\mathbf{F}_{56}^{N6}$, we have to study link 5. We can determine $\mathbf{F}_{65}^{N6}$ with the equation for moments equilibrium. We calculate the moments about point B (Eq. 4.67) considering torques with counterclockwise direction to be positive and also considering that force $\mathbf{F}_{65}^{N6}$ is headed downwards (Fig. 4.36c) so that $\sum_i \mathbf{M}_i^B = 0$:

$$-M_5^{In} + F_5^{In} h_5^{In} + F_{65}^{N6} h_{65}^{N6} + F_{65}^{T6} h_{65}^{T6} = 0 \qquad (4.67)$$

Fig. 4.37 a Polygon of forces acting on link 6. b Force polygon for forces on link 5

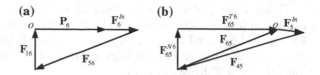

Clearing F_{65}^{N6}, we find its magnitude and direction. Force $\mathbf{F}_{65}$ (Eq. 4.38) is, consequently, fully defined as:

$$\mathbf{F}_{65} = \mathbf{F}_{65}^{N6} + \mathbf{F}_{65}^{T6} \tag{4.68}$$

Back to link 6, we can verify that all forces acting on it are now defined since $\mathbf{F}_{56} = -\mathbf{F}_{65}$ and $\mathbf{F}_{16} = \mathbf{F}_{65}^{N6}$ (Fig. 4.37a).

We formulate $\sum_i \mathbf{F}_i = 0$ in link 5. This way, we can find $\mathbf{F}_{45}$, the only force acting on it that remains undetermined. Figure 4.37b shows the force polygon of this link.

The next step is to examine links 3 and 4 and calculate the force transmitted between them. As stated in Sect. 4.1.1, this force has to be perpendicular to link 4 (Fig. 4.1b).

First consider the equilibrium of link 4 on which forces $\mathbf{F}_4^{In}$ and $\mathbf{F}_{54}$, of known magnitude and direction, are exerted. Forces $\mathbf{F}_{34}$, of known direction but unknown location of its line of action, and $\mathbf{F}_{14}$, totally unknown, are also acting on this link (Fig. 4.38a).

Hence, we have four unknowns. The point of application of force $\mathbf{F}_{34}$ depends on the force distribution in the contact area between slider 3 and link 4.

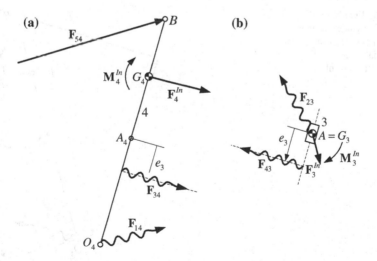

Fig. 4.38 a Free-body diagram of link 4 with the line of action of force $\mathbf{F}_{34}$ located at distance e_3 from G_3. b Free-body diagram of link 3

To solve this problem, we are going to study the equilibrium of link 3. Since there is an inertia torque acting on it (Fig. 4.38b), the reaction of link 4 on link 3, $\mathbf{F}_{43}$, has to be located in such a way that this force balances the torque (Eq. 4.69) since force $\mathbf{F}_{23}$ passes through point A and produces no torque about this point. Thus, $\mathbf{F}_{43}$ is at distance e_3 from G_3:

$$F_{43}e_3 = I_{G_3}\alpha_3 \tag{4.69}$$

Torque $F_{43}e_3$ opposes $I_{G_3}\alpha_3$. However, because the magnitude of $\mathbf{F}_{43}$ remains unknown, we cannot determine the value of e_3. We know that $\mathbf{F}_{34}$ has to produce a torque of equal value and opposite direction to the one produced by $\mathbf{F}_{43}$. In other words, it has the same magnitude and direction as the inertia torque of link 3. This way, we can assume that a torque is applied to link 4 with value $I_{G_3}\alpha_3$ and that $\mathbf{F}_{34}$ acts at point A, which coincides with G_3 (Fig. 4.39).

Considering that $\sum\limits_i \mathbf{M}_i^{O_4} = 0$ (Eq. 4.70), we can determine $\mathbf{F}_{34}$ (Eq. 4.71):

$$F_4^{In}h_4 + F_{54}h_{54} - F_{34}h_{34} + I_{G_4}\alpha_4 + I_{G_3}\alpha_3 = 0 \tag{4.70}$$

Thus:

$$F_{34} = \frac{F_4^{In}h_4 + F_{54}h_{54} + I_{G_4}\alpha_4 + I_{G_3}\alpha_3}{h_{34}} \tag{4.71}$$

The force polygon of link 4 gives us the magnitude and direction of $\mathbf{F}_{14}$ (Fig. 4.40a).

Fig. 4.39 Free-body diagram of link 4 considering that force $\mathbf{F}_{34}$ acts at point G_3 and the inertial torque of link 3 acts directly on link 4

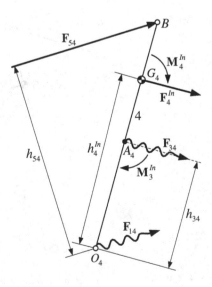

Fig. 4.40 **a** Force polygon of link 4. **b** Force polygon of link 3

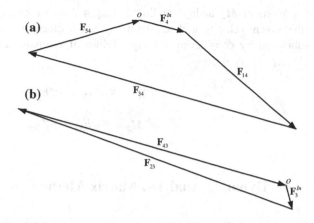

Back to link 3, we can now calculate the value of distance e_3 (Eq. 4.72) that defines the location of the line of action of force $\mathbf{F}_{43}$:

$$e_3 = \frac{I_{G_3}\alpha_3}{F_{43}} \tag{4.72}$$

Furthermore, the force polygon that is shown in Fig. 4.40b yields the value of $\mathbf{F}_{23}$.

Last, (Fig. 4.41a) shows the forces and torques acting on link 2. The direction and magnitude of $\mathbf{F}_{12}$ are obtained through the force polygon (Fig. 4.41b).

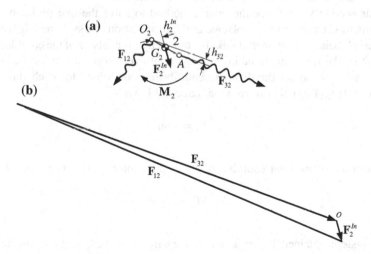

Fig. 4.41 **a** Free-body diagram of link 2. **b** Polygon of forces acting on link 2

Moment M_2 acting on link 2 (Eq. 4.74), for the mechanism to move with the given velocity and acceleration, has a clockwise direction and its value is obtained by considering the equilibrium of moments about point O_2 (Eq. 4.73), $\sum_i \mathbf{M}_i^{O_2} = 0$, thus:

$$-M_2 + F_{32}h_{32} - F_2^{In}h_2^{In} = 0 \qquad (4.73)$$

$$M_2 = F_{32}h_{32} - F_2^{In}h_2^{In} \qquad (4.74)$$

4.3 Dynamic Analysis. Matrix Method

Graphical methods of force analysis in planar mechanisms discussed so far have clear positive aspects because of their great explanatory clarity and great educational value, which they have in common, among other things, with graphic kinematic analysis methods of mechanisms. Similarly to the latter, there are analytical methods that can be employed in the dynamic analysis of a planar mechanism, which we will look into below.

There are two types of dynamic problems. In the first, the movement of all the links is defined in terms of the motion of the input link and we are asked to find the forces and moments associated with this movement. In other words, given the motion variables, we have to find the forces that have to be applied to produce it. This is what we call the inverse problem. The second type is the direct problem. Given a set of forces, we have to obtain the motion that occurs through the integration of a system of nonlinear differential equations.

In this section we will use the matrix method to solve the first problem.

This method can be briefly explained as the formulation of a system with $3(N-1)$ equations (N being the number of links) obtained by the analysis of the equilibrium of each link of the mechanism, taking into account every force acting on it. For this purpose, we have to set the dynamic equilibrium conditions for each link in the mechanism (Eq. 4.75), that is, Newton's Second Law:

$$\sum_i \mathbf{F}_i = m\mathbf{a}_G \qquad (4.75)$$

As well as the moment equilibrium about the center of gravity G (Eq. 4.76):

$$\sum_i \mathbf{M}_i^G = I_G\boldsymbol{\alpha} \qquad (4.76)$$

The system obtained is linear and can easily be solved, yielding the reaction forces at the joints and the input torque, provided we have already performed the acceleration analysis.

Fig. 4.42 Four-bar
mechanism in movement with
all the parameters involved in
its dynamic analysis

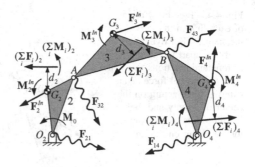

To explain and develop this method, we will make use of the four-bar mechanism in Fig. 4.42.

The following nomenclature will be used to define the characteristics of given link j:

- m_j Mass.
- I_{G_j} Moment of inertia about the center of gravity G_j.
- θ_j Angular position.
- ω_j Angular velocity.
- α_j Angular acceleration.

Forces acting at joints are given as $\mathbf{F}_{jk}$, which represents the force exerted by link j on link k. Also, the system can also be under the action of external forces and torques applied to each link, $(\sum \mathbf{F})_j$ and $(\sum \mathbf{M})_j$, different to motor torque $\mathbf{M}_0$ applied to the motor link.

Figure 4.42 shows all the parameters involved in the dynamic analysis of the four-bar mechanism. Only forces exerted by link $j+1$ on link j are displayed. Once these forces have been found, forces exerted by link j on link $j+1$ can easily be found as $\mathbf{F}_{jk} = -\mathbf{F}_{kj}$. This way, only forces $\mathbf{F}_{21}$, $\mathbf{F}_{32}$, $\mathbf{F}_{43}$ and $\mathbf{F}_{14}$ will be considered unknowns.

Distance d_j defines the position of a point on the line of action of $(\sum \mathbf{F})_j$ with respect to G_j (center of gravity of link j). Vectors $\mathbf{q}_j = \mathbf{r}_{(j-1)G_j}$ and $\mathbf{p}_j = \mathbf{r}_{(j)G_j}$ define the position of the center of gravity of link j with respect to joints with links $j-1$ and j respectively.

Figure 4.43 shows the free body diagram of each link showing the forces and parameters involved at the same time.

The motion equations of link j can be formulated (Eqs. 4.77–4.78) as:

$$\mathbf{F}_{(j+1)j} - \mathbf{F}_{j(j-1)} + \left(\sum_i \mathbf{F}_i\right)_j = m_j \mathbf{a}_{G_j} \qquad (4.77)$$

Fig. 4.43 Free-body diagram of each link in the mechanism. For dynamic analysis, apart from the actions and reactions, we will consider inertial force as an external force acting on the center of gravity and the inertia torque as an external torque acting on the link

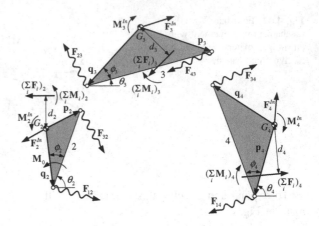

These generic equations can be applied to any of the links in the four-bar mechanism. Hence, we obtain the following equations:

$$\mathbf{r}_{(j)G_j} \wedge \mathbf{F}_{(j+1)j} - \mathbf{r}_{(j-1)G_j} \wedge \mathbf{F}_{j(j-1)} + \mathbf{d}_j \wedge \left(\sum_i \mathbf{F}_i\right)_j + \left(\sum_i \mathbf{M}_i\right)_j$$

$$= \mathbf{p}_j \wedge \mathbf{F}_{(j+1)j} - \mathbf{q}_j \wedge \mathbf{F}_{j(j-1)} + \mathbf{d}_j \wedge \left(\sum_i \mathbf{F}_i\right)_j + \left(\sum_i \mathbf{M}_i\right)_j = I_{G_j}\alpha_j \quad (4.78)$$

These generic equations can be applied to any of the links in the four-bar mechanism. Hence, we obtain the following equations:

- Link 2, (Eqs. 4.79 and 4.80):

$$\mathbf{F}_{32} - \mathbf{F}_{21} = m_2\mathbf{a}_{G_2} - \left(\sum_i \mathbf{F}_i\right)_2 \quad (4.79)$$

$$\mathbf{r}_{(2)G_2} \wedge \mathbf{F}_{32} - \mathbf{r}_{(1)G_2} \wedge \mathbf{F}_{21} + \mathbf{M}_0$$

$$= \mathbf{p}_2 \wedge \mathbf{F}_{32} - \mathbf{q}_2 \wedge \mathbf{F}_{21} + \mathbf{M}_0$$

$$= I_{G_2}\alpha_2 - \mathbf{d}_2 \wedge \left(\sum_i \mathbf{F}_i\right)_2 - \left(\sum_i \mathbf{M}_i\right)_2 \quad (4.80)$$

- Link 3, (Eqs. 4.81 and 4.82):

$$\mathbf{F}_{43} - \mathbf{F}_{32} = m_3\mathbf{a}_{G_3} - \left(\sum_i \mathbf{F}_i\right)_3 \quad (4.81)$$

$$\mathbf{r}_{(3)G_3} \wedge \mathbf{F}_{43} - \mathbf{r}_{(2)G_3} \wedge \mathbf{F}_{32}$$

$$= \mathbf{p}_3 \wedge \mathbf{F}_{43} - \mathbf{q}_3 \wedge \mathbf{F}_{32}$$

$$= I_{G_3} \alpha_3 - \mathbf{d}_3 \wedge \left(\sum_i \mathbf{F}_i \right)_3 - \left(\sum_i \mathbf{M}_i \right)_3 \qquad (4.82)$$

- Link 4, (Eqs. 4.83 and 4.84):

$$\mathbf{F}_{14} - \mathbf{F}_{43} = m_4 \mathbf{a}_{G_4} - \left(\sum_i \mathbf{F}_i \right)_4 \qquad (4.83)$$

$$\mathbf{r}_{(4)G_4} \wedge \mathbf{F}_{14} - \mathbf{r}_{(3)G_4} \wedge \mathbf{F}_{43}$$

$$= \mathbf{p}_4 \wedge \mathbf{F}_{14} - \mathbf{q}_4 \wedge \mathbf{F}_{43}$$

$$= I_{G_4} \alpha_4 - \mathbf{d}_4 \wedge \left(\sum_i \mathbf{F}_i \right)_4 - \left(\sum_i \mathbf{M}_i \right)_4 \qquad (4.84)$$

Where terms have been reorganized so that the unknowns and dimensional parameters are on the left side of the equations and the right side contains external forces and torques, which is known data.

The equations that define force equilibrium will be broken into their x and y components. We can also develop the vector product of vectors $\mathbf{r}$ and $\mathbf{F}$ (Eq. 4.85), both on the same plane:

$$\mathbf{r} \wedge \mathbf{F} = r_x F_y - r_y F_x \qquad (4.85)$$

Hence, we have (Eqs. 4.86–4.94):

$$F_{32_x} - F_{21_x} = m_2 a_{G_{2x}} - \left(\sum_i F_{i_x} \right)_2 \qquad (4.86)$$

$$F_{32_y} - F_{21_y} = m_2 a_{G_{2y}} - \left(\sum_i F_{i_y} \right)_2 \qquad (4.87)$$

$$(p_{2_x} F_{32_y} - p_{2_y} F_{32_x}) - (q_{2_x} F_{21_y} - q_{2_y} F_{21_x}) + M_0$$

$$= I_{G_2} \alpha_2 - \mathbf{d}_2 \wedge \left(\sum_i \mathbf{F}_i \right)_2 \cdot \hat{\mathbf{k}} - \left(\sum_i \mathbf{M}_i \right)_2 \qquad (4.88)$$

$$F_{43_x} - F_{32_x} = m_3 a_{G_{3x}} - \left(\sum_i F_{i_x} \right)_3 \qquad (4.89)$$

$$F_{43_y} - F_{32_y} = m_3 a_{G_{3y}} - \left(\sum_i F_{i_y} \right)_3 \tag{4.90}$$

$$(p_{3_x} F_{43_y} - p_{3_y} F_{43_x}) - (q_{3_x} F_{32_y} - q_{3_y} F_{32_x})$$
$$= I_{G_3} \alpha_3 - \mathbf{d}_3 \wedge \left(\sum_i \mathbf{F}_i \right)_3 \cdot \hat{\mathbf{k}} - \left(\sum_i M_i \right)_3 \tag{4.91}$$

$$F_{14_x} - F_{43_x} = m_4 a_{G_{4x}} - \left(\sum_i F_{i_x} \right)_4 \tag{4.92}$$

$$F_{14_y} - F_{43_y} = m_4 a_{G_{4y}} - \left(\sum_i F_{i_y} \right)_4 \tag{4.93}$$

$$(p_{4_x} F_{14_y} - p_{4_y} F_{14_x}) - (q_{4_x} F_{43_y} - q_{4_y} F_{43_x})$$
$$= I_{G_4} \alpha_4 - \mathbf{d}_4 \wedge \left(\sum_i \mathbf{F}_i \right)_4 \cdot \hat{\mathbf{k}} - \left(\sum_i M_i \right)_4 \tag{4.94}$$

Components p_{j_x}, p_{j_y}, q_{j_x} and q_{j_y} can easily be calculated knowing the distances between the centers of gravity and the joints as well as angles ϕ_j and θ_j of each link.

This way, we have nine equations to find the nine unknowns of the problem: F_{21_x}, F_{21_y}, F_{32_x}, F_{32_y}, F_{43_x}, F_{43_y}, F_{14_x}, F_{14_y} and M_0.

The system of equations can be written as a matrix multiplication and expressed in matrix form (Eq. 4.95):

$$\begin{pmatrix} -1 & 0 & 1 & 0 & 0 & 0 & 0 & 0 & 0 \\ 0 & -1 & 0 & 1 & 0 & 0 & 0 & 0 & 0 \\ q_{2_y} & -q_{2_x} & -p_{2_y} & p_{2_x} & 0 & 0 & 0 & 0 & 1 \\ 0 & 0 & -1 & 0 & 1 & 0 & 0 & 0 & 0 \\ 0 & 0 & 0 & -1 & 0 & 1 & 0 & 0 & 0 \\ 0 & 0 & q_{3_y} & -q_{3_x} & -p_{3_y} & p_{3_x} & 0 & 0 & 0 \\ 0 & 0 & 0 & 0 & -1 & 0 & 1 & 0 & 0 \\ 0 & 0 & 0 & 0 & 0 & -1 & 0 & 1 & 0 \\ 0 & 0 & 0 & 0 & q_{4_y} & -q_{4_x} & -p_{4_y} & p_{4_x} & 0 \end{pmatrix} \begin{pmatrix} F_{21_x} \\ F_{21_y} \\ F_{32_x} \\ F_{32_y} \\ F_{43_x} \\ F_{43_y} \\ F_{14_x} \\ F_{14_y} \\ M_0 \end{pmatrix} = [L]\mathbf{q}$$
$$= \mathbf{F}$$

$$\tag{4.95}$$

with:

- **F**: External force and torque vector, also including inertia forces and moments (Eq. 4.96).

$$
\mathbf{F} =
\begin{pmatrix}
m_2 a_{G_{2x}} - \left(\sum_i F_{i_x} \right)_2 \\[2mm]
m_2 a_{G_{2y}} - \left(\sum_i F_{i_y} \right)_2 \\[2mm]
I_{G_2} \alpha_2 - \mathbf{d}_2 \wedge \left(\sum_i \mathbf{F}_i \right)_2 \cdot \hat{\mathbf{k}} - \left(\sum_i M_i \right)_2 \\[2mm]
m_3 a_{G_{3x}} - \left(\sum_i F_{i_x} \right)_3 \\[2mm]
m_3 a_{G_{3y}} - \left(\sum_i F_{i_y} \right)_3 \\[2mm]
I_{G_3} \alpha_3 - \mathbf{d}_3 \wedge \left(\sum_i \mathbf{F}_i \right)_3 \cdot \hat{\mathbf{k}} - \left(\sum_i M_i \right)_3 \\[2mm]
m_4 a_{G_{4x}} - \left(\sum_i F_{i_x} \right)_4 \\[2mm]
m_4 a_{G_{4y}} - \left(\sum_i F_{i_y} \right)_4 \\[2mm]
I_{G_4} \alpha_4 - \mathbf{d}_4 \wedge \left(\sum_i \mathbf{F}_i \right)_4 \cdot \hat{\mathbf{k}} - \left(\sum_i M_i \right)_4
\end{pmatrix}
\tag{4.96}
$$

• $[L]$: Matrix of geometric parameters of the mechanism (Eq. 4.97).

$$
[L] =
\begin{pmatrix}
-1 & 0 & 1 & 0 & 0 & 0 & 0 & 0 & 0 \\
0 & -1 & 0 & 1 & 0 & 0 & 0 & 0 & 0 \\
q_{2_y} & -q_{2_x} & -p_{2_y} & p_{2_x} & 0 & 0 & 0 & 0 & 1 \\
0 & 0 & -1 & 0 & 1 & 0 & 0 & 0 & 0 \\
0 & 0 & 0 & -1 & 0 & 1 & 0 & 0 & 0 \\
0 & 0 & q_{3_y} & -q_{3_x} & -p_{3_y} & p_{3_x} & 0 & 0 & 0 \\
0 & 0 & 0 & 0 & -1 & 0 & 1 & 0 & 0 \\
0 & 0 & 0 & 0 & 0 & -1 & 0 & 1 & 0 \\
0 & 0 & 0 & 0 & q_{4_y} & -q_{4_x} & -p_{4_y} & p_{4_x} & 0
\end{pmatrix}
\tag{4.97}
$$

• $\mathbf{q}$: Vector of unknown forces and moments (Eq. 4.98).

$$
\mathbf{q} =
\begin{pmatrix}
F_{21_x} \\
F_{21_y} \\
F_{32_x} \\
F_{32_y} \\
F_{43_x} \\
F_{43_y} \\
F_{14_x} \\
F_{14_y} \\
M_0
\end{pmatrix}
\tag{4.98}
$$

Hence, we can solve the unknown matrix (Eq. 4.99) as:

$$\mathbf{q} = [L]^{-1}\mathbf{F} \tag{4.99}$$

The system is linear so it can easily be solved by Gaussian elimination, for example.

It may be interesting to obtain the combination of the reactions on link 1 (Eq. 4.100). Remember that the value of this resultant force, $\mathbf{F}_S$, is:

$$\mathbf{F}_S = \mathbf{F}_{21} - \mathbf{F}_{14} \tag{4.100}$$

It can also be obtained through its components (Eq. 4.101):

$$\left.\begin{array}{l} F_{S_x} = F_{21_x} - F_{14_x} \\ F_{S_y} = F_{21_y} - F_{14_y} \end{array}\right\} \tag{4.101}$$

Example 13 In the crank-shaft mechanism in Fig. 4.44 we know the geometry of the mechanism as well as its mass and the moment of inertia of every link. We also know the acceleration of the links, the location of the centers of gravity and the value of force $\mathbf{P}_4$ acting on the piston. We want to find the reactions at the joints and input torque $\mathbf{M}_0$.

Free-body diagrams of the links are shown in Fig. 4.45a–c.

Considering the equilibrium in the links:

- Link 2, (Eqs. 4.102 and 4.103):

$$\mathbf{F}_{32} - \mathbf{F}_{21} = m_2\mathbf{a}_{G_2} \tag{4.102}$$

$$\begin{aligned} \mathbf{r}_{(2)G_2} \wedge \mathbf{F}_{32} - \mathbf{r}_{(1)G_2} \wedge \mathbf{F}_{21} + \mathbf{M}_0 \\ = \mathbf{p}_2 \wedge \mathbf{F}_{32} - \mathbf{q}_2 \wedge \mathbf{F}_{21} + \mathbf{M}_0 = I_{G_2}\boldsymbol{\alpha}_2 \end{aligned} \tag{4.103}$$

- Link 3, (Eqs. 4.104 and 4.105):

$$\mathbf{F}_{43} - \mathbf{F}_{32} = m_3\mathbf{a}_{G_3} \tag{4.104}$$

Fig. 4.44 Crank-shaft mechanism with required data to carry out its dynamic analysis

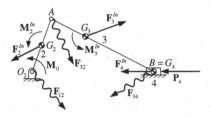

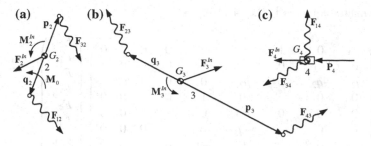

Fig. 4.45 Free-body diagram of the links **a** 2, **b** 3 and **c** 4 with external forces and reactions

$$\mathbf{r}_{(3)G_3} \wedge \mathbf{F}_{43} - \mathbf{r}_{(2)G_3} \wedge \mathbf{F}_{32}$$
$$= \mathbf{p}_3 \wedge \mathbf{F}_{43} - \mathbf{q}_3 \wedge \mathbf{F}_{32} = I_{G_3}\alpha_3 \tag{4.105}$$

- Link 4, (Eqs. 4.106 and 4.107):

$$\mathbf{P}_4 + \mathbf{F}_{14} - \mathbf{F}_{43} = m_4\mathbf{a}_{G_4} \tag{4.106}$$

$$\mathbf{r}_{(4)G_4} \wedge \mathbf{F}_{14} - \mathbf{r}_{(3)G_4} \wedge \mathbf{F}_{43}$$
$$= \mathbf{p}_4 \wedge \mathbf{F}_{14} - \mathbf{q}_4 \wedge \mathbf{F}_{43} = I_{G_4}\alpha_4 - \mathbf{d}_4 \wedge \mathbf{P}_4 \tag{4.107}$$

Torque equilibrium (Eq. 4.107) is not considered in link 4, as all forces converge at the same point, and $\alpha_4 = 0$. Developing these equations and breaking them into their x and y components, we obtain (Eqs. 4.108–4.115):

$$F_{32_x} - F_{21_x} = m_2 a_{G_{2x}} \tag{4.108}$$

$$F_{32_y} - F_{21_y} = m_2 a_{G_{2y}} \tag{4.109}$$

$$(p_{2_x}F_{32_y} - p_{2_y}F_{32_x}) - (q_{2_x}F_{21_y} - q_{2_y}F_{21_x}) + M_0 = I_{G_2}\alpha_2 \tag{4.110}$$

$$F_{43_x} - F_{32_x} = m_3 a_{G_{3x}} \tag{4.111}$$

$$F_{43_y} - F_{32_y} = m_3 a_{G_{3y}} \tag{4.112}$$

$$(p_{3_x}F_{43_y} - p_{3_y}F_{43_x}) - (q_{3_x}F_{32_y} - q_{3_y}F_{32_x}) = I_{G_3}\alpha_3 \tag{4.113}$$

$$F_{14_x} - F_{43_x} = m_4 a_{G_{4x}} - P_{4_x} \tag{4.114}$$

$$F_{14_y} - F_{43_y} = m_4 a_{G_{4y}} - P_{4_y} \tag{4.115}$$

In Eqs. (4.114) and (4.115), we know that $F_{14_x} = P_{4_y} = a_{G_{4y}} = 0$.

The matrix equation is (Eq. 4.116):

$$
\begin{pmatrix}
-1 & 0 & 1 & 0 & 0 & 0 & 0 & 0 & 0 \\
0 & -1 & 0 & 1 & 0 & 0 & 0 & 0 & 0 \\
q_{2_y} & -q_{2_x} & -p_{2_y} & p_{2_x} & 0 & 0 & 0 & 0 & 1 \\
0 & 0 & -1 & 0 & 1 & 0 & 0 & 0 & 0 \\
0 & 0 & 0 & -1 & 0 & 1 & 0 & 0 & 0 \\
0 & 0 & q_{3_y} & -q_{3_x} & -p_{3_y} & p_{3_x} & 0 & 0 & 0 \\
0 & 0 & 0 & 0 & -1 & 0 & 1 & 0 & 0 \\
0 & 0 & 0 & 0 & 0 & -1 & 0 & 1 & 0
\end{pmatrix}
\begin{pmatrix}
F_{21_x} \\
F_{21_y} \\
F_{22_x} \\
F_{22_y} \\
F_{43_x} \\
F_{43_y} \\
F_{14_x} \\
F_{14_y} \\
M_0
\end{pmatrix}
$$

$$
=
\begin{pmatrix}
m_2 a_{G_{2x}} \\
m_2 a_{G_{2y}} \\
I_{G_2}\alpha_2 \\
m_3 a_{G_{3x}} \\
m_3 a_{G_{3y}} \\
I_{G_3}\alpha_3 \\
m_4 a_{G_{4x}} - P_{4_x} \\
m_4 a_{G_{4y}} - P_{4_y}
\end{pmatrix}
\tag{4.116}
$$

Components p_{j_x}, p_{j_y}, q_{j_x} and q_{j_y} are calculated from known distances and angle θ_j of each link (Eqs. 4.117–4.124). In this example, these values are:

$$q_{2_x} = \overline{O_2 G_2}\cos(\theta_2 + 180°) \tag{4.117}$$

$$q_{2_y} = \overline{O_2 G_2}\sin(\theta_2 + 180°) \tag{4.118}$$

$$p_{2_x} = \overline{AG_2}\cos\theta_2 \tag{4.119}$$

$$p_{2_y} = \overline{AG_2}\sin\theta_2 \tag{4.120}$$

$$q_{3_x} = \overline{AG_3}\cos(\theta_3 + 180°) \tag{4.121}$$

$$q_{3_y} = \overline{AG_3}\sin(\theta_3 + 180°) \tag{4.122}$$

$$p_{3_x} = \overline{BG_3}\cos\theta_3 \tag{4.123}$$

$$p_{3_y} = \overline{BG_3}\sin\theta_3 \tag{4.124}$$

There are eight equations left to find the following eight unknowns: F_{21_x}, F_{21_y}, F_{32_x}, F_{32_y}, F_{43_x}, F_{43_y}, F_{14_y} and M_0. Therefore, we can find the value of the unknowns by solving the system.

Fig. 4.46 Forces and torques
acting on a four-link
mechanism with a slider
rotating about a fixed point

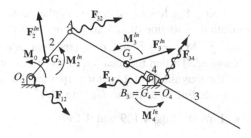

Example 14 Carry out a dynamic analysis of the mechanism in Fig. 4.46. Consider
the acceleration of the links and the location of the centers of gravity to be known.
No external forces are applied. Point B_3 of link 3 coincides at the instant studied
with the center of rotation of link 4, O_4.

We start by drawing the free-body diagram of each link (Fig. 4.47) and
formulating the equilibrium equations (Eqs. 4.125–4.128):

- Link 2, (Eqs. 4.125 and 4.126):

$$\mathbf{F}_{32} - \mathbf{F}_{21} = m_2\mathbf{a}_{G_2} \tag{4.125}$$

$$\mathbf{r}_{(2)G_2} \wedge \mathbf{F}_{32} - \mathbf{r}_{(1)G_2} \wedge \mathbf{F}_{21} + \mathbf{M}_0$$
$$= \mathbf{p}_2 \wedge \mathbf{F}_{32} - \mathbf{q}_2 \wedge \mathbf{F}_{21} + \mathbf{M}_0 = I_{G_2}\alpha_2 \tag{4.126}$$

- Link 3, (Eqs. 4.127 and 4.128):

$$\mathbf{F}_{43} - \mathbf{F}_{32} = m_3\mathbf{a}_{G_3} \tag{4.127}$$

$$\mathbf{r}_{(3)G_3} \wedge \mathbf{F}_{43} - \mathbf{r}_{(2)G_3} \wedge \mathbf{F}_{32}$$
$$= (\mathbf{r}_{B_3G_3} + \mathbf{e}_4) \wedge \mathbf{F}_{43} - \mathbf{q}_3 \wedge \mathbf{F}_{32} = I_{G_3}\alpha_3 \tag{4.128}$$

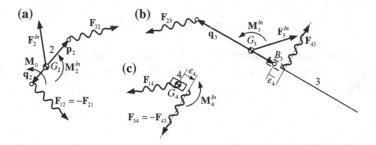

Fig. 4.47 Free-body diagram of each link of the mechanism in Fig. 4.46

Force $\mathbf{F}_{43}$ has to be perpendicular to link 3. However, we do not know the location of its line of action. It cannot go through point B_3 since this would mean that $\sum_i \mathbf{M}_i^{G_4} = 0$ is not met, since the inertia moment of link 4 would not be balanced. To find the location of force $\mathbf{F}_{43}$, we have to introduce a new unknown, e_4, the magnitude of a vector perpendicular to the force that measures the distance from its line of action to G_4 (Fig. 4.47).

- Link 4, (Eqs. 4.129 and 4.130):

$$\mathbf{F}_{14} - \mathbf{F}_{43} = m_4 \mathbf{a}_{G_4} = 0 \rightarrow \mathbf{F}_{14} = \mathbf{F}_{43} \tag{4.129}$$

Because G_4 is a steady point, its acceleration is zero.

$$0 \wedge \mathbf{F}_{14} - \mathbf{e}_4 \wedge \mathbf{F}_{43} = I_{G_4} \alpha_4 = I_{G_4} \alpha_3 \tag{4.130}$$

Since the angular acceleration of link 4 is the same as the one of link 3.

Moreover, we have to add the condition of $\mathbf{F}_{43}$ being perpendicular to link 3. This condition is imposed with dot product $\mathbf{e}_4 \cdot \mathbf{F}_{43} = 0$, or, in other words, $\hat{\mathbf{u}} \cdot \mathbf{F}_{43} = 0$, with: $\mathbf{e}_4 = e_4 \hat{\mathbf{u}}$. That is, $\hat{\mathbf{u}}$ is a unit vector with the same direction as $\mathbf{e}_4$.

We can break the previous equations into their x and y terms following the same procedure as in the previous example. Thus, we obtain nine equations to which we have to add this last one, making a total of ten equations to solve the ten unknowns: F_{21_x}, F_{21_y}, F_{32_x}, F_{32_y}, F_{43_x}, F_{43_y}, F_{14_x}, F_{14_y}, e_4 and M_0. Nevertheless, we can simplify:

- Since $\mathbf{a}_{G_4} = 0$, from the force equilibrium equation in link 4 (Eq. 4.129), $\mathbf{F}_{14} = \mathbf{F}_{43}$.
- The moment equilibrium equation (Eq. 4.128) on link 3 can be written as:

$$\mathbf{r}_{B_3 G_3} \wedge \mathbf{F}_{43} + \mathbf{e}_4 \wedge \mathbf{F}_{43} - \mathbf{q}_3 \wedge \mathbf{F}_{32} = I_{G_3} \alpha_3 \tag{4.131}$$

Remember the moment equilibrium equation of link 4 (Eq. 4.130), then the equilibrium equation (Eq. 4.132) changes to:

$$\mathbf{r}_{B_3 G_3} \wedge \mathbf{F}_{43} - \mathbf{q}_3 \wedge \mathbf{F}_{32} = I_{G_3} \alpha_3 + I_{G_4} \alpha_3 \tag{4.132}$$

This way, we eliminate unknown e_4 which, together with $\hat{\mathbf{u}} \cdot \mathbf{F}_{43} = 0$, allows us to reduce the number of unknowns to seven. The matrix equation, consequently, remains (Eq. 4.133):

$$
\begin{pmatrix}
-1 & 0 & 1 & 0 & 0 & 0 & 0 & 0 & 0 \\
0 & -1 & 0 & 1 & 0 & 0 & 0 & 0 & 0 \\
q_{2_y} & -q_{2_x} & -p_{2_y} & p_{2_x} & 0 & 0 & 0 & 0 & 1 \\
0 & 0 & -1 & 0 & 1 & 0 & 0 & 0 & 0 \\
0 & 0 & 0 & -1 & 0 & 1 & 0 & 0 & 0 \\
0 & 0 & q_{3_y} & -q_{3_x} & -r_{B_3 G_3 y} & r_{B_3 G_3 x} & 0 & 0 & 0 \\
0 & 0 & 0 & 0 & 0 & 0 & u_x & u_y & 0
\end{pmatrix}
\begin{pmatrix}
F_{21_x} \\
F_{21_y} \\
F_{22_x} \\
F_{22_y} \\
F_{43_x} \\
F_{43_y} \\
M_0
\end{pmatrix}
=
\begin{pmatrix}
m_2 a_{G_{2x}} \\
m_2 a_{G_{2y}} \\
I_{G_2} \alpha_2 \\
m_3 a_{G_{3x}} \\
m_3 a_{G_{3y}} \\
(I_{G_3} + I_{G4}) \alpha_3 \\
0
\end{pmatrix}
$$

$$\tag{4.133}$$

where the x and y components of vectors $\mathbf{p}_2$, $\mathbf{q}_2$, $\mathbf{r}_{B_3 G_3}$, $\mathbf{q}_3$ and $\hat{\mathbf{u}}$ can be calculated as (Eqs. 4.134–4.143):

$$q_{2_x} = \overline{O_2 G_2} \cos(\theta_2 + 180°) \tag{4.134}$$

$$q_{2_y} = \overline{O_2 G_2} \sin(\theta_2 + 180°) \tag{4.135}$$

$$p_{2_x} = \overline{AG_2} \cos \theta_2 \tag{4.136}$$

$$p_{2_y} = \overline{AG_2} \sin \theta_2 \tag{4.137}$$

$$q_{3_x} = \overline{AG_3} \cos \theta_3 \tag{4.138}$$

$$q_{3_y} = \overline{AG_3} \sin \theta_3 \tag{4.139}$$

$$r_{B_3 G_{3x}} = \overline{BG_3} \cos(\theta_3 + 180°) \tag{4.140}$$

$$r_{B_3 G_{3y}} = \overline{BG_3} \sin(\theta_3 + 180°) \tag{4.141}$$

$$u_x = \cos(\theta_3 + 180°) \tag{4.142}$$

$$u_y = \sin(\theta_3 + 180°) \tag{4.143}$$

Substituting these values in the matrix, we can clear the unknowns.

Example 15 In the quick-return mechanism in (Fig. 4.48), we know the geometry data as well as the acceleration of all the links and centers of gravity. Force $\mathbf{P}_6$ that acts on piston is also known. Find the forces at the joints and input torque M_0 that is acting on link 2.

Fig. 4.48 Quick-return
mechanism with force $\mathbf{P}_6$
acting on the piston and an
input torque M_0 acting on
link 2

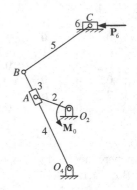

Figure 4.49 shows the free body diagrams of every link. We can write the equilibrium equations of each one.

- Link 2, (Eqs. 4.144 and 4.145):

$$\mathbf{F}_{32} - \mathbf{F}_{21} = m_2 \mathbf{a}_{G_2} \qquad (4.144)$$

$$\mathbf{r}_{(2)G_2} \wedge \mathbf{F}_{32} - \mathbf{r}_{(1)G_2} \wedge \mathbf{F}_{21} + \mathbf{M}_0$$
$$= \mathbf{p}_2 \wedge \mathbf{F}_{32} - \mathbf{q}_2 \wedge \mathbf{F}_{21} + \mathbf{M}_0 = I_{G_2} \alpha_2 \qquad (4.145)$$

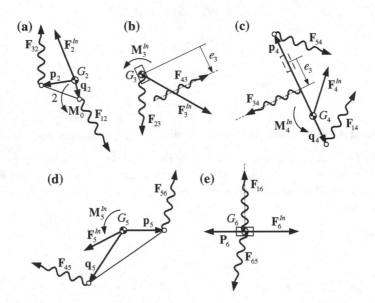

Fig. 4.49 Free-body diagram of the links of the mechanism in Fig. 4.48

- Link 3, (Eqs. 4.146 and 4.147):

$$\mathbf{F}_{43} - \mathbf{F}_{32} = m_3 \mathbf{a}_{G_3} \tag{4.146}$$

$$\mathbf{r}_{(3)G_3} \wedge \mathbf{F}_{43} - \mathbf{r}_{(2)G_3} \wedge \mathbf{F}_{32} \\ = \mathbf{e}_3 \wedge \mathbf{F}_{43} = I_{G_3} \alpha_3 \tag{4.147}$$

where distance e_3 is unknown.

- Link 4, (Eqs. 4.148 and 4.149):

$$\mathbf{F}_{54} - \mathbf{F}_{43} + \mathbf{F}_{14} = m_4 \mathbf{a}_{G_4} \tag{4.148}$$

$$\mathbf{r}_{(5)G_4} \wedge \mathbf{F}_{54} - \mathbf{r}_{(3)G_4} \wedge \mathbf{F}_{43} + \mathbf{r}_{(1)G_4} \wedge \mathbf{F}_{14} \\ = \mathbf{p}_4 \wedge \mathbf{F}_{54} - (\mathbf{r}_{A_4 G_4} + \mathbf{e}_3) \wedge \mathbf{F}_{43} + \mathbf{q}_4 \wedge \mathbf{F}_{14} = I_{G_4} \alpha_4 \tag{4.149}$$

Vector $\mathbf{r}_{A_4 G_4}$ is known, since it depends on the position of the slider which has to be determined before calculating velocities and accelerations in the mechanism.

- Link 5, (Eqs. 4.150–4.151):

$$\mathbf{F}_{65} - \mathbf{F}_{54} = m_5 \mathbf{a}_{G_5} \tag{4.150}$$

$$\mathbf{r}_{(6)G_5} \wedge \mathbf{F}_{65} - \mathbf{r}_{(4)G_5} \wedge \mathbf{F}_{54} \\ = \mathbf{p}_5 \wedge \mathbf{F}_{65} - \mathbf{q}_5 \wedge \mathbf{F}_{54} = I_{G_5} \alpha_5 \tag{4.151}$$

- Link 6, (Eq. 4.152):

$$\mathbf{F}_{16} - \mathbf{F}_{65} + \mathbf{P}_6 = m_6 \mathbf{a}_{G_6} \tag{4.152}$$

There is no torque equation, since all the forces go through G_6 and also $\alpha_6 = 0$.

These equations yield a system of 14 equations and 16 unknowns. We need two additional equations so that the system can be solved. As we know that the reaction force on the piston only has a vertical component, F_{16_x}, hence, we have our first equation.

The second equation (Eq. 4.153) is obtained by making force $\mathbf{F}_{43}$, exerted by link 4 on 3, perpendicular to link 3. That is:

$$\mathbf{e}_4 \cdot \mathbf{F}_{43} = 0 \tag{4.153}$$

Equation (Eq. 4.153) can be substituted by $\hat{\mathbf{u}} \cdot \mathbf{F}_{43} = 0$, with $\mathbf{e}_4 = e_4 \hat{\mathbf{u}}$. Again, $\hat{\mathbf{u}}$ is a unit vector in the direction of $\mathbf{e}_4$, which is known as it is the same as link 4.

Thus, as in previous examples, we decompose the equations into their components and develop the vector products in the moment equations. We also substitute the moment equation of link 3 (Eq. 4.147) in the one of link 4 so that the torque equation of link 4 (Eq. 4.149) becomes (remember that $\alpha_3 = \alpha_4$):

$$\mathbf{p}_4 \wedge \mathbf{F}_{54} - \mathbf{r}_{A_4 G_4} \wedge \mathbf{F}_{43} + \mathbf{q}_4 \wedge \mathbf{F}_{14} = I_{G_3}\alpha_3 + I_{G_4}\alpha_4 \qquad (4.154)$$

We obtain the following system (Eq. 4.155):

$$\left.\begin{array}{r}
F_{32_x} - F_{21_x} = m_2 a_{G_{2x}} \\
F_{32_y} - F_{21_y} = m_2 a_{G_{2y}} \\
(p_{2_x}F_{32_y} - p_{2_y}F_{32_x}) - (q_{2_x}F_{21_y} - q_{2_y}F_{21_x}) + M_0 = I_{G_2}\alpha_2 \\
F_{43_x} - F_{32_x} = m_3 a_{G_{3x}} \\
F_{43_y} - F_{32_y} = m_3 a_{G_{3y}} \\
F_{54_x} + F_{14_x} - F_{43_x} = m_4 a_{G_4 x} \\
F_{54_y} + F_{14_y} - F_{43_y} = m_4 a_{G_4 y} \\
(p_{4_x}F_{54_y} - p_{4_y}F_{54_x}) - (r_{A_4 G_4 x}F_{43_y} - r_{A_4 G_4 y}F_{43_x}) + (q_{4_x}F_{14_y} - q_{4_y}F_{14_x}) = (I_{G_3} + I_{G_4})\alpha_4 \\
F_{65_x} - F_{54_x} = m_5 a_{G_{5x}} \\
F_{65_y} - F_{54_y} = m_5 a_{G_{5y}} \\
(p_{5_x}F_{65_y} - p_{5_y}F_{65_x}) - (q_{5_x}F_{54_y} - q_{5_y}F_{54_x})_0 = I_{G_5}\alpha_5 \\
-F_{65_x} = m_6 a_{G_{6x}} - P_{6_x} \\
F_{16_y} - F_{65_y} = m_6 a_{G_{6y}} = 0
\end{array}\right\} \qquad (4.155)$$

We have to add equation $\hat{\mathbf{u}} \cdot \mathbf{F}_{43} = 0$, that is to say: $u_x F_{43_x} + u_y F_{43_y} = 0$.

Writing the equation system in its matrix form, it remains (Eqs. 4.156–4.159):

$$[L]\mathbf{q} = \mathbf{F} \qquad (4.156)$$

where:

$$[L] = \begin{bmatrix}
-1 & 0 & 1 & 0 & 0 & 0 & 0 & 0 & 0 & 0 & 0 & 0 & 0 & 0 \\
0 & -1 & 0 & 1 & 0 & 0 & 0 & 0 & 0 & 0 & 0 & 0 & 0 & 0 \\
q_{2_y} & -q_{2_x} & -p_{2_y} & p_{2_x} & 0 & 0 & 0 & 0 & 0 & 0 & 0 & 0 & 0 & 1 \\
0 & 0 & -1 & 0 & 1 & 0 & 0 & 0 & 0 & 0 & 0 & 0 & 0 & 0 \\
0 & 0 & 0 & -1 & 0 & 1 & 0 & 0 & 0 & 0 & 0 & 0 & 0 & 0 \\
0 & 0 & 0 & 0 & -1 & 0 & 1 & 0 & 1 & 0 & 0 & 0 & 0 & 0 \\
0 & 0 & 0 & 0 & 0 & -1 & 0 & 1 & 0 & 1 & 0 & 0 & 0 & 0 \\
0 & 0 & 0 & 0 & r_{A_4 G_4 y} & -r_{A_4 G_4 x} & -p_{4_y} & p_{4_x} & -q_{4_y} & q_{4_x} & 0 & 0 & 0 & 0 \\
0 & 0 & 0 & 0 & 0 & 0 & -1 & 0 & 0 & 0 & 1 & 0 & 0 & 0 \\
0 & 0 & 0 & 0 & 0 & 0 & 0 & -1 & 0 & 0 & 0 & 1 & 0 & 0 \\
0 & 0 & 0 & 0 & 0 & 0 & q_{5_y} & q_{5_x} & 0 & 0 & -p_{5_y} & p_{5_x} & 0 & 0 \\
0 & 0 & 0 & 0 & 0 & 0 & 0 & 0 & 0 & 0 & -1 & 0 & 0 & 0 \\
0 & 0 & 0 & 0 & 0 & 0 & 0 & 0 & 0 & 0 & 0 & -1 & 1 & 0 \\
0 & 0 & 0 & 0 & u_x & u_y & 0 & 0 & 0 & 0 & 0 & 0 & 0 & 0
\end{bmatrix}$$

$$(4.157)$$

$$\mathbf{q} = \begin{bmatrix} F_{21_x} \\ F_{21_y} \\ F_{32_x} \\ F_{32_y} \\ F_{43_x} \\ F_{43_y} \\ F_{54_x} \\ F_{54_y} \\ F_{14_x} \\ F_{14_y} \\ F_{65_x} \\ F_{65_y} \\ F_{16_y} \\ M_0 \end{bmatrix} \tag{4.158}$$

$$\mathbf{F} = \begin{bmatrix} m_2 a_{G_{2x}} \\ m_2 a_{G_{2y}} \\ I_{G_2}\alpha_2 \\ m_3 a_{G_{3x}} \\ m_3 a_{G_{3y}} \\ m_4 a_{G_4x} \\ m_4 a_{G_4y} \\ (I_{G_3}+I_{G_4})\alpha_4 \\ m_5 a_{G_{5x}} \\ m_5 a_{G_{5y}} \\ I_{G_5}\alpha_5 \\ m_6 a_{G_{6x}} - P_{6_x} \\ 0 \\ 0 \end{bmatrix} \tag{4.159}$$

Once values p_{j_x}, p_{j_y}, q_{j_x}, q_{j_y}, $r_{A_4G_4x}$, $r_{A_4G_4y}$, u_x and u_y have been calculated, we can solve the system (Eq. 4.156) to obtain the reaction forces at the joints and torque M_0 that acts on link 2.

4.4 Exercises with Solutions

Example 16 Find the value of the input torque on the crank-shaft mechanism in Fig. 4.50. The centers of gravity of link 2, 3 and 4 are at G_2, G_3 and G_4 respectively. The masses and moments of inertia of the links are: $m_2 = 2$ kg, $m_3 = 0.5$ kg,

Fig. 4.50 Crank-shaft
mechanism moving with
known velocity and
acceleration

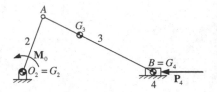

$m_4 = 0.5\,\text{kg}$, $I_{G_2} = 0.001\,\text{kg}\,\text{m}^2$, $I_{G_3} = 0.01\,\text{kg}\,\text{m}^2$ and $I_{G_4} = 0.002\,\text{kg}\,\text{m}^2$. It is
known that: $\overline{O_2A} = 3\,\text{cm}$, $\overline{AB} = 7\,\text{cm}$, $\overline{AG_3} = 2\,\text{cm}$, $\theta_2 = 60°$, $\omega_2 = -20\,\text{rad/s}$,
$\alpha_2 = -100\,\text{rad/s}^2$ and that external force $\mathbf{P}_4$ has a magnitude of 98 N.

In Example 2 in Chap. 3, we did the kinematic analysis of this mechanism yielding
$\theta_3 = 338.2°$ and the following accelerations for links 3 and 4:

$$\mathbf{a}_{G_3} = -250.26\hat{\mathbf{i}} - 849.45\hat{\mathbf{j}}\,\text{cm/s}^2 \quad \alpha_3 = 174.44\,\text{rad/s}^2$$
$$\mathbf{a}_{G_4} = -25.45\hat{\mathbf{i}}\,\text{cm/s}^2 \qquad\qquad \alpha_4 = 0$$

Next, we write the equilibrium equation for each link so that we obtain a system
of equations as we did in Example 13 in this chapter. It yields the matrix system
(Eq. 4.116), where values p_i and q_i are:

$$q_{2_x} = \overline{O_2G_2}\cos(\theta_2 + 180°) = 0$$
$$q_{2_y} = \overline{O_2G_2}\sin(\theta_2 + 180°) = 0$$
$$p_{2_x} = \overline{AG_2}\cos\theta_2 = 1.5\,\text{cm}$$
$$p_{2_y} = \overline{AG_2}\sin\theta_2 = 2.6\,\text{cm}$$
$$q_{3_x} = \overline{AG_3}\cos(\theta_3 + 180°) = -1.86\,\text{cm}$$
$$q_{3_y} = \overline{AG_3}\sin(\theta_3 + 180°) = 0.74\,\text{cm}$$
$$p_{3_x} = \overline{BG_3}\cos\theta_3 = 4.64\,\text{cm}$$
$$p_{3_y} = \overline{BG_3}\sin\theta_3 = -1.86\,\text{cm}$$

Hence, the system remains fully determined as:

$$
\begin{pmatrix}
-1 & 0 & 1 & 0 & 0 & 0 & 0 & 0 \\
0 & -1 & 0 & 1 & 0 & 0 & 0 & 0 \\
0 & 0 & -2.6\times10^{-2} & 1.5\times10^{-2} & 0 & 0 & 0 & 1 \\
0 & 0 & -1 & 0 & 1 & 0 & 0 & 0 \\
0 & 0 & 0 & -1 & 0 & 1 & 0 & 0 \\
0 & 0 & 0.74\times10^{-2} & 1.86\times10^{-2} & 1.86\times10^{-2} & 4.64\times10^{-2} & 0 & 0 \\
0 & 0 & 0 & 0 & -1 & 0 & 0 & 0 \\
0 & 0 & 0 & 0 & 0 & -1 & 1 & 0
\end{pmatrix}
\begin{pmatrix}
F_{21_x} \\
F_{21_y} \\
F_{22_x} \\
F_{22_y} \\
F_{43_x} \\
F_{43_y} \\
F_{14_y} \\
M_0
\end{pmatrix}
=
\begin{pmatrix}
0 \\
0 \\
-0.1 \\
-1.25 \\
-4.25 \\
1.74 \\
98 \\
0
\end{pmatrix}
$$

The solution of the matrix equation yields the following values:

$$\begin{pmatrix} F_{21_x} \\ F_{21_y} \\ F_{22_x} \\ F_{22_y} \\ F_{43_x} \\ F_{43_y} \\ F_{14_y} \\ M_0 \end{pmatrix} = \begin{pmatrix} -96.75\,\text{N} \\ 68.86\,\text{N} \\ -96.75\,\text{N} \\ 68.86\,\text{N} \\ -98.0\,\text{N} \\ 64.61\,\text{N} \\ 64.61\,\text{N} \\ -3.65\,\text{Nm} \end{pmatrix}$$

Example 17 Figure 4.51 shows a mixing machine whose kinematic analysis was carried out in Example 10 in Chap. 2. Find the value of the input torque that acts on link 2. Consider the external force applied on point C to have a magnitude of 100 N and have an opposite direction to the velocity. The mass of the spatula is 10 kg and its moment of inertia is 0.1 kg m^2. We assume that the spatula is the only link that has mass and a moment of inertia. The mass of the rest of the links is considered negligible.

In the first place, we need the values of the angular acceleration of the different links as well as the linear acceleration of their centers of gravity. The results of the kinematic analysis give us the following values:

$$\begin{array}{ccc} & \mathbf{a}_{G_2} = 0 & \alpha_2 = 0 \\ \theta_3 = 122.47° \quad \overline{B_3G_3} = 13.04\,\text{cm} & \mathbf{a}_{G_3} = -700\hat{i}\,\text{cm/s}^2 & \alpha_3 = -19.3\,\text{rad/s}^2 \\ & \mathbf{a}_{G_4} = 0 & \alpha_4 = -19.3\,\text{rad/s}^2 \end{array}$$

$$\mathbf{v}_C = 51.03\hat{i} + 102.45\hat{j} = 114.4\,\text{cm/s}\,\angle 63.52°$$

$$\mathbf{P}_3 = 100(\cos(63.52° + 180°)\hat{i} + \sin(63.52° + 180°)\hat{j}) = -44.59\hat{i} - 89.51\hat{j}\,\text{N}$$

Moreover, the dynamic equilibrium of the mechanism was solved in Example 15 in this chapter (Eqs. 4.156–4.159). However, in this case, we have an external force applied on link 3. Thus, the equilibrium equations of link 3 remain as follows:

$$\mathbf{P}_3 + \mathbf{F}_{43} - \mathbf{F}_{32} = m_3\mathbf{a}_{G_3}$$

$$\mathbf{r}_{CG_3} \wedge \mathbf{P}_3 + (\mathbf{r}_{B_3G_3} + \mathbf{e}_4) \wedge \mathbf{F}_{43} - \mathbf{q}_3 \wedge \mathbf{F}_{32} = I_{G_3}\alpha_3$$

Hence, the matrix equation remains:

$$\begin{bmatrix} -1 & 0 & 1 & 0 & 0 & 0 & 0 \\ 0 & -1 & 0 & 1 & 0 & 0 & 0 \\ q_{2y} & -q_{2x} & -p_{2y} & p_{2x} & 0 & 0 & 1 \\ 0 & 0 & -1 & 0 & 1 & 0 & 0 \\ 0 & 0 & 0 & -1 & 0 & 1 & 0 \\ 0 & 0 & q_{3x} & -q_{3y} & -r_{B_3G_3y} & r_{B_3G_3x} & 0 \\ 0 & 0 & 0 & 0 & u_x & u_y & 0 \end{bmatrix} \begin{bmatrix} F_{21_x} \\ F_{21_y} \\ F_{32_x} \\ F_{32_y} \\ F_{43_x} \\ F_{43_y} \\ M_0 \end{bmatrix} = \begin{bmatrix} m_2a_{G_{2x}} \\ m_2a_{G_{2y}} \\ I_{G_2}\alpha_2 \\ m_3a_{G_{3x}} \\ m_3a_{G_{3y}} \\ (I_{G_3} + I_{G_4})\alpha_3 - (r_{CG_3x}F_{3y} - r_{CG_3y}F_{3x}) \\ 0 \end{bmatrix}$$

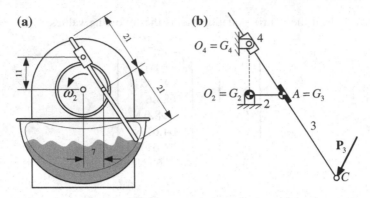

Fig. 4.51 a Mixing machine. **b** Kinematic skeleton with force that acts on link 3

where x and y components of vectors $\mathbf{p}_2$, $\mathbf{q}_2$, $\mathbf{r}_{CG_3}$, $\mathbf{r}_{B_3G_3}$ and $\mathbf{q}_3$ can be computed as:

$$q_{2_x} = \overline{O_2G_2} \cos(\theta_2 + 180°) = 0$$
$$q_{2_y} = \overline{O_2G_2} \sin(\theta_2 + 180°) = 0$$
$$p_{2_x} = \overline{AG_2} \cos \theta_2 = 7 \, \text{cm}$$
$$p_{2_y} = \overline{AG_2} \sin \theta_2 = 0 \, \text{cm}$$
$$r_{CG_3x} = \overline{CG_3} \cos(\theta_3 + 180°) = 11.27 \, \text{cm}$$
$$r_{CG_3y} = \overline{CG_3} \sin(\theta_3 + 180°) = -17.71 \, \text{cm}$$
$$r_{B_3G_3x} = \overline{B_3G_3} \cos \theta_3 = -7 \, \text{cm}$$
$$r_{B_3G_3y} = \overline{B_3G_3} \sin \theta_3 = 11 \, \text{cm}$$
$$q_{3_x} = \overline{AG_3} \cos \theta_3 = 0 \, \text{cm}$$
$$q_{3_y} = \overline{AG_3} \sin \theta_3 = 0 \, \text{cm}$$
$$u_x = \cos(\theta_3 + 180°) = 0.54$$
$$u_y = \sin(\theta_3 + 180°) = -0.84$$

Once these values have been obtained, the matrix system is:

$$
\begin{bmatrix}
-1 & 0 & 1 & 0 & 0 & 0 & 0 \\
0 & -1 & 0 & 1 & 0 & 0 & 0 \\
0 & 0 & 0 & 7 \times 10^{-2} & 0 & 0 & 1 \\
0 & 0 & -1 & 0 & 1 & 0 & 0 \\
0 & 0 & 0 & -1 & 0 & 1 & 0 \\
0 & 0 & 0 & 0 & -11 \times 10^{-2} & -7 \times 10^{-2} & 0 \\
0 & 0 & 0 & 0 & 0.54 & -0.84 & 0
\end{bmatrix}
\begin{bmatrix}
F_{21_x} \\
F_{21_y} \\
F_{32_x} \\
F_{32_y} \\
F_{43_x} \\
F_{43_y} \\
M_0
\end{bmatrix}
=
\begin{bmatrix}
0 \\
0 \\
0 \\
-25.41 \, \text{N} \\
89.51 \, \text{N} \\
16.05 \, \text{Nm} \\
0
\end{bmatrix}
$$

The solution of this matrix equation yields the following results:

$$\begin{bmatrix} F_{21_x} \\ F_{21_y} \\ F_{32_x} \\ F_{32_y} \\ F_{43_x} \\ F_{43_y} \\ M_0 \end{bmatrix} = \begin{bmatrix} -78.14\,\text{N} \\ -156.08\,\text{N} \\ -78.14\,\text{N} \\ -156.08\,\text{N} \\ -103.55\,\text{N} \\ -66.57\,\text{N} \\ 10.93\,\text{Nm} \end{bmatrix}$$

Chapter 5
Balancing of Machinery

Abstract In the previous chapter, we have studied forces that act on rigid bodies due to accelerations. We named them inertia forces and studied the effect they have on a mechanism. In most situations, these forces have a negative effect on the machine, as they cause parasite time-periodic stress with negative consequences for the machine. Vibration is an important issue that will be addressed further ahead in this book. Usually, the best way to handle vibrations in a machine is to equilibrate inertia forces and torques with other forces. The process of studying and making changes in a machine in order to reduce or eliminate shaking forces and torques is called balancing. A link in a planar linkage can have two basic types of motion, rotation and translation, which give raise to two different balancing processes: the balancing of rotating masses and the balancing of masses with reciprocating linear motion. Both methods will be analyzed in this chapter.

5.1 Rotor Balancing

A rotor is a rigid body that rotates about a fixed axis. The balancing of rotors is extremely important when designing and building any machine.

In order to check the importance of this process, we will see an example where the force generated by a slight imbalance is significant when the rotation speed is high.

The flywheel of a machine (Fig. 5.1) has a mass of 5 kg and rotates at 6.000 rpm about a shaft which is mounted with an eccentricity of 1 mm due to a manufacturing defect. We will calculate the resultant of the inertial forces.

When the flywheel rotates with constant angular velocity, there is no inertia torque ($M^{In} = I_G\alpha = 0$). We can assume that the inertia force (centrifugal force) acts on its center of mass with the same line of action as the normal acceleration component, $\mathbf{a}_G^n$, but with opposite direction. If the center of mass is located on the axis of rotation, this force is null.

© Springer International Publishing Switzerland 2016
A. Simón Mata et al., *Fundamentals of Machine Theory and Mechanisms*,
Mechanisms and Machine Science 40, DOI 10.1007/978-3-319-31970-4_5

Fig. 5.1 Flywheel rotating
with its center of gravity at
1 mm away from the center of
rotation

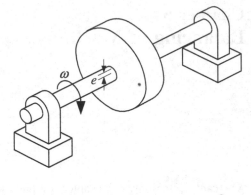

Fig. 5.2 Schematic
representation of the mass of
the flywheel in (Fig. 5.1)

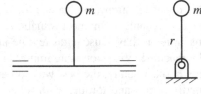

Figure 5.2 shows a schematic representation of the main parameters involved in
the imbalance of the flywheel. Its mass is represented as a sphere rotating with a
radius equal to the eccentricity.

As explained before, the acceleration of the center of mass only has a normal
component. The magnitude of the inertia force generated by this acceleration is
originated by the eccentricity of the center of mass. This force (Eq. 5.1) can be
computed as:

$$F^{In} = ma^n_G = mr\omega^2 = 5 \cdot 0.001 \cdot \left(\frac{2\pi}{60}6000\right)^2 \cong 2000N \qquad (5.1)$$

Note that this force is about 40 times higher than the weight of the flywheel and,
as a consequence, powerful reactions, whose directions change continuously, will
be originated at the supports. We should avoid these shaking forces or at least
reduce their magnitude as much as possible.

5.1.1 Static Balance

A rotor is statically balanced when its center of gravity is located on its revolution
axis. When this happens, it will remain immobile in any angular position, since it is
balanced.

Figure 5.3 represents a rotor with two masses, m_1 and m_2, that rotate about the
AB axis with eccentricity r_1 and r_2 respectively. The first condition of static

Fig. 5.3 A rotor with two masses, m_1 and m_2, which rotate about the AB axis with radius r_1 and r_2 respectively

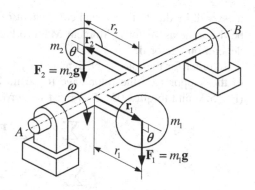

equilibrium is that the resultant force must be zero. The only force that acts in a static equilibrium state is the weight of the rotor itself. In this case the weight of the two masses will be equilibrated by the reactions at supports A and B. The second condition is that the net torque must be zero. In Fig. 5.3, the weight of each mass generates a torque about the AB axis with magnitude $rF \sin \theta = rmg \sin \theta$, where θ is the angle between vectors $\mathbf{r}$ (eccentricity or radius) and $\mathbf{F} = mg$ (weight). As all masses rotate together, the second condition of static equilibrium of a rotor can be written as:

$$\sum_{j=1}^{n} m_j g \mathbf{r}_j = g \left(\sum_{j=1}^{n} m_j \mathbf{r}_j \right) = 0 \rightarrow \sum_{j=1}^{n} m_j \mathbf{r}_j = 0 \qquad (5.2)$$

where n is the number of masses.

Equation (5.2) is met when the center of mass of the system formed by masses m_1 and m_2 is located on rotation axis AB.

5.1.2 Dynamic Balance

Figure 5.4 shows a schematic representation of the rotor in Fig. 5.3. Assuming that the masses rotate with constant angular velocity, inertia force $\mathbf{F}^{In}$ of each rotating

Fig. 5.4 Moments $\mathbf{M}_1^{In}$ and $\mathbf{M}_2^{In}$ produced by inertia forces $\mathbf{F}_1^{In}$ and $\mathbf{F}_2^{In}$ about the Y-axis

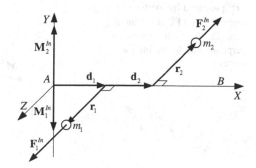

mass will be due to the normal component of its acceleration and its direction will be the same as vector **r**. Moment $\mathbf{M}^{In}$ produced by force $\mathbf{F}^{In}$ about the Y-axis will depend on vector **d**.

In order for the rotor to be dynamically balanced, the shaking forces transmitted to the supports by the rotor have to be null. That is, the sum of inertial forces (Eq. 5.3) and torques (Eq. 5.4) must be zero:

$$\sum_j \mathbf{F}_j = \sum_j \mathbf{F}_j^{In} = 0 \tag{5.3}$$

$$\sum_j \mathbf{M}_j = \sum_j \mathbf{M}_j^{In} = 0 \tag{5.4}$$

where $\mathbf{F}_j^{In}$ is the inertia force of mass j while $\mathbf{M}_j^{In}$ is the torque produced by this force about the Y-axis.

Now, let us imagine a rotor like the one in Fig. 5.5 with two masses located in different rotation planes.

To balance the rotor we will need two balancing masses as any number of unbalanced masses can be equilibrated with just two counterweights. To verify this statement, it is necessary to remember that any system of forces can be reduced to a combination of one force (Eq. 5.5) and one torque (Eq. 5.6):

$$\mathbf{R} = \sum_j^n \mathbf{F}_j^{In} = \mathbf{F}_1^{In} + \mathbf{F}_2^{In} + \cdots + \mathbf{F}_n^{In} \tag{5.5}$$

$$\mathbf{R} = \sum_j^n \mathbf{F}_j^{In}$$

$$\mathbf{M}_R = \sum_j^n \mathbf{M}_j^{In} = \mathbf{M}_1^{In} + \mathbf{M}_2^{In} + \cdots + \mathbf{M}_n^{In} \tag{5.6}$$

Fig. 5.5 Moments $\mathbf{M}_1^{In}$ and $\mathbf{M}_2^{In}$ produced by inertia forces $\mathbf{F}_1^{In}$ and $\mathbf{F}_2^{In}$ about the Y-axis. The moment vectors are perpendicular to the inertia forces

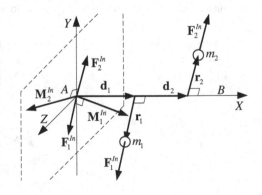

Hence, masses m_{B1} and m_{B2} producing a torque equal to $-\mathbf{M}_R$ and a force equal to $-\mathbf{R}$ are enough to equilibrate the system.

Inertia forces and torques (Eq. 5.7) about the Y-axis, originated by the rotation masses could be calculated as:

$$\left. \begin{array}{l} \mathbf{F}_j^{In} = \omega^2 m_j \mathbf{r}_j \\ \mathbf{M}_j^{In} = \mathbf{d}_j \wedge \mathbf{F}_j^{In} = (\omega^2 m_j) \mathbf{d}_j \wedge \mathbf{r}_j \end{array} \right\} \tag{5.7}$$

- Force equilibrium:

$$\sum_j \mathbf{F}_j = \sum_j \mathbf{F}_j^{In} = \sum_j \omega^2 m_j \mathbf{r}_j = \omega^2 \sum_j m_j \mathbf{r}_j = 0 \tag{5.8}$$

- Torque equilibrium:

$$\sum_j \mathbf{M}_j = \sum_j \mathbf{M}_j^{In} = \sum_j (\omega^2 m_j) \mathbf{d}_j \wedge \mathbf{r}_j = 0 \tag{5.9}$$

We can write Eq. (5.9) with vectors that have the same direction as radius $\mathbf{r}_j$. These vectors are perpendicular to the moment vectors and have proportional magnitude (Eq. 5.10):

$$\omega^2 \sum_j d_j m_j \mathbf{r}_j = 0 \tag{5.10}$$

The system would be dynamically balanced if Eqs. (5.8) and (5.10) were met. Therefore, the dynamic equilibrium conditions in Eqs. (5.3) and (5.4) can be written as Eqs. (5.11) and (5.12):

$$\sum_j m_j \mathbf{r}_j = 0 \tag{5.11}$$

$$\sum_j d_j m_j \mathbf{r}_j = 0 \tag{5.12}$$

Note that the first condition Eq. (5.11) a rotor has to meet in order to be dynamically balanced is the same as Eq. (5.2), that is, the rotor has to be in static balance.

The magnitude of the vectors in Eq. (5.11) is proportional to the magnitude of the inertia forces.

We need two polygons to solve this problem graphically: one proportional to the force polygon and another proportional to the torque polygon. The directions of both polygons are given by vectors $\mathbf{r}_j$.

In Eq. (5.12), d_j is the distance of mass m_j to a transverse reference plane that contains the Y-axis. When there are masses on both sides of the reference plane, we

Fig. 5.6 A rotor with a single
concentrated mass m and
radius r can be dynamically
balanced with mass m_B in the
same transverse plane with
radius $\mathbf{r}_B$ at $180°$ from $\mathbf{r}$

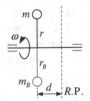

will have positive and negative values for d_j. The negative value will give us a
direction for vector $d_j m_j \mathbf{r}_j$ opposite vector $\mathbf{r}_j$.

Example 1 Balancing a single rotating mass.

Assume that mass m belongs to a rotor that is spinning at constant angular speed
(Fig. 5.6). The distance from the center of gravity of such a mass to the rotation axis
is denominated r. When the rotor is dynamically balanced, we verify Eqs. (5.11)
and (5.12).

First, we have to place a new mass, m_B, so that it verifies (Eq. 5.11), that is,
$m\mathbf{r} + m_B\mathbf{r}_B = 0$. The simplest solution would be to place a mass of the same value
and eccentricity, with $\mathbf{r}_B$ at $180°$ from $\mathbf{r}$. To comply with Eq. (5.12), the new mass
has to be added in the same transverse plane as m so that both masses are at the
same distance d. In Fig. 5.6 the equilibrating mass has been drawn with a thin line.

Example 2 Balancing two rotating masses in two given transverse planes with an
angle of $180°$ between their radiuses.

Figure 5.7 represents a rotor with masses m_1 and m_2 lying in the same longi-
tudinal plane. If we assume that the rotor is statically balanced with m_1 and m_2,
masses m_{B1} and m_{B2}, which have to be added to achieve dynamic balance, have to
yield (Eqs. 5.13–5.14):

$$m_1\mathbf{r}_1 + m_2\mathbf{r}_2 + m_{B1}\mathbf{r}_{B1} + m_{B2}\mathbf{r}_{B2} = 0 \qquad (5.13)$$

$$d_1 m_1\mathbf{r}_1 + d_2 m_2\mathbf{r}_2 + d_{B1}m_{B1}\mathbf{r}_{B1} + d_{B2}m_{B2}\mathbf{r}_{B2} = 0 \qquad (5.14)$$

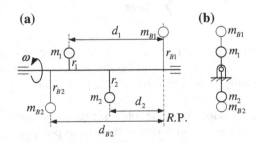

Fig. 5.7 a Front view of a rotor with two single concentrated masses m_1 and m_2. Original masses
are drawn with continuous lines while equilibrating masses, m_{B1} and m_{B2}, are represented with a
thin line **b** lateral view of the same rotor

In other words, the torque produced by m_{B1} and m_{B2} has to have the same magnitude as the one produced by m_1 and m_2 but with opposite direction.

If m_1 and m_2 are not initially statically balanced, the system could be fully equilibrated with one mass, for example m_{B2}. In this case, Eq. (5.14) is valid, but instead of Eq. (5.13), the first condition of the dynamic equilibrium will give us Eq. (5.15):

$$m_1\mathbf{r}_1 + m_2\mathbf{r}_2 + m_{B2}\mathbf{r}_{B2} = 0 \qquad (5.15)$$

Example 3 Balancing several rotating masses in the same transverse plane

Figure 5.8 shows a rotor with four concentrated masses all of them lying in the same plane of rotation. As all the lines of action of the inertia forces intersect at one point, if the system is statically balanced then it will also be dynamically balanced. So, if the rotor is not statically balanced, a single mass, m_B, located in the same plane of rotation is enough to balance it not only statically but also dynamically.

For static balance (Eq. 5.16), the sum of the inertia forces due to the four original masses and added mass m_B must be null:

$$m_1\mathbf{r}_1 + m_2\mathbf{r}_2 + m_3\mathbf{r}_3 + m_4\mathbf{r}_4 + m_B\mathbf{r}_B = 0 \qquad (5.16)$$

In Fig. 5.9 the problem has been graphically solved by means of vectors that represent $m_j\mathbf{r}_j$. By closing the polygon we find the magnitude of vector $m_B\mathbf{r}_B$. To balance the rotor we can either choose a small mass at a large radius or a large mass at a small radius.

Fig. 5.8 a Front view of a rotor with four concentrated masses all of them lying in the same transverse plane **b** Lateral view with the equilibrating mass, m_B, drawn with a *thin line*

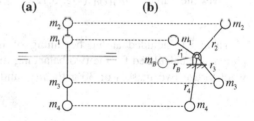

(a)   **(b)**

Fig. 5.9 The polygon of the inertia forces in the rotor including the added mass m_B has to be closed

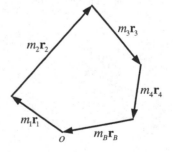

Fig. 5.10 a Front view of an
unbalanced rotor, with three
rotating masses, that has to be
balanced with two new
masses, located in planes P_{B1}
and P_{B2} **b** lateral view of the
same rotor

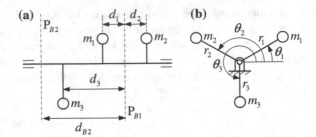

Example 4 Balance a rotor with three rotating masses, m_1, m_2 and m_3, in different
transverse planes. Use two masses, m_{B1} and m_{B2}, lying in planes P_{B1} and P_{B2}
(Fig. 5.10). The parameters of the rotor are included in Table 5.1. Distances d_j have
been considered to be positive on the left of plane P_{B1}.

As seen at the beginning of this section, we will need two polygons to solve this
problem graphically: one proportional to the force polygon and another proportional to the torque polygon. The steps to be followed are:

1. Build a polygon with the vectors in Eq. (5.12). Value d_j is the distance from
   mass m_j to plane P_{B1} (Fig. 5.10). We consider this distance to be positive for the
   masses on the left of the plane and negative on the right. The vector which
   closes the polygon is $d_{B2}m_{B2}\mathbf{r}_{B2}$ (Fig. 5.11). Its direction gives us the direction
   of $\mathbf{r}_{B2}$, the radius of balancing mass m_{B2}. Dividing the magnitude of vector
   $d_{B2}m_{B2}\mathbf{r}_{B2}$ by d_{B2}, we obtain value $m_{B2}r_{B2}$.
2. Build a polygon with the vectors in Eq. (5.11) including vector $m_{B2}\mathbf{r}_{B2}$. Vector
   $m_{B1}\mathbf{r}_{B1}$, which closes the polygon, yields the direction of $\mathbf{r}_{B1}$ and its magnitude
   gives the value of $m_{B1}r_{B1}$ (Fig. 5.12).

Table 5.1 included at the beginning of this example can be filled with the
parameters calculated for the two balancing masses (Table 5.2). Assuming that we
want to use two masses of 30 g for m_{B1} and 16 g for m_{B2}, we can calculate the

Fig. 5.11 Closed polygon
with the inertia torque vectors

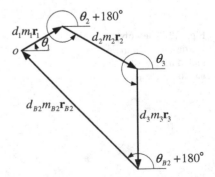

Fig. 5.12 Closed polygon
with the inertia force vectors

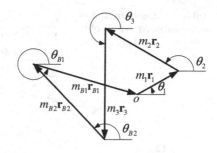

Table 5.1 Geometrical data
and unbalanced masses of the
rotor in (Fig. 5.10)

j	m_j	r_j	d_j	θ_j
1	15	4	4	30°
2	30	4	−4	150°
3	15	5	8	270°
B1	?	?	0	?
B2	?	?	10	?

Table 5.2 Masses, radius
and angular position
calculated for rotor in
example 4

j	m_j	r_j	d_j	θ_j	$m_j r_j$	$d_j m_j r_j$
1	15	4	4	30°	60	240
2	30	4	−4	150°	120	−480
3	15	5	8	270°	75	600
B1	30	4.88	0	324°	146.5	0
B2	16	6	10	131°	96	960

needed radius r_{B1} and r_{B2}. We could also decide to use a certain value for the radius
and calculate the needed mass.

Example 5 Assuming that the rotor in Fig. 5.13 rotates at constant speed, find
masses m_{B1} and m_{B2} at a 2 cm radius and their angular positions to balance it
dynamically. The two balancing masses have to be located in transverse planes P_{B1}
and P_{B2}. Table 5.3 includes the parameters of the rotor. Distances d_j have been
considered to be positive on the right of plane P_{B1}. The table also includes the
magnitudes of vectors $m_j \mathbf{r}_j$ and $d_j m_j \mathbf{r}_j$ of the known masses.

Fig. 5.13 a Front view of a
rotor with three rotating
masses that has to be balanced
with two new masses located
in planes P_{B1} and P_{B2} **b** lateral
view

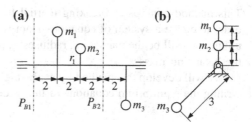

Table 5.3 Geometrical data and unbalanced masses of the rotor in (Fig. 5.13)

j	m_j	r_j	d_j	θ_j	$m_j r_j$	$d_j m_j r_j$
1	200	2	2	90°	400	800
2	400	1	4	90°	400	1600
3	300	3	8	225°	900	7200
B1	?	?	0	?		
B2	?	?	6	?		

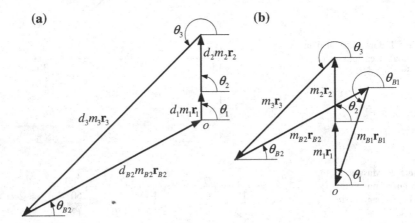

(a)   **(b)**

Fig. 5.14 a Closed polygon of inertia torque vectors **b** closed polygon of inertia force vectors

Table 5.4 Balancing masses and their angular position

j	m_j	r_j	d_j	θ_j	$m_j r_j$	$d_j m_j r_j$
B1	325	2	0	250°	650	0
B2	479	2	6	28°	958	5748

We start by drawing the polygon of inertia torques to obtain the unknown parameters for mass m_{B2} (Fig. 5.14a). Then, we draw the inertia force polygon (Fig. 5.14b) to obtain the unknown parameters of mass m_{B1} (Table 5.4).

5.1.3 Analytical Method

This method consists of breaking inertial forces into their x and y components in order to create a system of equations from where we can find the unknowns. Such unknowns will be the masses m_j, radiuses r_j, distances d_j and angular positions θ_j of the balancing masses.

We will develop this method by analyzing the rotor in Figs. 5.15a, b. We start by writing the equilibrium equations of the force x and y components including those

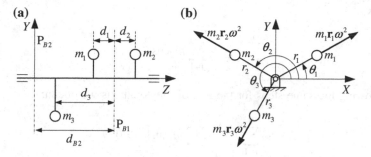

Fig. 5.15 **a** Front view of the rotor with two balancing masses, drawn with a *thin line*, located in two predetermined planes, P_{B1} and P_{B2} **b** lateral view of a rotor with three masses with the x and y components of their inertia forces

forces generated by the two balancing masses, m_{B1} and m_{B2}. This yields Eqs. (5.17 and 5.18):

$$m_1 r_1 \cos \theta_1 + m_2 r_2 \cos \theta_2 + m_3 r_3 \cos \theta_3 + m_{B1} r_{B1} \cos \theta_{B1} + m_{B2} r_{B2} \cos \theta_{B2} = 0$$
(5.17)

$$m_1 r_1 \sin \theta_1 + m_2 r_2 \sin \theta_2 + m_3 r_3 \sin \theta_3 + m_{B1} r_{B1} \sin \theta_{B1} + m_{B2} r_{B2} \sin \theta_{B2} = 0$$
(5.18)

We define a new X' — Y' Cartesian coordinate system in one of the two planes that contain a balancing mass. In Fig. 5.15a, this system is defined in plane P_{B2}. Then we write the equilibrium equation of the moments (Eqs. 5.19 and 5.20) generated by the x and y component of the inertia forces about the Y' and X' axis respectively:

$$d_1 m_1 r_1 \cos \theta_1 + d_2 m_2 r_2 \cos \theta_2 + d_3 m_3 r_3 \cos \theta_3 + d_{B1} m_{B1} r_{B1} \cos \theta_{B1} + d_{B2} m_{B2} r_{B2} \cos \theta_{B2} = 0$$
(5.19)

$$d_1 m_1 r_1 \sin \theta_1 + d_2 m_2 r_2 \sin \theta_2 + d_3 m_3 r_3 \sin \theta_3 + d_{B1} m_{B1} r_{B1} \sin \theta_{B1} + d_{B2} m_{B2} r_{B2} \sin \theta_{B2} = 0$$
(5.20)

We can simplify the equations, so that they depend on single common radius R, $(m_j r_j = m'_j R)$. This leads to the following changes in Eqs. (5.21–5.24).

$$R(m'_1 \cos \theta_1 + m'_2 \cos \theta_2 + m'_3 \cos \theta_3 + m'_{B1} \cos \theta_{B1} + m'_{B2} \cos \theta_{B2}) = 0$$

$$\sum_{j=1}^{n} F_{j_x}^{In} = 0 = R\omega^2 \sum_{j=1}^{n} m'_j \cos \theta_j$$

Thus:

$$\sum_{j=1}^{n} F_{j_x}^{In} = \sum_{j=1}^{n} m'_j \cos \theta_j = 0 \qquad (5.21)$$

Following the same steps for the rest of the equations we obtain:

$$\sum_{j=1}^{n} F_{j_y}^{In} = \sum_{j=1}^{n} m'_j \sin \theta_j = 0 \qquad (5.22)$$

$$\sum_{j=1}^{n} M_{j_y}^{In} = \sum_{j=1}^{n} d_j m'_j \cos \theta_j = 0 \qquad (5.23)$$

$$\sum_{j=1}^{n} M_{j_x}^{In} = \sum_{j=1}^{n} d_j m'_j \sin \theta_j = 0 \qquad (5.24)$$

Example 6 Balance the rotor in Fig. 5.16 assuming that rotating velocity ω is constant and knowing that the value of the masses is $m_1 = 200\,g$, $m_2 = 400\,g$ and $m_3 = 300\,g$. Use two balancing masses located in planes P_{B1} and P_{B2}.

We will proceed as in the previous example and unify the radius value. We will consider the common radius to be $R = 2\,cm$, hence, the new values for the masses are:

$$m_1 r_1 = m'_1 R \Rightarrow m'_1 = \frac{r_1}{R} m_1 = 200\,g$$

$$m_2 r_2 = m'_2 R \Rightarrow m'_2 = \frac{r_2}{R} m_2 = 200\,g$$

$$m_3 r_3 = m'_3 R \Rightarrow m'_3 = \frac{r_3}{R} m_3 = 450\,g$$

Equilibrium equations (Eqs. 5.21–5.24) will allow us to calculate the four unknowns: the two balancing masses and their angular position.

Fig. 5.16 a Front view of a rotor with three rotating masses and planes P_{B1} and P_{B2} where the balancing masses have to be located **b** lateral view

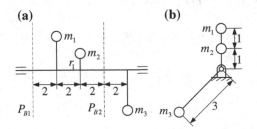

For moment equations (Eqs. 5.23 and 5.24) we will use distance d_j from each mass to plane P_{B1}. This way the moment for mass m_{B1} will be zero and this unknown will not appear in the moment equations.

Plugging the new mass values and distances into the equilibrium equations we obtain:

$$200\cos 90\degree + 200\cos 90\degree + 450\cos 225\degree + m_{B1}\cos\theta_{B1} + m_{B2}\cos\theta_{B2} = 0 \quad (5.25)$$

$$200\sin 90\degree + 200\sin 90\degree + 450\sin 225\degree + m_{B1}\sin\theta_{B1} + m_{B2}\sin\theta_{B2} = 0 \quad (5.26)$$

$$2\cdot 200\cos 90\degree + 4\cdot 200\cos 90\degree + 8\cdot 450\cos 225\degree + 6\cdot m_{B2}\cos\theta_{B2} = 0 \quad (5.27)$$

$$2\cdot 200\sin 90\degree + 4\cdot 200\sin 90\degree + 8\cdot 450\sin 225\degree + 6\cdot m_{B2}\sin\theta_{B2} = 0 \quad (5.28)$$

Operating in Eqs. (5.27 and 5.26), we get Eqs. (5.29 and 5.30):

$$450\frac{\sqrt{2}}{2} = m_{B1}\cos\theta_{B1} + m_{B2}\cos\theta_{B2} \quad (5.29)$$

$$450\frac{\sqrt{2}}{2} - 400 = m_{B1}\sin\theta_{B1} + m_{B2}\sin\theta_{B2} \quad (5.30)$$

From Eq. (5.27) we can clear m_{B2} (Eq. 5.31):

$$8\cdot 450\frac{\sqrt{2}}{2} = 6m_{B2}\cos\theta_{B2} \Rightarrow m_{B2} = \frac{8\cdot 450\dfrac{\sqrt{2}}{2}}{6\cos\theta_{B2}} \quad (5.31)$$

Operating in Eq. (5.28) we obtain Eq. (5.32):

$$8\cdot 450\frac{\sqrt{2}}{2} - 1200 = 6\,m_{B2}\sin\theta_{B2} \quad (5.32)$$

Plugging m_{B2} Eqs. (5.31) into (5.32), operating and simplifying we calculate unknown θ_{B2}:

$$\tan\theta_{B2} = \frac{8\cdot 450\dfrac{\sqrt{2}}{2} - 1200}{8\cdot 450\dfrac{\sqrt{2}}{2}} \Rightarrow \theta_{B2} = 27.86\degree$$

Operating in Eq. (5.31) with θ_{B2} we obtain mass $m_{B2} = 479.88$ g. Plugging these two values into Eqs. (5.29 and 5.30), we calculate $m_{B1} = 323.81$ g and $\theta_{B1} = 250.88°$.

Hence, the rotor can be dynamically equilibrated with two masses m_{B1} and m_{B2} located in transverse planes P_{B1} and P_{B2}.

5.2 Inertia Balancing of Single and Multi-cylinder Engines

In this section we will study the inertia forces and torques, first in single-cylinder engines and then in multicylinder ones. The main goal is to know how to reduce shaking forces and torques.

5.2.1 One-Cylinder Engines

To carry out the balancing of a single-cylinder engine we have to previously know the inertial force generated by each of the slider-crank mechanism links. We will start by studying the piston. Then, we will move on to the connecting rod and will finish by studying the crankshaft. As usual, we consider link 2 to be the input link and moving with constant angular velocity ω (Fig. 5.17).

5.2.1.1 The Inertia Force of the Piston

Since the piston in an engine mechanism has pure translational motion, its inertia force is $\mathbf{F}^{In} = -m\mathbf{a}$.

We will find the equation that defines position x of the slider or piston (Eq. 5.34) in terms of angle θ of the crank (Fig. 5.18). Differentiating the position equation

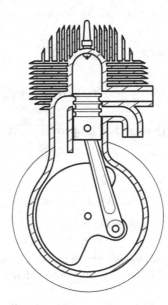

Fig. 5.17 Single-cylinder-two-stroke engine with a counterweight on the crankshaft to balance the inertia forces

Fig. 5.18 Kinematic skeleton of a slider-crank mechanism with the parameters required to calculate the position of the piston using the trigonometric method

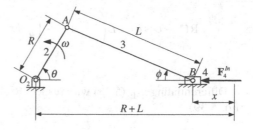

with respect to time gives us velocity and acceleration and multiplying the latter by the piston mass yields the expression of its inertia force (Eq. 5.33).

$$x = f(\theta) \Rightarrow v = \frac{d}{dt}f(\theta) \Rightarrow a = \frac{d^2}{dt^2}f(\theta) \Rightarrow F^{In} = ma \qquad (5.33)$$

In Fig. 5.18 we can calculate x as:

$$
\begin{aligned}
x &= R + L - \overline{O_2B} = R + L - (\overline{O_2M} + \overline{MB}) \\
&= R + L - (R\cos\theta + L\cos\phi) \\
&= R(1 - \cos\theta) + L(1 - \cos\phi)
\end{aligned}
\qquad (5.34)
$$

Angle ϕ can be eliminated by establishing the relationship (Eq. 5.35):

$$\overline{AM} = R\sin\theta = L\sin\phi \Rightarrow \sin\phi = \frac{R}{L}\cos\theta \qquad (5.35)$$

Next, we find $\cos\phi$ (Eq. 5.36):

$$\cos\phi = \sqrt{1 - \sin^2\phi} = \sqrt{1 - \left(\frac{R}{L}\sin\theta\right)^2} \qquad (5.36)$$

This way, angle ϕ is expressed in terms of angle θ. A series expansion of the binomial term (Eq. 5.37) will simplify its derivative.

$$
\begin{aligned}
\sqrt{1 - \left(\frac{R}{L}\sin\theta\right)^2} = {}& 1 - \frac{1}{2}\left(\frac{R}{L}\sin\theta\right)^2 \\
& - \frac{1}{2\cdot 4}\left(\frac{R}{L}\sin\theta\right)^4 - \frac{3}{2\cdot 4\cdot 6}\left(\frac{R}{L}\sin\theta\right)^6 + \cdots
\end{aligned}
\qquad (5.37)
$$

We just take the first two terms of the series and plug them into Eq. (5.34). The rest of the terms can be neglected, the error made being very small.

$$x = R(1 - \cos\theta) + L\left(1 - 1 + \frac{1}{2}\left(\frac{R}{L}\sin\theta\right)^2\right) = R(1 - \cos\theta) + \frac{R^2}{2L}\sin^2\theta$$

$$(5.38)$$

Differentiating Eq. (5.38) with respect to time, we find the velocity of the piston (Eq. 5.39):

$$v = \frac{dx}{dt} = R\sin\theta\frac{d\theta}{dt} + \frac{R^2}{L}\sin\theta\cos\theta\frac{d\theta}{dt}$$

$$= R\omega\left(\sin\theta + \frac{R}{L}\sin\theta\cos\theta\right)$$

$$(5.39)$$

Likewise, the velocity equation can be time-differentiated so that it yields the acceleration of the piston in terms of angle θ (Eq. 5.40). Remember that we have considered ω to be constant.

$$a = \frac{dv}{dt} = R\omega\left(\cos\theta\frac{d\theta}{dt} + \frac{R}{L}(\cos^2\theta - \sin^2\theta)\frac{d\theta}{dt}\right)$$

$$= R\omega^2\left(\cos\theta + \frac{R}{L}\sin 2\theta\right)$$

$$(5.40)$$

Thus, once the acceleration of the piston has been calculated, we can compute its inertia force (Eq. 5.41) for any position of the crank:

$$F^{In} = ma = mR\omega^2\left(\cos\theta + \frac{R}{L}\sin 2\theta\right)$$

$$= \underbrace{mR\omega^2\cos\theta}_{F_p^{In}} + \underbrace{m\frac{R^2}{L}\omega^2\sin 2\theta}_{F_s^{In}}$$

$$(5.41)$$

Where m is the mass of the piston, R is the radius of the crank, L is the length of the connecting rod, ω is the rotating velocity of the crank and θ is its angular position.

As we can see in Eq. (5.41), the inertia force of the piston consists of two terms. We call the first one the primary inertia force F_p^{In} whose value varies following a simple harmonic motion law with angular frequency ω (angular velocity of the crank). The second term is called secondary inertia force F_s^{In}. Its frequency is 2ω and its magnitude is R/L times the primary inertia force. We can represent these forces by means of a system with two masses rotating with different angular velocities ω and 2ω (Fig. 5.19).

In the diagram in Fig. 5.20, we can observe that the amplitude of the secondary force is proportional to the amplitude of the primary one, its maximum being less than 50 % of the primary inertia force, since L is more than twice as much as R.

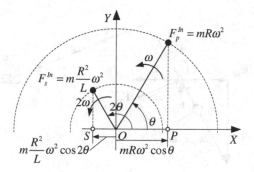

Fig. 5.19 Horizontal projections $\overline{OP}$ and $\overline{OS}$ represent the primary and secondary inertia force respectively. Radius F_p^{In} rotates at velocity ω while F_s^{In} rotates at 2ω

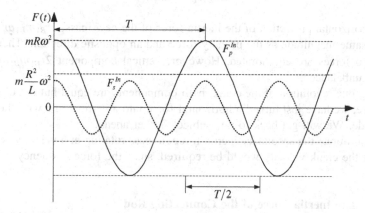

Fig. 5.20 Primary and secondary force diagram with respect to the crank position

As observed, at positions $\theta = 0°$ and $\theta = 180°$, the primary and secondary inertia forces reach their maximum magnitudes. As we can observe, at position $\theta = 180°$ (inferior dead-point), the primary force is partially equilibrated by the secondary force and at position $\theta = 0°$ (superior dead-point) both forces add their effects.

In Fig. 5.21 we can see how the primary inertia force is transmitted to the frame of the engine. $\mathbf{F}_{21}$ and $\mathbf{F}_{41}$ are the forces exerted by links 2 and 4 on the frame. The horizontal component of force $\mathbf{F}_{21}$, $\mathbf{F}_{21_x}$, is equal to the primary force $\mathbf{F}_p^{In}$ while its vertical component, $\mathbf{F}_{21_y}$, has the same magnitude and opposite direction to $\mathbf{F}_{41}$. This gives way to the idea of locating mass $2m_{B1}$ with rotation radius r_{B1} at the extension of the crank (Eq. 5.42), so that:

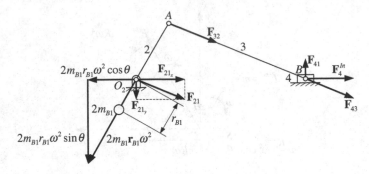

Fig. 5.21 Forces $\mathbf{F}_{21}$ and $\mathbf{F}_{41}$ transmit the primary force to the frame of the engine. Mass $2m_{B1}$ is added to the crank to balance the primary force

$$2m_{B1}r_{B1} = mR \qquad (5.42)$$

The horizontal projection of the inertia force of the new mass, $2r_{B1}m_{B1}\omega^2 \cos\theta$, has the same magnitude as the primary force and an opposite direction. Therefore, these two forces are equilibrated. However, vertical component $2r_{B1}m_{B1}\omega^2 \sin\theta$ remains unbalanced.

Along one revolution of the crank, both components are equal but out of phase. Therefore, we have just turned a horizontal force into a vertical one with the same magnitude. We will get back to this subject further ahead.

To equilibrate the secondary inertia force, a mass rotating at a velocity twice as much as the crank velocity would be required, since the force frequency is 2ω.

5.2.1.2 The Inertia Force of the Connecting Rod

After studying the inertia force of the piston, we will now go on to study the inertia force of the connecting rod, link 3. Since the movement of the connecting rod combines rotation and translation, we will simplify the problem by reducing the rod to a dynamically equivalent system consisting of two concentrated masses. The first one, m_B, located at the piston pin (B) and the second one, m_E, located at a point (E) located between the center of mass (G) and the crankpin (A).

For these two concentrated masses to be dynamically equivalent to the original rod, the following conditions have to be met (Eqs. 5.43–5.45):

• Mass m_3 does not change:

$$m_3 = m_B + m_E \qquad (5.43)$$

Fig. 5.22 Connecting rod

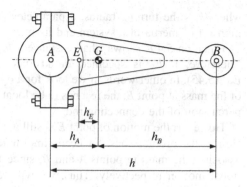

- The center of mass does not change:

$$0 = m_B h_B - m_E h_E \qquad (5.44)$$

h_B and h_E being the distances of masses m_B and m_E to the center of mass G respectively.
- Moment of inertia I_G does not change (Fig. 5.22):

$$I_G = m_B h_B^2 + m_E h_E^2 \qquad (5.45)$$

From Eqs. (5.43 and 5.44) we have Eq. (5.46):

$$m_B = (m_3 - m_B)\frac{h_E}{h_B} \Rightarrow m_B(h_B + h_E) = m_3 h_E \qquad (5.46)$$

As point E is quite close to point A, we can simplify Eq. (5.46) by considering that $h = h_B + h_E$ (Eq. 5.47):

$$m_B = \frac{h_E}{h_B + h_E} m_3 \approx \frac{h_E}{h} m_3 \qquad (5.47)$$

We can work out m_E (Eq. 5.48) the same way:

$$m_E = \frac{h_B}{h_B + h_E} m_3 \approx \frac{h_B}{h} m_3 \qquad (5.48)$$

Once we have obtained m_B and m_E, we plug their values into Eq. (5.45):

$$I_G = m_3 k^2 = m_3 \frac{h_B}{h} h_E^2 + m_3 \frac{h_E}{h} h_B^2$$
$$= m_3 h_B h_E \frac{h_B + h_E}{h} \simeq m_3 h_B h_E \rightarrow k^2 \sim h_B h_E \qquad (5.49)$$

where k is the turning radius, a parameter that depends on the ratio between the moment of inertia of a system and its mass.

Equation 5.49 shows that distances h_B and h_E are mutually dependent. Therefore, if h_B is specified in advance, distance h_E can be determined by Eq. (5.45). In other words, if we look for a dynamically equivalent system with part of the mass at point B, the rest has to be located at point E, which is called center of percussion of the connecting rod.

However, the motion of point E is still a combination of rotation and translation. Hence, the problem has not been simplified. The ideal thing to do would be to associate the mass to points A and B, since they have pure rotational and translational motion respectively. Thus, we will reformulate the original system fixing point B and distance h_B and moving point E to the location of point A, consequently using distance h_A instead of h_E. This way, there is no need to use Eq. (5.45) in the system. Consequently, there will be an error, since the moment of inertia of the new system is different to the one of the connecting rod, which means that the inertia torque will be slightly different.

The approximation is nonetheless acceptable, as the distance between point E and A is small and, therefore, the error as well.

Also, the shape of connecting rods made by different manufacturers is usually similar, since they are designed to endure high traction and compression (buckling) loads. This causes the position of the center of mass to be similar in most connecting rods, being located approximately at 0.7 h of point B and 0.3 h of point A. Considering these values, the concentrated masses at point A and B (Eq. 5.50) can be obtained in terms of the mass of the connecting rod:

$$\left.\begin{array}{c} m_3 = m_A + m_B \\ 0 = m_A h_A + m_B h_B \end{array}\right\} \Rightarrow \left.\begin{array}{c} m_A = m_3 \dfrac{h_B}{h} = 0.7 m_3 \\ m_B = m_3 \dfrac{h_A}{h} = 0.3 m_3 \end{array}\right\} \tag{5.50}$$

In short, to simplify the analysis of the inertia forces on the mechanism, the connecting rod will be replaced by a dynamically equivalent system which consists of two concentrated masses:

- Mass m_A is approximately 70 % of the mass of the connecting rod, moves with pure rotation and is located at the crankpin.
- Mass m_B, located at the piston pin, is approximately 30 % of the connecting rod mass and has rectilinear translation.

5.2.1.3   The Inertia Force of the Crankshaft

Figure 5.23 shows the crankshaft of the engine with its mass m_2 concentrated at its center of mass G_2. Mass m_A of the connecting rod obtained in the last section can be considered part of the crank. These two masses, m_2 and m_A, rotate with constant

Fig. 5.23 Crankshaft with
two m_{B2} masses that balance
the inertia forces of masses m_2
and m_A

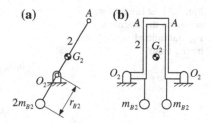

radiuses, r_{G_2} and R. Therefore, their centrifugal forces can be fully equilibrated with
two masses, m_{B2}, located on an extension of the crank (Eq. 5.51), so that:

$$2m_{B2}r_{B2}\omega^2 = m_2r_{G_2}\omega^2 + m_A R\omega^2 \qquad (5.51)$$

5.2.1.4   Inertia Balancing of the Engine

To balance the engine, we have to consider the inertia forces of each of the parts
previously studied. First, we will add m_B to the piston mass m_4. Remember that m_B
is a concentrated mass located at the piston pin that is part of a system dynamically
equivalent to the connecting rod. This way, the magnitude of the primary inertia
force of the piston (Eq. 5.52) will be:

$$F_p^{In} = (m_4 + m_B)R\omega^2 \cos\theta \qquad (5.52)$$

We will not consider the secondary inertia force, since its frequency is twice the
engine rotation frequency and it cannot be balanced with a mass that rotates at the
primary frequency. Apart from that, as we have already mentioned, the magnitude
of this force is less than 50 % of the magnitude of the primary inertia force.

Figure 5.24 represents the inertia forces acting on the engine. Masses m_{B1} are
used to balance the primary inertia force of the piston (due to m_4 and m_B) and
masses m_{B2} are added to equilibrate the inertia force of the crankshaft (due to m_2
and m_A).

Therefore, to equilibrate the whole engine, we have to locate two pairs of
masses, one of them meant to balance the inertia force of the crank and the inertia of
the mass of the connecting rod at the crankpin ($2m_{B2}r_{B2}$), and another one
($2m_{B1}r_{B1}$) that equilibrates the inertia force of the piston and the inertia of the mass
of the connecting rod at the piston pin (Eq. 5.53). The equilibrium equation is:

$$2m_{B1}r_{B1}\omega^2 + 2m_{B2}r_{B2}\omega^2 = m_2r_{G_2O_2}\omega^2 + m_A R\omega^2 + (m_4 + m_B)R\omega^2 \qquad (5.53)$$

This way we equilibrate F_{21_x} but F_{21_y} increases its value. What we have really
done here is to turn the horizontal component of the reaction force 90°. In order to

Fig. 5.24 a Inertia forces acting on the engine **b** force F_{21} transmitted to the frame by the crank

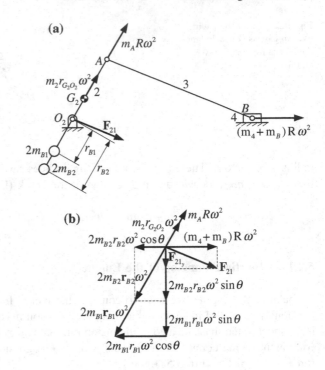

reduce the vertical component, we do not completely equilibrate F_{21_x} (Eq. 5.54). We take a value for between $2/3$ and $1/2$ of $(m_4 + m_B)R$:

$$\frac{2}{3}(m_4 + m_B)R \leq 2m_{B1}r_{B1} \leq \frac{1}{2}(m_4 + m_B)R \qquad (5.54)$$

Figure 5.25 represents the shaking force in the mechanism for a complete turn of the crank-shaft in the following three cases:

1. The mechanism is not balanced.
2. The mass of the counterweights balances $2/3$ of the primary force (including m_B) and the inertia force of the crank-shaft (including m_A).
3. The horizontal component of the primary force at O_2 is fully equilibrated (including the mass of the connecting rod at B) as well as the inertia force of the crank-shaft (including the mass of the connecting rod at A).

The forces considered to draw the polar diagram in Fig. 5.25 are the following:

- The inertia of the revolving mass of the crank (including m_A):
  We will call this force $\mathbf{F}_c$. Analysing the polar diagram in Fig. 5.25 at position θ, we can observe that straight line $\overline{oa} = F_c = m_2 r_{G_2}\omega^2 + m_A R\omega^2$ represents this force.

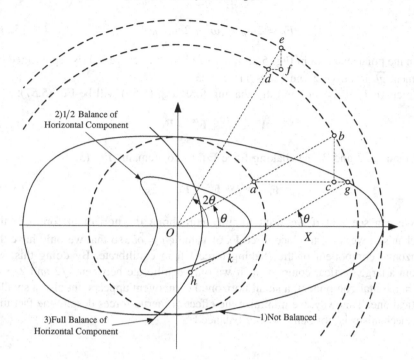

Fig. 5.25 Three polar diagrams of the shaking-force on the crankshaft bearing along a cycle depending on the counterweight mass

- The primary inertia force of the piston including mass m_B:
  The maximum value for this force is represented by straight line $\overline{ab}$, its value being $\overline{ab} = (m_4 + m_B)R\omega^2$. However, we are interested in its projection, represented by $\overline{ac}$, its value being $\overline{ac} = F_p^{In} = (m_4 + m_B)R\omega^2|\cos\theta|$.
- The secondary inertia force of the piston including mass m_B:
  In order to obtain this force graphically, we have to trace $\overline{od}$ with an angle of 2θ. After obtaining $\overline{de} = (m_4 + m_B)R\omega^2\frac{R}{L}$ (its maximum value), we consider projection $\overline{df} = F_s^{In} = (m_4 + m_B)R\omega^2\frac{R}{L}|\cos 2\theta|$. We draw segment $\overline{df}$ to the extreme of point c, yielding segment $\overline{cg}$. g is a point of the unbalanced curve for the shaft angle equal to θ.
- The centrifugal force of the counterweight mass (Eq. 5.55):

  We will call this force $\mathbf{F}_b$.

$$\mathbf{F}_b = F_b \angle \theta + 180^\circ \qquad (5.55)$$

Where magnitude F_b is Eq. (5.56):

$$F_b = 2m_{B1}r_{B1}\omega^2 + 2m_{B2}r_{B2}\omega^2 \tag{5.56}$$

In the polar diagram in Fig. 5.25 this force is null in curve 1. It is represented by segment $\overline{gh}$ in curve 2 and by $\overline{gk}$ in curve 3.

In curve 1, $\mathbf{F}_b$ is zero and the shaking force Eq. (5.56) will be Eq. (5.57):

$$\mathbf{F}_c + \mathbf{F}_p^{In} + \mathbf{F}_s^{In} = \mathbf{F}_S \tag{5.57}$$

In curves 2 and 3, the shaking force Eq. (5.56) remains Eq. (5.58):

$$\mathbf{F}_c + \mathbf{F}_p^{In} + \mathbf{F}_s^{In} + \mathbf{F}_b = \mathbf{F}_S \tag{5.58}$$

We can see that if we want to fully equilibrate the horizontal forces in the mechanism, we have to place force $\mathbf{F}_b$ of value $\overline{gh} = \overline{ob}$, so that we only have the horizontal component of the shaking force left to equilibrate. By doing this, we obtain a large vertical component. If we only equilibrate between 1/2 and 2/3 of the horizontal component, a small horizontal component appears but also a smaller vertical one. This way, we minimize the effect of inertia forces despite the fact that the mechanism is not being totally balanced.

5.2.2 Multi-cylinder in-Line Engines

As demonstrated before, it is not possible to equilibrate the inertia force of the piston in a single-cylinder engine along one revolution. We can add counterweights to decrease the shaking force but we cannot eliminate it.

However, when we have more than one piston, we can study the distribution of the crankpins, so that the inertia forces cancel each other in a way that completely equilibrates the engine. In Fig. 5.26 a four-cylinder in-line engine can be seen. In this case, the crankpins of the first and forth cylinders have an angle of 180° with the crankpins of the second and third cylinders. In this section we will learn how to choose the best relative angles for the crankpins of different cylinders.

To illustrate that it is possible to equilibrate the inertia forces of the piston in multi-cylinder engines, Fig. 5.27 shows a two-cylinder engine layout in which the cylinders form a V at an angle of 180°. The primary and secondary inertia forces of the two pistons have the same magnitude and opposite direction at any position of the crankshaft.

In Fig. 5.28 a diagram with the values of these forces shows that the resultant inertia force is null along the whole cycle. However, in Fig. 5.27 we can see that an unbalanced torque with a moment arm of value d appears.

In this chapter, we will study in-line engines with any number of cylinders N_C, so that we can compare the balance obtained for different configurations of the

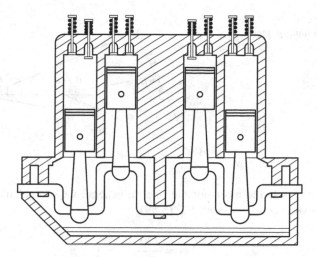

Fig. 5.26 Four-cylinder in-line engine

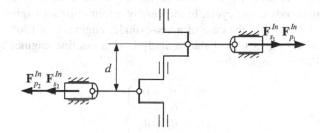

Fig. 5.27 Primary and secondary inertia forces in a two-cylinder 180-degree V-engine

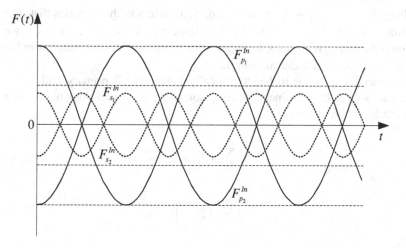

Fig. 5.28 Diagram of the primary and secondary inertia forces along a cycle in the two-cylinder 180-degree V-engine in (Fig. 5.27)

Fig. 5.29 Angles of the
crankpins in an inline engine
with 4 cylinders

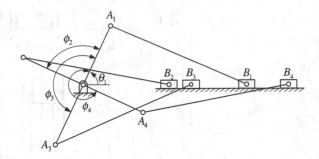

crankpins and choose the most suitable one. In general, we can calculate the best
angle (ϕ) between crankpins with Eq. (5.59).

$$\phi = \frac{720^\circ}{N_C} \tag{5.59}$$

Number 720° is used because this is the angle for a four-stroke engine to
complete its thermodynamic cycle. In the case of a four-cylinder engine, this angle
is 180° (Fig. 5.26) and in the case of a six-cylinder engine, it is 120°.

In order to perform a force balance analysis in an in-line engine, we need to
define the following angles (Fig. 5.29):

$$\left.\begin{array}{l} \theta_1 = \theta_1 + 0^\circ \\ \theta_2 = \theta_1 + \phi_2 \\ \theta_3 = \theta_1 + \phi_3 \\ \vdots \end{array}\right\} \tag{5.60}$$

In general, $\theta_j = \theta_1 + \phi_j$ where θ_j (Eq. 5.60) refers to those angles that give the
angular position of every crankpin with respect to a fixed line (in this case the
piston path) and ϕ_j represents the angles that define its relative position with respect
to the crankpin of cylinder number 1.

In order for a motor to be dynamically balanced, all primary and secondary
forces as well as their moments have to sum zero. The conditions that have to be
met are Eqs. (5.61–5.64):

$$\sum_{j=1}^{N_C} F_{p_j}^{In} = 0 \tag{5.61}$$

$$\sum_{j=1}^{N_C} F_{s_j}^{In} = 0 \tag{5.62}$$

$$\sum_{j=1}^{N_C} M_{Pj}^{In} = 0 \tag{5.63}$$

$$\sum_{j=1}^{N_C} M_{Sj}^{In} = 0 \tag{5.64}$$

We will develop these equations using the following nomenclature:

- θ_j is the angle between crankpin j and the horizontal line.
- ϕ_j is the angle between crankpin j and the first one.
- d_j is the axial distance between crankpin j and the first one.

We begin by developing the equilibrium of the primary forces (Eq. 5.65):

$$\sum_{j=1}^{N_C} F_{Pj}^{In} = \sum_{j=1}^{N_C} m_j R_j \omega^2 \cos \theta_j = \omega^2 \sum_{j=1}^{N_C} m_j R_j \cos(\theta_1 + \phi_j) = 0 \tag{5.65}$$

We develop the cosine in Eq. (5.66):

$$\cos(\theta_1 + \phi_j) = \cos \theta_1 \cos \phi_j - \sin \theta_1 \sin \phi_j \tag{5.66}$$

Hence, we obtain Eq. (5.67):

$$\omega^2 \sum_{j=1}^{N_C} m_j R_j (\cos \theta_1 \cos \phi_j - \sin \theta_1 \sin \phi_j) = \omega^2 \cos \theta_1 \sum_{j=1}^{N_C} m_j R_j \cos \phi_j$$

$$- \omega^2 \sin \theta_1 \sum_{j=1}^{N_C} m_j R_j \sin \phi_j = 0 \tag{5.67}$$

In order to comply with Eq. (5.67), the conditions Eqs. (5.68 and 5.69) have to be met:

$$\omega^2 \cos \theta_1 \sum_{j=1}^{N_C} m_j R_j \cos \phi_j = 0 \tag{5.68}$$

$$\omega^2 \sin \theta_1 \sum_{j=1}^{N_C} m_j R_j \sin \phi_j = 0 \tag{5.69}$$

Otherwise, considering all the masses and radius to have the same value, $m_j = m$ and $R_j = R$, we can write Eqs. (5.68 and 5.69) as Eqs. (5.70 and 5.71):

$$mR\omega^2 \cos\theta_1 \sum_{j=1}^{N_C} \cos\phi_j = 0 \tag{5.70}$$

$$mR\omega^2 \sin\theta_1 \sum_{j=1}^{N_C} \sin\phi_j = 0 \tag{5.71}$$

This finally yields Eqs. (5.72 and 5.73):

$$\sum_{j=1}^{N_C} \cos\phi_j = 0 \tag{5.72}$$

$$\sum_{j=1}^{N_C} \sin\phi_j = 0 \tag{5.73}$$

The sum of moments due to primary forces with respect to an arbitrary reference plane (Eq. 5.74) also has to be zero. In this case we will use the rotation plane of the crankpin of the first cylinder as the reference plane.

$$\sum_{j=1}^{N_C} M_{Pj}^{In} = \sum_{j=1}^{N_C} d_j F_{Pj}^{In} = 0 \tag{5.74}$$

Following a procedure similar to the one described above we reach (Eqs. 5.75 and 5.76):

$$mR\omega^2 \cos\theta_1 \sum_{j=1}^{N_C} d_j \cos\phi_j = 0 \tag{5.75}$$

$$mR\omega^2 \sin\theta_1 \sum_{j=1}^{N_C} d_j \sin\phi_j = 0 \tag{5.76}$$

This yields Eqs. (5.77 and 5.78):

$$\sum_{j=1}^{N_C} d_j \cos\phi_j = 0 \tag{5.77}$$

$$\sum_{j=1}^{N_C} d_j \sin\phi_j = 0 \tag{5.78}$$

We will continue studying the conditions for the equilibrium of the secondary inertia forces (Eq. 5.79):

$$\sum_{j=1}^{N_C} F_{Sj}^{In} = \sum_{j=1}^{N_C} m_j \frac{R_j^2}{L_j} \omega^2 \cos 2\theta_j = 0 \qquad (5.79)$$

Substituting Eq. (5.80):

$$\omega^2 \sum_{j=1}^{N_C} m_j \frac{R_j^2}{L_j} \cos(2\theta_1 + 2\phi_j) = 0 \qquad (5.80)$$

Again, we apply trigonometry and develop the angle-sum cosine in Eq. (5.81):

$$\omega^2 \sum_{j=1}^{N_C} m_j \frac{R_j^2}{L_j} (\cos 2\theta_1 \cos 2\phi_j - \sin 2\theta_1 \sin 2\phi_j) = 0 \qquad (5.81)$$

In order to comply with this expression, Eqs. (5.82 and 5.83) have to be met:

$$\omega^2 \cos 2\theta_1 \sum_{j=1}^{N_C} m_j \frac{R_j^2}{L_j} \cos 2\phi_j = 0 \qquad (5.82)$$

$$\omega^2 \sin 2\theta_1 \sum_{j=1}^{N_C} m_j \frac{R_j^2}{L_j} \sin 2\phi_j = 0 \qquad (5.83)$$

Reorganizing and considering $m_j = m$, $R_j = R$ and $L_j = L$, we obtain Eqs. (5.84 and 5.85):

$$m \frac{R^2}{L} \omega^2 \cos 2\theta_1 \sum_{j=1}^{N_C} \cos 2\phi_j = 0 \qquad (5.84)$$

$$m \frac{R^2}{L} \omega^2 \sin 2\theta_1 \sum_{j=1}^{N_C} \sin 2\phi_j = 0 \qquad (5.85)$$

This finally yields Eqs. (5.86 and 5.87):

$$\sum_{j=1}^{N_C} \cos 2\phi_j = 0 \qquad (5.86)$$

$$\sum_{j=1}^{N_C} \sin 2\phi_j = 0 \qquad (5.87)$$

Finally, the sum of moments of the secondary inertia forces relative to an arbitrary plane (Eq. 5.88) has to be zero:

$$\sum_{j=1}^{N_C} M_{S_j}^{In} = \sum_{j=1}^{N_C} d_j F_{S_j}^{In} = 0 \qquad (5.88)$$

This yields Eqs. (5.89 and 5.90):

$$\omega^2 \cos 2\theta_1 \sum_{j=1}^{N_C} d_j m_j \frac{R_j^2}{L_j} \cos 2\phi_j = 0 \qquad (5.89)$$

$$\omega^2 \sin 2\theta_1 \sum_{j=1}^{N_C} d_j m_j \frac{R_j^2}{L_j} \sin 2\phi_j = 0 \qquad (5.90)$$

Eventually Eqs. (5.91 and 5.92):

$$\sum_{j=1}^{N_C} d_j \cos 2\phi_j = 0 \qquad (5.91)$$

$$\sum_{j=1}^{N_C} d_j \sin 2\phi_j = 0 \qquad (5.92)$$

Therefore, Eqs. (5.72, 5.73, 5.77, 5.78, 5.86, 5.87, 5.91, 5.92) have to be met in order for the engine to be balanced. Whenever these eight conditions are verified, we can assert that the engine is fully balanced. If any of these equations is different to zero, there will be some imbalance in the system. As stated previously in this section, primary inertia forces have a bigger effect than secondary ones. Thus, the imbalance of the system will have a greater dependence on primary forces. The same happens when comparing the effect of forces with the effect of moments in the system. The first ones have a bigger effect on balance.

Example 7 Two-cylinder inline engine.

Figures 5.30a, b shows two possible configurations for the crankshaft of a two-cylinder inline engine.

To check which of the two configurations is better, we make a table with the eight conditions for the engine to be fully balanced.

As we can see in Table 5.5, the second configuration (crankpins at 180°) succeeds in balancing the primary inertia forces while in the first configuration none of

Fig. 5.30 a Crankshaft of a two-cylinder inline engine with 360° between the two crankpins **b** the same engine with 180° between the crankpins

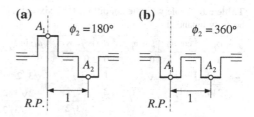

Table 5.5 Above, the eight conditions for a two-cylinder engine with 180° between the two crankpins to be fully balanced (Fig. 5.30 a). Below, the same conditions for a two-cylinder engine with 360° between the two crankpins (Fig. 5.30 b)

j	ϕ_j	$\cos\phi_j$	$\sin\phi_j$	d_j	$d_j\cos\phi_j$	$d_j\sin\phi_j$	$\cos 2\phi_j$	$\sin 2\phi_j$	$d_j\cos 2\phi_j$	$d_j\sin 2\phi_j$
1	$0°$	1	0	0	0	0	1	0	0	0
2	$180°$	−1	0	1	−1	0	1	0	1	0
$\sum\limits_{j=1}^{2}$		0	0		−1	0	2	0	1	0
1	$0°$	1	0	0	0	0	1	0	0	0
2	$360°$	1	0	1	1	0	1	0	1	0
$\sum\limits_{j=1}^{2}$		2	0		1	0	2	0	1	0

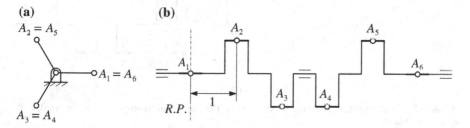

(a)
$A_2 = A_5$
$A_1 = A_6$
$A_3 = A_4$

(b)
A_2 A_5
A_1 A_6
R.P. A_3 A_4

Fig. 5.31 a Lateral and **b** front views of the crankshaft of a six-cylinder inline engine with the position of the crankpin of each cylinder

the inertia forces and torques are balanced. Therefore, the second configuration will give us a better balanced engine.

Example 8 Six-cylinder inline engine.

Figure 5.31 shows lateral and front views of the crankshaft of a six-cylinder inline engine with 120° between the crankpins.

As seen in Table 5.6, all inertia forces and torques are balanced in this engine. This is, therefore, the best possible configuration for an inline multi-cylinder engine.

Table 5.6 Table with the conditions for a six-cylinder engine with 120° between the crankpins to be fully balanced

j	ϕ_j	$\cos\phi_j$	$\sin\phi_j$	d_j	$d_j\cos\phi_j$	$d_j\sin\phi_j$	$\cos 2\phi_j$	$\sin 2\phi_j$	$d_j\cos 2\phi_j$	$d_j\sin 2\phi_j$
1	0°	1	0	0	0	0	1	0	0	0
2	120°	$-1/2$	$\sqrt{3}/2$	1	$-1/2$	$\sqrt{3}/2$	$-1/2$	$-\sqrt{3}/4$	$-1/2$	$-\sqrt{3}/4$
3	240°	$-1/2$	$-\sqrt{3}/2$	2	$-2/2$	$-2\sqrt{3}/2$	$-1/2$	$\sqrt{3}/4$	$-2/2$	$2\sqrt{3}/4$
4	240°	$-1/2$	$-\sqrt{3}/2$	3	$-3/2$	$-3\sqrt{3}/2$	$-1/2$	$\sqrt{3}/4$	$-3/2$	$3\sqrt{3}/4$
5	120°	$-1/2$	$\sqrt{3}/2$	4	$-4/2$	$4\sqrt{3}/2$	$-1/2$	$-\sqrt{3}/4$	$-4/2$	$-4\sqrt{3}/4$
6	0°	1	0	5	5	0	1	0	5	0
$\sum_{j=1}^{6}$		0	0		0	0	0	0	0	0

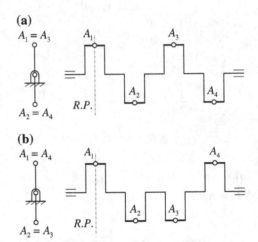

Fig. 5.32 a A configuration for the crankshaft of a four-cylinder inline engine with the position of the crankpin of each cylinder **b** a different configuration for the same engine

Table 5.7 Table with the balancing conditions for the four-cylinder engine with the crankshaft shown in (Fig. 5.32a)

j	ϕ_j	$\cos\phi_j$	$\sin\phi_j$	d_j	$d_j\cos\phi_j$	$d_j\sin\phi_j$	$\cos 2\phi_j$	$\sin 2\phi_j$	$d_j\cos 2\phi_j$	$d_j\sin 2\phi_j$
1	0°	1	0	0	0	0	1	0	0	0
2	180°	-1	0	1	-1	0	1	0	1	0
3	0°	1	0	2	2	0	1	0	2	0
4	180°	-1	0	3	-3	0	1	0	3	0
$\sum_{j=1}^{4}$		0	0		-2	0	4	0	6	0

Example 9 Four-cylinder inline engine.

Figure 5.32a, b shows two possible configurations for the crankshaft of a four-cylinder inline engine (Tables 5.7 and 5.8).

Table 5.8 Table with the balancing conditions for an engine with the crankshaft in (Fig. 5.32b)

j	ϕ_j	$\cos\phi_j$	$\sin\phi_j$	d_j	$d_j\cos\phi_j$	$d_j\sin\phi_j$	$\cos 2\phi_j$	$\sin 2\phi_j$	$d_j\cos 2\phi_j$	$d_j\sin 2\phi_j$
1	$0°$	1	0	0	0	0	1	0	0	0
2	$180°$	−1	0	1	−1	0	1	0	1	0
3	$180°$	−1	0	2	−2	0	1	0	2	0
4	$0°$	1	0	3	3	0	1	0	3	0
$\sum_{j=1}^{4}$		0	0		0	0	4	0	6	0

In both cases the primary forces are balanced but only in the second case the moments of the primary forces are also balanced. Hence, the latter is the best solution for a four-cylinder inline engine.

5.3 Problems with Solutions

Example 10 Figure 5.33a shows a wheel balancer that has two force sensors at supports 1 and 2. In Fig. 5.33b the graph shows the values of the vertical component of the force on each support with respect to the angular position of the wheel. The maximum value of this component gives us the magnitude of the force vector. The following results are obtained:

- Support 1:
  - Force magnitude: 10 N
  - Phase angle: $\theta = 45°$
- Support 2:
  - Force magnitude: 20 N
  - Phase angle: $\theta = 90°$

Find the value of the required balancing masses as well as their angular position. These masses have to be located on the balancing planes, their rotation radius has to be the same as the wheel rim, in this case 15 cm. The angular velocity is $\omega = 1500$ rpm. The values for distances in Fig. 5.33a are: $D = 20$ cm, $d_1 = 20$ cm and $d_2 = 30$ cm.

To solve the problem, Fig. 5.34a, b shows a scheme in which the forces on supports 1 and 2 have been replaced by a dynamically equivalent system. This system consists of centrifugal forces F_1 and F_2, generated by masses m_1 and m_2, which have the same magnitude and angle as the original forces on the supports. To equilibrate the wheel, we will use two masses, m_{B1} and m_{B2}, located in the balancing planes, one on each face of the wheel rim.

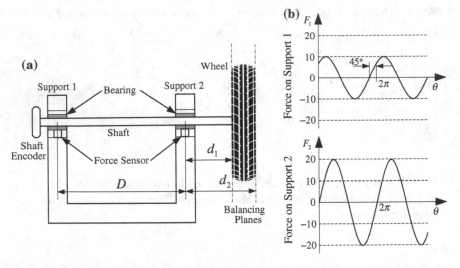

Fig. 5.33 **a** Wheel balancer **b** force graph showing the vertical component of the forces on the supports vs. angle

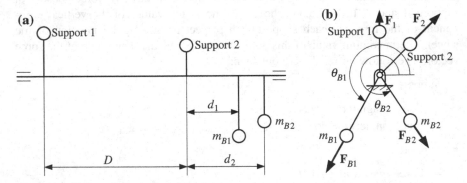

Fig. 5.34 **a** Front and **b** lateral views of the rotor with two centrifugal forces, $\mathbf{F}_1$ and $\mathbf{F}_2$, with the same magnitude and angle as the measured forces at the supports and the centrifugal forces of the unknown balancing masses, $\mathbf{F}_{B1}$ and $\mathbf{F}_{B2}$

We have to calculate the magnitudes of the centrifugal forces of the balancing masses, $\mathbf{F}_{B1}$ and $\mathbf{F}_{B2}$, as well as their angles, θ_{B1} and θ_{B2}. These angles will give us the location of the equilibrating masses.

We use the force and moment equilibrium equations (Eq. 5.93):

$$\left.\begin{array}{r} 10\cos 45^\circ + 20 + F_{B1}\cos\theta_{B1} + F_{B2}\cos\theta_{B2} = 0 \\ 10\sin 45^\circ + F_{B1}\sin\theta_{B1} + F_{B2}\sin\theta_{B2} = 0 \\ 10\cdot 50\sin 45^\circ + 10F_{B1}\sin\theta_{B1} = 0 \\ 10\cdot 50\cos 45^\circ + 20\cdot 30 + 10F_{B1}\cos\theta_{B1} = 0 \end{array}\right\} \qquad (5.93)$$

The solution to the system yields the following values for the unknowns:

$$\mathbf{F}_{B1} = 101.7 N \angle 200.34°$$
$$\mathbf{F}_{B2} = 73.91 N \angle 22.49°$$

To find the value of the balancing masses (Eqs. 5.94 and 5.95) that have to be placed on each side of the wheel rim, we compute:

$$m_{B1} = \frac{F_{B1}}{r\omega^2} = \frac{101.7}{0.15\left(1,500\dfrac{2\pi}{60}\right)^2} = 0.027\,\text{kg} \qquad (5.94)$$

$$m_{B2} = \frac{F_{B2}}{r\omega^2} = \frac{73.91}{0.15\left(1,500\dfrac{2\pi}{60}\right)^2} = 0.020\,\text{kg} \qquad (5.95)$$

Example 11 The device shown in Fig. 5.35 is used to equilibrate the inertia force of the piston in a slider-crank mechanism. The device consists of a link (5) which rotates at the same angular speed as the crank (2), but with opposite direction. Moreover, the angles formed by links 2 and 5, measured as shown in the figure, are the same along a cycle. We know the mass of the crank, $m_2 = 2$ kg, and assume its center of gravity to be at point A. The masses of the connecting rod (3) and the piston (4) are $m_3 = 1$ kg and $m_4 = 1$ kg, with their centers of mass at points G_3 and $G_4 = B$ respectively. We also know $\overline{O_2A} = \overline{O_2M_2} = \overline{O_2M_5} = \overline{AG_3} = 3$ cm and $\overline{AB} = 12$ cm. Find the value of masses m_{B5} and m_{B2}, so that the mechanism is completely balanced. Consider the secondary component of the inertia force on the piston to be negligible.

First of all, we calculate two concentrated masses m_A and m_B (Eq. 5.96) that will replace the connecting rod for a dynamically equivalent system.

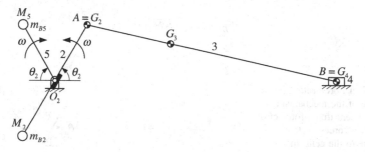

Fig. 5.35 Slider-crank mechanism with one balancing mass, m_{B5}, which rotates at the same velocity as link 2 but with opposite direction and another mass, m_{B2}, which rotates with the crank

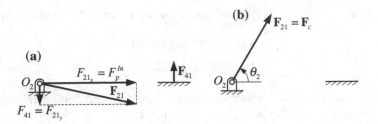

Fig. 5.36 a Forces acting on the frame of the mechanism due to the primary force of the piston **b** forces on the frame due to the centrifugal force of the crankshaft

$$\left.\begin{array}{c} 1 = m_A + m_B \\ 0 = 3m_A - 9m_B \end{array}\right\} \Rightarrow \left.\begin{array}{c} m_A = 0.75\,\text{kg} \\ m_B = 0.25\,\text{kg} \end{array}\right\} \tag{5.96}$$

The primary inertia force of the piston (Eq. 5.97), considering m_B, will have a magnitude of:

$$F_p^{In} = (1 + 0.25)3\omega^2 \cos\theta_2 \tag{5.97}$$

And the centrifugal force of the crank (Eq. 5.98), considering m_A, will be:

$$F_c = (2 + 0.75)3\omega^2 \tag{5.98}$$

Figure 5.36 shows the effects produced by these two forces on the supports of the mechanism.

In Fig. 5.36b, we can check that the vertical components of the action of the primary force on the frame, $\mathbf{F}_{21_y}$ and $\mathbf{F}_{41}$ have the same magnitude and opposite direction. Therefore, their sum is zero and they will not be considered in the equilibrium equations.

The forces that have to be equilibrated are: the horizontal component of the primary force and the horizontal and vertical components of the centrifugal force of the crankshaft. To do so, we use two counterweights with masses m_{B2} and m_{B5}, which produce the following forces (Eqs. 5.99 and 5.100) (Fig. 5.37):

$$F_{B2} = m_{B2}\overline{O_2 M_2}\omega^2 \tag{5.99}$$

Fig. 5.37 a Force acting on the frame of the mechanism due to the centrifugal force of mass m_{B2} **b** forces on the frame due to the centrifugal force of mass m_{B5}

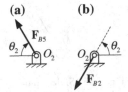

$$F_{B5} = m_{B5}\overline{O_2M_5}\omega^2 \tag{5.100}$$

The equilibrium of horizontal and vertical forces remains (Eq. 5.101):

$$\left.\begin{array}{l} ((m_4 + m_B)\overline{O_2A} + (m_2 + m_A)\overline{O_2A} - m_{B2}\overline{O_2M_2} - m_{B5}\overline{O_2M_5})\omega^2\cos\theta_2 = 0 \\ ((m_2 + m_A)\overline{O_2A} - m_{B2}\overline{O_2M_2} + m_{B5}\overline{O_2M_5})\omega^2\sin\theta_2 = 0 \end{array}\right\} \tag{5.101}$$

Operating with the known values:

$$\left.\begin{array}{l} 1.25 + 2.75 - m_{B2} - m_{B5} = 0 \\ 2.75 - m_{B2} + m_{B5} = 0 \end{array}\right\}$$

By solving these equations, we find the mass of the required counterweights:

$$\left.\begin{array}{l} m_{B2} = 3.375\,\text{kg} \\ m_{B5} = 0.625\,\text{kg} \end{array}\right\}$$

Note that 2.75 kg of m_{B2} will equilibrate the inertial force of the crankshaft (2 kg) and 75 % of the mass of the connecting rod (0.75 kg). The rest of mass m_{B2} (0.625 kg) together with mass m_{B5} will equilibrate the primary force.

Chapter 6
Flywheel Calculations

Abstract Oscillations in magnitude of the torque applied to an axle along a kinematic cycle produce variations in turning speed, which can cause machine malfunction. A flywheel is a rotating mass that is used to store energy, so that it reduces the aforesaid oscillations. It stores mechanical energy when the input torque is higher than the requirement and releases it when the resistant torque is more than the input one. The permissible variation in speed depends on the machine. In this chapter, we will study how to design a flywheel that absorbs such torque oscillations to achieve the desired regularity.

6.1 Forces and Torques in Mechanisms

Forces and torques applied to different parts of mechanisms can be classified in three groups:

- Internal forces and moments.
  Since we are considering links as rigid bodies that keep their shape, we will not consider these forces in the energy balance, as they are already balanced and produce no work.
- External forces and torques.
  External forces directly applied to the links of a mechanism can be divided into:

  - Weight of the links: Generally, this force is negligible when we compare it with the rest of external forces. The reason for this is that links are usually designed to have the minimum mass possible in order to reduce inertia. Machines with links that have high dimensions work at lower velocities and, consequently, the weight of the links might be considerable.
  - Motor force and torque: These are the forces and torques that originate motion of the links and produce a positive amount of work.
  - Resistant forces and torques: These are the ones which resist motion of the links and produce a negative amount of work. They can be classified as follows:

© Springer International Publishing Switzerland 2016
A. Simón Mata et al., *Fundamentals of Machine Theory and Mechanisms*,
Mechanisms and Machine Science 40, DOI 10.1007/978-3-319-31970-4_6

- Active forces and pairs: These are used to produce useful work (the kind of work the machine is designed for).
- Passive forces: These are energy losses due to friction. Usually machine links work in lubricated environments, so they are usually negligible.

- Inertia forces and moments: As we have already studied in this book, these forces are caused by acceleration in the links and they can be big enough to be taken into consideration. They produce positive work in a part of the kinematic cycle and negative work in the rest, its energy balance at the end of the kinematic cycle being null. That is to say, they neither deliver nor consume energy along the whole cycle.

6.2   General Equation of Mechanism Motion

A mechanism is a system of masses to which we can apply the theorem of live forces: "In a system of moving masses to which a set of external forces is applied, work done by these forces is equal to the variation of kinetic energy in the system".

In other words, work W_{12} (Eq. 6.1) done by external forces between instants 1 and 2 of the motion is:

$$W_{12} = E_{K_2} - E_{K_1} \tag{6.1}$$

where:

- E_{K_2} is the kinetic energy of the mechanism at instant 2.
- E_{K_1} is the kinetic energy of the mechanism at instant 1.

The following terms are included in W_{12} (Eq. 6.2):

- Motor work (W_M). Work produced by those forces and torques applied to the mechanism in order to make it move.
- Resistant work (W_R). Work created by resistant forces and moments. We can distinguish two types.

  - Active work (W_A), which is produced by those forces that have to be overcome so that the machine or mechanism can carry out its mission.
  - Passive work (W_P), work needed to overcome passive resistance.

Hence, we have:

$$W_{12} = W_M - W_R = W_M - W_A - W_P \tag{6.2}$$

As for kinetic energy, it is obtained as the sum of energy produced by masses in their translational and rotational motion (Eq. 6.3).

$$E_K = \frac{1}{2} \sum_i m_i v_i^2 + \frac{1}{2} \sum_i I_i \omega_i^2 \qquad (6.3)$$

where m_i are the masses with translational motion, v_i are their velocities, I_i are the mass moments of inertia about the rotation axis of masses with rotational motion and ω_i are their angular velocity.

Equation (6.3) can be simplified by using the reduced mass moment of inertia of the machine with respect to one axle (Eq. 6.4). In other words, the reduced moment of inertia will be equivalent to the mass inertia of the complete machine.

Let I_R be the reduced mass moment of inertia with respect one axle and ω_R its angular velocity. It has to be verified that:

$$\frac{1}{2} I_R \omega_R^2 = \frac{1}{2} \sum_i m_i v_i^2 + \frac{1}{2} \sum_i I_i \omega_i^2 \qquad (6.4)$$

Clearing I_R form this equation, we find the reduced moment of inertia (Eq. 6.5):

$$I_R = \sum_i m_i \left(\frac{v_i}{\omega_R} \right)^2 + \sum_i I_i \left(\frac{\omega_i}{\omega_R} \right)^2 \qquad (6.5)$$

The kinetic energy of the machine (Eq. 6.6) can be expressed as:

$$E_K = \frac{1}{2} I_R \omega_R^2 \qquad (6.6)$$

Hence, the work done between time instant 1 and 2 (Eq. 6.7) can be written as:

$$W_{12} = W_M - W_R = \frac{1}{2} I_R \omega_{R_2}^2 - \frac{1}{2} I_R \omega_{R_1}^2 \qquad (6.7)$$

From now on, we will use I and ω when referring to the moment of inertia of a machine reduced to the axle of the flywheel and the angular velocity reduced to the same axle (Eq. 6.8). This way, we write:

$$W_{12} = W_M - W_R = \frac{1}{2} I \omega_2^2 - \frac{1}{2} I \omega_1^2 \qquad (6.8)$$

6.3   Working Periods of a Cyclic Machine

Cyclic machines are commonly used in real-life applications, since this system allows maintaining work indefinitely by the repetition of consecutive cycles.

A machine is said to be cyclic when the different positions of its links are repeated periodically. In a cyclic machine, position diagrams with respect to time are formed by sections that are repeated. Each one of these sections represents the positions in a kinematic cycle. Usually, kinematic cycles coincide with working cycles. However, in some machines, like the four-stroke engine, a working cycle comprises more than one kinematic cycle. When velocity and acceleration diagrams repeat themselves over time, the velocity at the beginning of the cycle is the same as at the end. Hence, kinetic energy does not change. In this case, the machine works in a steady-state regime. Other operating regimes of a machine are start and stop. These regimes have the following conditions.

- Starting period:
  The initial velocity is $\omega_1 = 0$ and after some time, the machine reaches velocity $\omega_2 = \omega$. From Eq. (6.8), we obtain Eqs. (6.9) and (6.10):

$$W_{12} = W_M - W_R = \frac{1}{2}I\omega_2^2 - 0 = \frac{1}{2}I\omega^2 \tag{6.9}$$

$$W_M = W_R + \frac{1}{2}I\omega^2 \tag{6.10}$$

Therefore, the motor work is equal to the resistant work plus a supplementary amount of work equal to $1/2I\omega^2$, which is stored by the machine as it increases the velocity of its links. This work, which will be returned during the stopping period, is called inertia work and has to be considered when computing W_M. This way, during the starting period, it is verified that (Eq. 6.11):

$$W_M > W_R \tag{6.11}$$

- Steady state:
  The velocity is constant, $\omega_1 = \omega_2 = \omega$. Using Eq. (6.8) again, we find Eq. (6.12):

$$W_{12} = W_M - W_R = \frac{1}{2}I\omega^2 - \frac{1}{2}I\omega^2 = 0 \tag{6.12}$$

Hence, along this period, motor work and resistant work are equal (Eq. 6.13):

$$W_M = W_R \tag{6.13}$$

- Stopping period:
  The initial velocity is $\omega_1 = \omega$ and the final velocity is $\omega_2 = 0$. By using Eq. (6.8) again, we reach Eqs. (6.14) and (6.15):

$$W_{12} = W_M - W_R = 0 - \frac{1}{2}I\omega^2 = -\frac{1}{2}I\omega^2 \tag{6.14}$$

$$W_M = W_R - \frac{1}{2}I\omega^2 \tag{6.15}$$

During this period, the kinetic energy stored during the starting process of the machine, is returned. It is verified (Eq. 6.16):

$$W_M < W_R \qquad\qquad (6.16)$$

6.4 Steady State

As we already mentioned at the beginning of this chapter, forces applied to a machine can be classified in two groups: motor forces and resistant ones. Motor forces add energy to the system. Therefore, their work is computed as positive. On the other hand, resistant forces consume energy, so their work is computed as negative.

In order to study the dynamic behavior of a machine, we will display the diagrams corresponding to its motor and resistant forces during one kinematic cycle. In general, we show the input torque, T_M, exerted on the input axle and the resistant torque, T_R, exerted on the output axle.

There are numerous machines in which one of these diagrams is constant and the other varies along time. For instance, consider an internal combustion engine (T_M varies) that moves an electric generator (T_R can be considered constant and equal to the absolute value of the average motor torque). Conversely, consider an electric motor (T_M can be considered constant) that moves a press or an air compressor (T_R is variable).

Figure 6.1 shows the values, along a working cycle, of instant motor torque T_M and average motor torque $T_{M (average)}$ versus crankshaft angle in a single-cylinder engine.

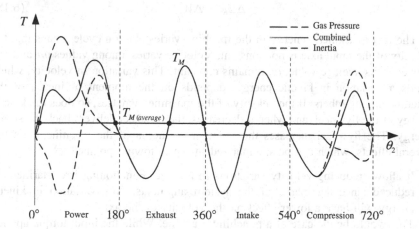

Fig. 6.1 Instant motor torque curve (*solid line*) obtained by addition of the pressure and inertia torque curves (*dashed lines*) in a single-cylinder combustion engine. The average motor torque is represented by means of a horizontal solid line

In Fig. 6.1, $T_{M\,(average)}$ is positive and constant, while T_M varies with positive and negative values along the cycle. Variations in the torque value make the machine work at different velocities at different instants along the cycle. However, it is possible to make the average velocity constant during successive cycles. In this case, as stated before, the machine is considered to be working at steady state.

In order to verify this, every cycle has to be the same as the previous one, which means that the machine reaches the end of a cycle with the same kinetic energy as when it started. It has to be verified (Eq. 6.17):

$$(\Delta W_M)_{Cycle} + (\Delta W_R)_{Cycle} = 0 \tag{6.17}$$

We can calculate work at every instant of the cycle by computing the area between the torque curve and the X-axis.

In order to fulfill this steady condition, the sum of the areas enclosed by curves T_M and T_R has to be zero. To do so, we consider the area to be positive above the X-axis and negative below it.

If this condition is met, the machine starts every new cycle with the same kinetic energy, E_K. Therefore, it starts at the same velocity as it did at the beginning of the previous cycle. This way, the average velocity along successive cycles remains constant.

6.5 Flywheels

As seen in Sect. 6.2, the theorem of live forces (Eq. 6.18) states that an increment in work value equals the increment in energy:

$$\Delta E_K = \Delta W \tag{6.18}$$

The fact that kinetic energy in the machine varies along a cycle means that the velocity of the input axle is not constant. Velocity varies among values ω_1 and ω_2. However, its average value, ω, remains constant. This variation in velocity, which yields a variation in kinetic energy, depends on the moment of inertia of the machine among others. If the velocity of the machine increases, so does the kinetic energy of the flywheel, and when it decreases, the flywheel delivers back the stored energy. The flywheel increases the moment of inertia of the machine. This has several effects, which can be summarized in the following points:

- It allows reducing velocity variations in a cycle. Functioning irregularities are reduced since the inertia of the system augments. Any variation in kinetic energy will have a lower effect on the velocity of the axle.
- This would be the case of a punching machine. While the input torque applied by an electric motor is almost constant, the resistant torque has a strong increment at the instant the punch pierces the metallic sheet, being practically

zero during the rest of the time. The loss of kinetic energy that is produced at the punching instant is so strong that if there was no flywheel (with its corresponding stored energy) the motor would have to be much more powerful in order to prevent the machine from stopping.

- A flywheel reduces the maximum stress some of the elements of the machine are affected by. This happens when the irregularity is originated by an excess in motor or resistant torque. In general, to diminish transmission stresses, the flywheel has to be located as close to the irregularity source as possible. The only members, apart from the flywheel, with the capacity to absorb torque variations will be those elements joining it to the link where the irregular torque is applied. For instance, consider a vehicle; the flywheel is usually installed in at the motor exit axle in order to prevent it from transmitting possible irregularities to the gearbox. This way, a smoother functioning of the motor is also achieved when it is working unengaged. However, in the punching machine, it will be assembled as close to the punch as possible, so that the motor and the elements joining it to the flywheel are isolated form the resistant torque fluctuations.
- It enlarges the starting and stopping phases of the machine. This effect, provoked by the inertia increase in the system, can be either beneficial or harmful, depending on the case. It will be beneficial in those machines where we expect the stopping process not to be quick. For instance, consider a hydraulic turbine, where, a sudden pressure pulse can produce a water hummer with the potential to damage it. On the other side, a flywheel is harmful when you need high acceleration in a machine as the velocity changes will develop less quickly (for example, in an internal-combustion engine for a vehicle).

6.6 Application Examples of Flywheels

Irregularities in machine torque can be caused by:

- A constant motor torque versus a variable resistant torque.
- Oscillation in a motor torque versus a constant resistant torque.
- Irregularities due to variable motor and resistant torques.

When motor and resistant torques acting on a machine are constant, there will be no variations in velocity. Consequently, it will not be necessary to assemble a flywheel.

When a machine is powered by an electric motor, we can consider its input torque to be constant, whereas, when it is powered by an internal-combustion engine, it will oscillate. Examples of machines with variable resistant torques are punching machines, stamping machines and air compressors, while electric generators can be considered as machines with constant resistant torques.

Below, two examples of flywheel applications are described:

- Velocity variations in an electric generator powered by a single-cylinder four-stroke combustion engine will be noticeable, since there is only one power stroke for every four strokes of the piston while the resistant torque produced by the electric generator can be considered constant. Since the voltage output of the electric current depends on the turning speed of the electric generator, changes in velocity will cause changes in the voltage. As a result there will be a flicker in the lights. In this case, a flywheel can ensure a more regular turning velocity in the generator during each cycle. An example of the calculation of a flywheel used with an electric generator moved by a two stroke combustion engine is included in Example 3 at the end of this chapter.

- In a punching machine powered by and electric motor, the input torque can be considered constant while the resistant torque is variable. The resistant torque is zero along most of the cycle and has a sharp increase at the punching instant. This implies that the power at this instant has to be high to prevent the machine from stopping. However, if we install a flywheel, it will store energy during the time interval between pierces returning it at the punching instant, so a smaller motor will be needed.

Example 1 A punching machine that rotates at 30 rpm carries out piercing operation in $1/3$ of a second and the work needed to carry out every puncture is 1570.8 J.

If no flywheel is mounted, we will need a motor with enough power to carry out 1570.8 J of work in $1/3$ of a second. To simplify the problem we will assume this work is constant over time. Figure 6.2 shows the work curve with respect to time.

The power of the required motor (Eq. 6.19) will be:

$$P = \frac{W}{\Delta t} = \frac{1570.8\,\text{J}}{1/3\,\text{s}} = 4712.4\,\text{W} \tag{6.19}$$

Actually, the force with which the material resists piercing is bigger at the beginning, so the work produced by the motor at the beginning of the process has to be higher. We can estimate the instant power needed at the beginning, so that the machine does not stop, to be twice as much as the average power of the process.

On the other hand, if a flywheel is mounted, the work the motor needs to carry out is the same. However, now we will consider this work is produced along the whole cycle since the flywheel will store the excess of energy during the part of the

Fig. 6.2 Work curve with respect to time in the punching machine of Example 1 assuming that the work is constant

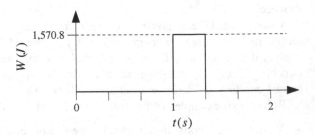

cycle where the motor does not have to overcome any resistance returning it at the piercing instant.

$$P = \frac{W}{\Delta t} = \frac{1570.8 \,\mathrm{J}}{2\,\mathrm{s}} = 785.4\,\mathrm{W} \tag{6.20}$$

Hence, a 785.4 W motor is enough (Eq. 6.20). The work accomplished by this motor in $1/3$ of a second is only 261.8 J, so the flywheel has to produce the remaining 1309 J.

6.7   Coefficient of Speed Fluctuation

We have seen that if a machine operates in steady state, there is no increase in work (and, therefore, kinetic energy) at the end of each cycle. However, the torque diagram in Fig. 6.1 shows that motor and resistant torques are not always even. Within a cycle, there will be positions of the machine where the motor torque is higher than the resistant torque and other positions where the resistant torque is higher. This produces velocity fluctuations in the cycle. The highest acceptable value of these oscillations will depend on the application of the machine. Following the previous examples, it is clear that we have to be more demanding with the regularity of the angular velocity in an electrical generator than in a punching machine.

When designing a machine flywheel, it is necessary to fix the maximum speed fluctuation allowed in a cycle. To do so, we define the speed fluctuation coefficient (Eq. 6.21) or irregularity factor, δ.

$$\delta = \frac{\omega_{max} - \omega_{min}}{\omega} \tag{6.21}$$

where ω_{max} is the maximum velocity of the flywheel during a cycle, ω_{min} is the minimum velocity and ω is the average velocity.

In those flywheels where the mass is concentrated in a rim, we can define this parameter (Eq. 6.22) as:

$$\delta = \frac{v_{max} - v_{min}}{v} \tag{6.22}$$

where v_{max}, v_{min} and v are the linear maximum, minimum and average velocity of a point in the rim respectively (it can actually be the velocity of any point in the flywheel as long as it is not zero).

The value of δ varies ranging from 0.2 for the punching machine and 0.002 for electricity generators. The list below shows the recommended δ values for some machines.

- Electric generator: 0.002
- Electric machine: 0.003
- Gear transmission: 0.02
- Machine tool: 0.03
- Stamping machine: 0.02.

6.8   Design of a Flywheel

The difference between motor work and resistant work at some instant of the cycle produces a change in velocity which affects the kinetic energy of the system. The theorem of live forces, described previously in this chapter, states that we can determine the variation in kinetic energy of a system (Eq. 6.23) based on its motor and resistant work:

$$\Delta E_K = \Delta W = \Delta W_M + \Delta W_R \tag{6.23}$$

We will consider time instants 1 and 2 of the cycle with a maximum ΔE_K between them. The kinetic energy at instant 1 (Eq. 6.24) attains its maximum value along the cycle and it can be calculated as:

$$E_{K_1} = \frac{1}{2} I \omega_{max}^2 \tag{6.24}$$

Instant 2 is the one with minimum kinetic energy (Eq. 6.25):

$$E_{K_2} = \frac{1}{2} I \omega_{min}^2 \tag{6.25}$$

Thus, the increment in kinetic energy (Eq. 6.26) is:

$$\begin{aligned} \Delta E_K &= \frac{1}{2} I \omega_{max}^2 - \frac{1}{2} I \omega_{min}^2 = \frac{1}{2} I (\omega_{max}^2 - \omega_{min}^2) \\ &= I \left(\frac{\omega_{max} + \omega_{min}}{2} \right) \omega \left(\frac{\omega_{max} - \omega_{min}}{\omega} \right) = I \omega^2 \delta \end{aligned} \tag{6.26}$$

Hence, the velocity fluctuation that will yield the variation in kinetic energy will depend on the moment of inertia of the machine reduced to one axle. For a kinetic energy variation, the bigger this moment is, the lower velocity fluctuation will be.

From Eq. (6.19) we can clear I (Eq. 6.27), the moment of inertia necessary to store energy ΔE_K at velocity ω so that oscillation in velocity is limited (δ).

$$I = \frac{\Delta E_K}{\omega^2 \delta} \tag{6.27}$$

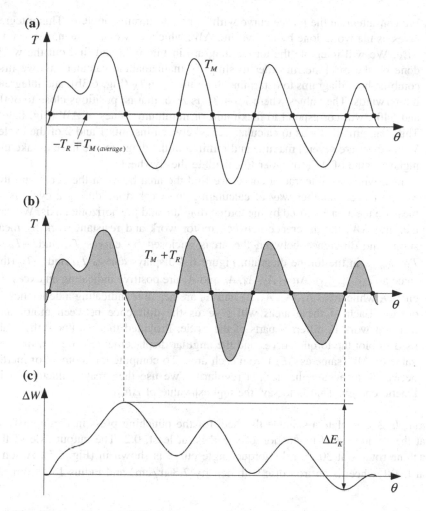

Fig. 6.3 a Motor and resistant torque curve versus axle angle **b** combination of the motor and resistant torque diagrams **c** integral curve of $T_M + T_R$

By use of Eq. (6.27) and knowing the moment of inertia of the machine, we can calculate the moment of inertia of the flywheel.

The highest ΔE_K that is produced along one cycle can be obtained from the motor and resistant torque diagrams (Fig. 6.3a). If the kinetic energy at the beginning and at the end of the cycle have to be the same, the average motor torque and the average resistant torque have to have the same magnitude and an opposite direction. In the diagram shown in Fig. 6.3a, the resistant torque is constant. Therefore, its absolute value along the cycle will be the same as the value of the average motor torque.

Basically, there are two methods of determining ΔE_K value.

- We can integrate the torque curve with respect to turning angle θ. This integral gives us the work done by the system, ΔW, which, as we have seen, is equal to ΔE_K. We will integrate the torque diagram in Fig. 6.3a and find out the work done by the machine. In order to simplify mathematical calculations, we first combine both diagrams to determine diagram $T_M + T_R$ (Fig. 6.3b), and integrate it afterwards. The values where $T_M + T_R$ is zero, that is, positions close to 360° and 540°, will correspond to maximum or minimum values of ΔW (Fig. 6.3c). This diagram allows us to calculate ΔE_K between instants 1 and 2 of the cycle. When there are several maxima and minima in the diagram, we have to take the highest value of ΔE_K in order to calculate the flywheel.
- In fact, when we integrate a curve, we find the area between the curve and the X-axis. Hence, another way of calculating the work done during a cycle is to measure the area enclosed by the torque diagram and the horizontal axis. We can measure ΔW, the difference between motor work and resistant work by measuring the difference between the areas enclosed by curves T_M and $-T_R = T_{M\,(average)}$ in the torque diagram. Figure 6.4 displays curves T_M and $-T_R$ (the same as $T_{M\,(average)}$). Areas A_2, A_4, A_6 and A_8 are positive indicating an excess in energy, while areas A_1, A_3, A_5, A_7 and A_9 are negative indicating a deficiency in energy. Each of these areas will give us the difference between motor and resistant work in different parts of the cycle. Provided that we know the scale used to plot the torque curves and the angular displacement, θ, we can find the value of ΔW (same as ΔE_K) from each area. To compute the moment of inertia necessary to ensure the desired regularity, we use the greatest fluctuation in kinetic energy. That is to say, the highest value of ΔE_K.

Example 2 Calculate a suitable flywheel for the punching press in Example 1, so that the coefficient of fluctuation of speed is, at least, 0.2. The output axle of the machine rotates at 30 rpm. Its torque-angle curve is shown in (Fig. 6.5). Assume that the flywheel is an iron disc with density $7.8\,\mathrm{gr/cm^3}$ and radius 1500 mm.

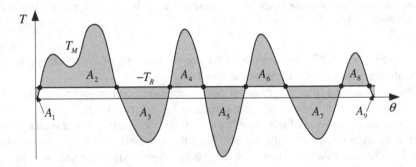

Fig. 6.4 Area between the motor torque curve and the average-torque line (same as the resistant torque line but different sign)

Fig. 6.5 Motor and resistant
torque-angle curves

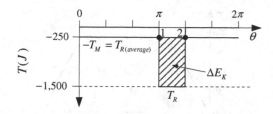

In Fig. 6.5 we can see that the motor torque is constant while the resistant torque
varies. When the press operates at steady-state, the average resistant torque and the
motor torque have to be the same. Between positions 1 and 2, the resistant torque is
higher and the machine slows down, so that the minimum velocity of the cycle is
reached at instant 2. During the rest of the cycle, there is no resistant torque and the
machine speeds up reaching its maximum velocity at instant 1.

To find the variation in kinetic energy (Eq. 6.28) produced between positions π
and $4\pi/3$, we calculate the area between the motor and resistant curves:

$$\Delta E_K = \frac{\pi}{3}250\,\text{J} - \frac{\pi}{3}1500\,\text{J} = -1309\,\text{J} \qquad (6.28)$$

where the negative sign in the expression indicates a decrease in velocity. During
the rest of the cycle, the resistant torque is zero and the kinetic energy increases the
same amount (Eq. 6.29), which means that the machine speeds up:

$$\Delta E_K = \frac{2\pi}{3}250\,\text{J} + \pi 250\,\text{J} = 1309\,\text{J} \qquad (6.29)$$

We can also compute ΔE_K by integrating the curve (Eq. 6.30). For example,
between positions π and $4\pi/3$ the variation in kinetic energy will be:

$$\Delta E_K = \int\limits_{\pi}^{4\pi/3} (T_M + T_R)d\theta = (250 - 1500)|_{\pi}^{4\pi/3} = -1309\,\text{J} \qquad (6.30)$$

Once we have the value of ΔE_K, we can find the moment of inertia that provides
the desired coefficient of fluctuation at the given speed (Eq. 6.31). To do so, in this
example we neglect the inertia of the links in the machine. Hence, the moment of
inertia of the flywheel is:

$$I = \frac{\Delta E_K}{\omega^2 \delta} = \frac{1309\,\text{J}}{(30\frac{2\pi}{60})\,\text{rad/s}\,0.2} = 667.15\,\text{kg}\,\text{m}^2 \qquad (6.31)$$

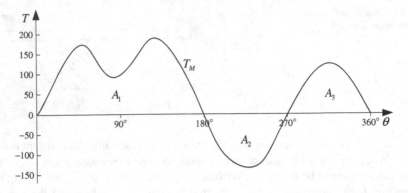

Fig. 6.6 Motor torque curve versus crank position in a two-stroke engine

Knowing that the moment of inertia of a disc (Eq. 6.32) with a mass equal to m and a radius equal to R is:

$$I = \frac{1}{2}mR^2 \qquad (6.32)$$

And also taking into account that the mass of a disc with a thickness equal to e and a density equal to ρ is $m = \pi R^2 e \rho$, we can clear the thickness of the flywheel in Eq. (6.33):

$$e = \frac{2I}{\pi R^4 \rho} = \frac{2 \cdot 667.15\,\text{kg}\,\text{m}^2}{\pi (0.75\,\text{m})^4\,7800\,\text{kg/m}^3} = 0.172\,\text{m} \qquad (6.33)$$

Example 3: Find a suitable flywheel for a two-stroke engine that, when working at a speed of 9000 rpm, has the motor torque-angle diagram shown in Fig. 6.6. The maximum fluctuation in speed admissible is 150 rpm when working at the aforesaid speed. Assume the flywheel to be a steel disc with a radius of 10 cm. The scale used to plot the torque diagram is $1\,\text{cm} = 37.23°$ for the horizontal axis and $1\,\text{cm} = 85.76\,\text{Nm}$ for the vertical axis.

First of all, we have to calculate the average motor torque. In order to do this, we have to measure the area enclosed by the motor curve and the horizontal axis (Eq. 6.34).

The total area enclosed by this curve is:

$$A_T = A_1 - A_2 + A_3 = 6.79 - 2.52 + 2.22 = 6.49\,\text{cm}^2 \qquad (6.34)$$

We can calculate the average torque by plotting a rectangle with a base of 9.65 cm (equivalent to 360° of θ) which encloses 6.49 cm^2 on the graph (Eq. 6.35).

$$A_T = bh \Rightarrow h = \frac{A_T}{b} = \frac{6.49 \, \text{cm}^2}{9.65 \, \text{cm}^2} = 0.672 \, \text{cm} \tag{6.35}$$

Since we know that the scale used in the vertical axis is 1 cm = 85.76 Nm, we can work out the average torque value (Eq. 6.36):

$$T_{M \, (average)} = 0.672 \, \text{cm} \cdot 85.76 \, \text{Nm/cm} = 57.67 \, \text{Nm} \tag{6.36}$$

Working at steady-state conditions, the areas between the horizontal axis and the motor and resistant torque diagrams have to be equal. Therefore, assuming that the resistant torque is constant, it will be the same as the average motor torque. Thus, the absolute values of average $T_{M \, (average)}$ and constant T_R are equal during the cycle (Fig. 6.7a).

To calculate the maximum fluctuation of kinetic energy, we have to find which instant has the greatest difference between T_M and $-T_R$ in the diagram. To do so, we compute the areas enclosed by curves T_M and $-T_R$ during time intervals 0–1, 1–2, 2–3, 3–4 and 4–0 and we check between which points the accumulated energy

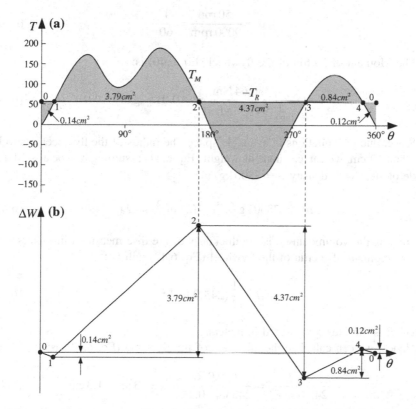

Fig. 6.7 **a** Areas between curves T_M and $-T_R$ during time intervals 0–1, 1–2, 2–3, 3–4 and 4–0 **b** accumulated kinetic energy along the same time intervals

is the highest. In this example, this happens between instants 2 and 3 and the area has a value of 4.37 cm², as seen in Fig. 6.7b. Graph diagrams are usually plotted with a computer program that allows the user to calculate areas with precision.

In order to find the value of ΔE_K during time interval 2–3 directly from the curve (Eq. 6.38), we have to use the scale defined in the wording of the problem (Eq. 6.37).

$$1\,cm^2 = 1\,cm \cdot 1\,cm = 37.23° \cdot 85.76\,Nm\frac{\pi}{180°} = 55.72\,Nm \qquad (6.37)$$

Thus,

$$\Delta E_K = 4.37\,cm^2\,55.72\,Nm/cm^2 = 243.52\,Nm \qquad (6.38)$$

This energy has to be stored in the flywheel with a variation in angular velocity lower than 150 rpm along the cycle.

Hence, the coefficient of speed fluctuation (Eq. 6.39) when the engine rotates at 9000 rpm is:

$$\delta = \frac{150\,rpm}{9000\,rpm} = \frac{1}{60} \qquad (6.39)$$

The Moment of Inertia of the flywheel (Eq. 6.40) is:

$$I = \frac{236.64\,Nm}{\frac{1}{60}\,(942.5\,rad/s)^2} = 0.016\,kg\,m^2 \qquad (6.40)$$

Since, due to limitations in available space, the radius of the flywheel has to be less than 10 cm, we can calculate its weight (Eq. 6.41) assuming it to be a solid disc made of steel with density $\rho = 7800\,kg/m^3$.

$$m = \rho V = 7800\,kg/m^3 e\pi(0.1\,m)^2 = e245.04\,kg \qquad (6.41)$$

where V is the volume and e is the thickness of the disc measured in meters.

The moment of inertia of the flywheel (Eq. 6.42) will be:

$$I = \frac{1}{2}(245.04e)R^2 \qquad (6.42)$$

where R is the radius measured in meters.

Hence, we can calculate the thickness of the flywheel (Eq. 6.43) as:

$$e = \frac{2 \cdot I}{245.04 \cdot R^2} = \frac{2 \cdot 0.016}{245.04 \cdot 0.1^2} = 0.013\,m = 1.3\,cm \qquad (6.43)$$

Chapter 7
Vibrations in Systems with One Degree of Freedom

Abstract A mechanical system can undergo mechanical solicitations that can either be constant or variable in time. The behavior of the system will change widely depending on what type of load is applied. Even loads of similar magnitude will have different effects. Traditionally, time variation of stress in a system was not considered and only calculations considering maximum stress were carried out. It was up to the designer and his experience to choose a higher or lower safety factor according to how variable the load was in order to anticipate any eventuality. Theoretical analysis methods gave the opportunity to study system characteristics and dynamical behavior with precision. These dynamical loads, which are variable in time, provoke deformations in elastic bodies that can give way to vibration processes. The study of vibrations is related to oscillatory motion of non-rigid bodies and the forces related to them. Any system with mass and elasticity can vibrate.

7.1 Introduction to Oscillatory Motion

In this section we will develop the basics of the oscillatory motion of a mechanical system.

7.1.1 Fundamental Concepts of Vibrations

We define a mechanical vibration as the oscillatory motion of an elastic body about its equilibrium position.

The first distinction that we have to make is to define the difference between motion in rigid and deformable bodies. This second kind of motion is the one that gives way to vibrations. Since a rigid body does not exist in reality as part of a mechanical system, we will use the terms elastic solid body and solid body

© Springer International Publishing Switzerland 2016
A. Simón Mata et al., *Fundamentals of Machine Theory and Mechanisms*,
Mechanisms and Machine Science 40, DOI 10.1007/978-3-319-31970-4_7

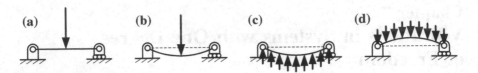

Fig. 7.1 An external force acts on an elastic body generating deformation. When it stops acting, an oscillation motion about its equilibrium position appears due to the action of internal forces

interchangeably. We could say that oscillations are small vibrations in deformable bodies.

In Fig. 7.1, an external force is applied to a body in order to produce a deformation and moves it away from its equilibrium position. When the external force disappears, internal forces try to bring it back to its equilibrium position. However, during this process, the body tends to exceed its equilibrium position due to its inertia and internal forces act again, in the opposite direction.

This way, when a deformable body is displaced from its equilibrium position, a vibration is generated due to the body trying to regain equilibrium by means of internal forces. When this process is repeated, the body keeps oscillating about an equilibrium position giving place to the phenomenon of vibration.

The time during which the deformable body runs a complete cycle of motion is called period. The number of cycles that take place per unit of time is called frequency. And, finally, the maximum displacement measured from the equilibrium position is called amplitude.

It is hard to avoid vibrations in real life applications since they are sometimes produced by unbalance in rotating elements, but other times they are caused by manufacturing tolerance and clearance.

Depending on the relationship between the frequency of the force and the frequency of the vibration, small unbalanced forces can trigger resonance in a system originating important stress and deformation.

However, vibrations do not always have a negative impact, since they can be used to carry out certain functions, such as the case of concrete vibrators, pneumatic hammers or vibrating feeders.

There are two main types of vibrations: free and forced. Free vibrations are those in which motion is maintained without any variable external force. Hence, internal forces are the only ones responsible for the motion of the deformable body. Depending on how well the system dissipates energy, vibration will be damped more quickly or slowly. At the same time, vibrations are denominated damped or undamped depending on whether we consider dissipative forces to be acting on the body or not. However, there is always damping since energy dissipation is an inherent element of any real mechanical system.

When a system with one degree of freedom undergoes free vibration, its vibration frequency is called natural frequency, which is a characteristic of a system that depends only on its mass and stiffness.

The vibration of a system due to external forces that vary in time is called forced vibration, which can be classified in:

- Harmonic excitation: forces vary in a sinusoidal waveform.
- Periodic excitation: forces are repeated at time intervals.
- Impulse excitation: forces of high intensity act during an infinitesimal time.
- General excitation: forces vary randomly.

The main difference that defines vibration versus oscillation is energy. In an oscillation, a pendulum for example, there are two types of energy: kinetic and gravitational potential energy. In a mechanical system vibration, deformation energy, elastic-potential energy or elastic-plastic strain energy have to take part.

The study of vibrations in systems makes it possible to make a suitable design of a system, so that we reduce internal dynamical stresses in it as well as energy losses and so that it has the proper stiffness.

7.1.2 Concept of Degree of Freedom (DOF)

In this book we have already defined the degree of freedom of a mechanical system as the number of independent parameters that define the position of all parts of that system completely at any instant of time. This way, a particle with a generic spatial motion has three degrees of freedom while a rigid body has six, three to define the position in space of one of its points and another three to define the required angles to know its orientation.

The dynamic analysis of physical systems by means of mathematical models that simulate its behavior along time requires the definition of these parameters.

Example 1 The system in Fig. 7.2 has two concentrated masses, m_2 and m_1, which represent the suspended and non-suspended masses of a car. Assuming that the masses are allowed to move only in vertical direction, the system will have as many

Fig. 7.2 Two degree of freedom system with two concentrated masses which represent the suspended and non-suspended masses of a car

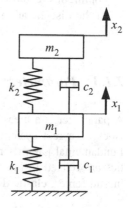

Fig. 7.3 Truss structure with
nine degrees of freedom

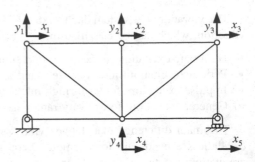

degrees of freedom as masses. Parameters x_2 and x_1 define the position of each mass
in terms of parameters m (mass), k (stiffness) and c (damping).

Example 2 In the truss structure in Fig. 7.3, there are as many degrees of freedom
as possible directions for the displacements of the nodes (joints) in the plane.

Example 3 The shape of a continuous elastic body will be defined by an infinite
number of degrees of freedom. The variables that define its position are a function
of the displacement and have to be computed by integration of the general equations of elasticity defined for that specific system, by using the inertia forces
considering them volume forces and completing the problem with the definition of
boundary conditions for the solid. The resulting problem is quite complex and an
analytical solution can only be found for some simple cases.

   Basically, there are two ways to find the solution to this problem:

1. Some parts of the body are defined as totally rigid, without mass and without
   any capacity to dissipate energy. This way, the real system is substituted by
   another one dynamically equivalent but with a finite number of degrees of
   freedom (concentrated parameter model). In fact, a big number of vibration
   problems can be computed with enough precision if the complete system is
   reduced to one with only a few degrees of freedom.
2. The system is defined as a discrete one, where each element has mass, is
   deformed and dissipates energy. This process is known as discretization of
   continuous features. This way, we can find a solution as exact as the refinement
   of the discretization (distributed parameter system).

7.1.3 Parameters of a Mechanical System

A parameter is a measurable factor that reflects physical properties and characteristics of a real system which is being studied by using a mathematical model.
Fundamental parameters are mass, stiffness and damping. These three characteristics are related to the three most common types of forces in vibration problems:
inertia forces, elastic deformation forces and energy dissipation forces.

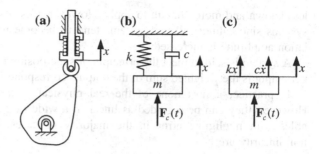

Fig. 7.4 a Disk cam with its follower. **b** Concentrated parameter model of the system. **c** Free body diagram of the system

These three parameters are usually known in problems of theoretical analysis of vibrations. We will assume that their values do not change with time or deformation (linear system).

In mathematical models (with a finite number of degrees of freedom) that are used to study real physical systems (with an infinite number of degrees of freedom), the following approximation is made: we suppose that the characteristics of the system are concentrated at certain elements or points. This is what we call concentrated parameter model.

Figure 7.4 shows a system composed of a disk cam and its follower, its concentrated parameter model and its free body diagram. The following points have been assumed:

- The whole capacity of the system to absorb elastic energy is attributed to an ideal spring that is mathematically defined by equation $F = kx$, where F is the force of the spring, k is the spring constant (factor that depends on its stiffness) and x is the spring displacement.
- Kinetic energy is assigned to non-deformable masses.
- The capacity of the system to dissipate energy is assigned to a viscous damper which is mathematically defined by equation $F = c\dot{x}$, where F is the force of the damper, c is the damping coefficient and $\dot{x}$ is the velocity (time derivative of the displacement).

Finally, another option is to assume that each element has mass, can be deformed and dissipates energy. Such models are known as distributed parameter models. This type of discrete system allows a more exact approximation of the physical system represented. An example of this discretization method is the Finite Elements Method (FEM).

7.1.4 Characterization of Oscillatory Systems

Oscillatory systems can be characterized as linear or non-linear. Linear systems meet the superposition principle, and mathematical techniques for their treatment are well developed. In contrast, techniques for the analysis of non-linear systems are less known and more difficult to apply. However, it is desirable to study non-linear systems since almost all real systems tend to become non-linear when the oscillation amplitude is increased.

A system is said to meet the principle of superposition and to be linear when the net output is the weighted sum of the outputs corresponding to each of these inputs.

In practice, almost none of the real physical systems have linear behavior. However, they can be regarded as linear in a wide range of their dependent variables with negligible error in the majority of cases. The principal causes of non-linearity are:

- Big deformations.
- Certain types of resistance or damping.
- Non-linearity in material behavior laws.

7.1.5   Harmonic Periodic Motion: Transient and Steady-State Regimes

Oscillatory motion can be repeated regularly, like the swinging of a pendulum in an antique clock, or it can show considerable irregularity as in the case of an earthquake. When motion is repeated at constant time intervals, T, it is said to be periodic. That time interval T is called period of oscillation and its inverse is called frequency, $f = 1/T$ If displacement x is defined as time function $x(t)$, then periodic motion has to fulfill relationship $x(t) = x(t + T)$.

A dynamic system works in a stationary-state regime when its time variation is periodic, that is, the variables of the problem repeat their values every T seconds, T being the period.

When a system does not have periodic behavior, then it is said to work in a transient regime.

The simplest form of periodic motion is the one called harmonic motion. This kind of motion can be simulated by means of a mass suspended from a spring. When the mass is displaced from its resting position and then released, it will oscillate up and down from its equilibrium position. If we plot its displacement on a strip of paper moving horizontally with constant velocity behind the mass (Fig. 7.5), the graph plotted by the mass during its motion can be expressed by Eq. (7.1):

Fig. 7.5 Displacement-time plot of a mass that oscillates suspended from a spring with time period T

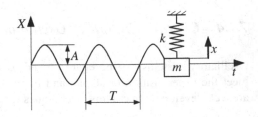

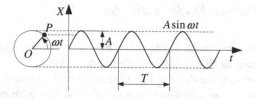

Fig. 7.6 Simple harmonic motion represented as the projection over a vertical line of point P that rotates with radius OP at constant angular velocity ω

$$x(t) = A \sin \frac{2\pi}{T} t \qquad (7.1)$$

where A is the amplitude of the oscillation measured from the equilibrium position of the mass and T is the period. The motion is repeated every $t = T$.

Harmonic motion is frequently represented as the projection over a vertical line of a point that moves at constant speed on the perimeter of a circle (Fig. 7.6). When the angular velocity of line OP is ω, displacement x (Eq. 7.2) can be expressed as:

$$x(t) = A \sin \omega t \qquad (7.2)$$

The value of ω is usually measured in radians per second and it is referred to as angular frequency. Since motion is repeated every 2π radians, we have (Eq. 7.3):

$$\omega = \frac{2\pi}{T} = 2\pi f \qquad (7.3)$$

where T and f are the period and frequency of the simple harmonic motion measured in seconds and cycles per second respectively.

7.2 Single Degree of Freedom (SDOF) Systems

Systems with one DOF are known as single degree of freedom systems. As we have already seen, these systems only require one coordinate to describe their position. The interest in their study is due to the following reasons.

- Many practical problems can be considered SDOF systems without losing much precision in the solution.
- They share many of their properties with systems with more than one DOF.
- Linear systems with n degrees of freedom can be solved by superposition of n systems with a single DOF.

7.2.1 Basic Discrete Model with One DOF

Figure 7.7 shows a concentrated parameter model of a mechanical system. This basic model is known as SDOF discrete parameter system.

In this model, we suppose that the kinetic energy of the system is stored in mass m, the elastic energy is concentrated in the massless spring with constant k and the energy dissipation capacity of the whole system is concentrated in the viscous damper, which moves at a velocity proportional to the force applied, c being the proportional constant.

The position of the system is defined by coordinate $x = x(t)$, the horizontal displacement of mass m. Since the system is considered linear, m, k and c have to be constants that do not depend on variable x.

The motion equation of the system in Fig. 7.7 can be obtained from its dynamic equilibrium (Eq. 7.4) (Fig. 7.8):

$$\sum_i F_i = ma \tag{7.4}$$

$$F(t) - kx - c\dot{x} = m\ddot{x} \tag{7.5}$$

$$m\ddot{x} + c\dot{x} + kx = F(t) \tag{7.6}$$

We can study this basic discrete system by using the superposition principle, considering only dynamic forces. Static forces, if there are any, can be studied separately by taking into account that they have to balance the elastic stress in the spring.

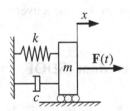

Fig. 7.7 Single-degree-of-freedom discrete parameter model

Fig. 7.8 Free body diagram
of the system in Fig. 7.7

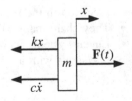

Differential equations are very common to linear SDOF systems (Eqs. 7.4–7.6) and are used to formulate and solve theoretical problems that are common to all of them.

7.3 Free Vibrations in SDOF Systems

Equation (7.6) represents the second-order ordinary differential equation (ODE) that defines linear systems with one DOF. This equation can be written as:

$$m\ddot{x}(t) + c\dot{x}(t) + kx(t) = F(t) \tag{7.7}$$

When the system moves without time varying external forces acting on it, $F(t) = 0$, its vibration is known as free vibration. This vibration is induced by initially disturbing the system from its equilibrium position.

The solution to Eq. (7.7) for a free vibration can be found assuming that we know the initial conditions (Eq. 7.8), that is, deformation x_0 and velocity $\dot{x}_0$ at initial instant $t = t_0$:

$$\left.\begin{array}{l} x_0 = x(t_0) \\ \dot{x}_0 = \dot{x}(t_0) \end{array}\right\} \tag{7.8}$$

We look for solutions (Eq. 7.9) of the following form:

$$x(t) = Ae^{st} \tag{7.9}$$

Deriving and substituting it in Eq. (7.7), we obtain (Eq. 7.10):

$$Ams^2 e^{st} + Acse^{st} + Ake^{st} = 0 \tag{7.10}$$

$$A(ms^2 + cs + k)e^{st} = 0 \tag{7.11}$$

The solutions that we look for ($x(t) = Ae^{st}$) will solve the differential equation for those s values that satisfy Eq. (7.11). These values are the roots of the characteristic equation (Eq. 7.12):

$$ms^2 + cs + k = 0 \tag{7.12}$$

Thus:

$$s = \frac{-c \pm \sqrt{c^2 - 4km}}{2m} \tag{7.13}$$

7.3.1 Undamped Free Vibrations of SDOF Systems

In this section we will study the lateral and torsional free vibrations in undamped systems with one degree of freedom.

7.3.1.1 Lateral Vibration

When a free vibration is not damped, $c = 0$, and substituting this value in Eq. (7.13), we obtain:

$$s = \pm\sqrt{\frac{-k}{m}} \tag{7.14}$$

Since k/m is a positive constant:

$$\omega^2 = \frac{k}{m} \tag{7.15}$$

where ω (Eq. 7.15) is called natural frequency. Hence, the s values are:

$$s = \pm\sqrt{-\omega^2} = \pm i\omega \tag{7.16}$$

Therefore, the general solution to the differential equation (Eq. 7.7) will be given by (Eq. 7.17):

$$x(t) = A_1 e^{i\omega t} + A_2 e^{-i\omega t} \tag{7.17}$$

where A_1 and A_2 are either real or complex constants.

Another way to express this solution is in its trigonometric form (Eq. 7.18), carrying out the change of variables of Eq. (7.19):

$$x(t) = A \cos \omega t + B \sin \omega t \tag{7.18}$$

where:

$$\left.\begin{array}{l} A = A_1 + A_2 \\ B = i(A_1 - A_2) \end{array}\right\} \tag{7.19}$$

Carrying out a new change of variables (Eq. 7.20):

$$\left.\begin{array}{l} A = X \cos \theta \\ B = X \sin \theta \end{array}\right\} \tag{7.20}$$

where:

$$X = \sqrt{A^2 + B^2} \tag{7.21}$$

$$\theta = \arctan \frac{B}{A} \tag{7.22}$$

Thus, the solution remains in the form (Eq. 7.23):

$$x(t) = X(\cos \theta \cos \omega t + \sin \theta \sin \omega t) \tag{7.23}$$

or (Eq. 7.24) using a trigonometric identity:

$$x(t) = X \cos(\omega t - \theta) \tag{7.24}$$

where, A, B, X, (Eq. 7.21) and θ (Eq. 7.22) are always real. Their values have to be determined from the initial conditions.

To determine A and B (Eq. 7.25), we impose the initial conditions for $t = 0$ in (Eq. 7.18) and in its time derivative:

$$\left. \begin{array}{l} x_0 = x(0) = A \cdot 1 + B \cdot 0 \quad \Rightarrow \quad A = x_0 \\ \dot{x}_0 = \dot{x}(0) = -A\omega \cdot 0 + B\omega \cdot 1 \quad \Rightarrow \quad B = \frac{\dot{x}_0}{\omega} \end{array} \right\} \tag{7.25}$$

Hence, the solution to the problem is (Eq. 7.26):

$$x(t) = x_0 \cos \omega t + \frac{\dot{x}_0}{\omega} \sin \omega t \tag{7.26}$$

The maximum vibration amplitude (Eq. 7.27) is:

$$X = \sqrt{x_0^2 + \left(\frac{\dot{x}_0}{\omega}\right)^2} \tag{7.27}$$

Thus, substituting the calculated values in Eq. (7.24), we can express the solution as (Eq. 7.28):

$$x(t) = \sqrt{x_0^2 + \left(\frac{\dot{x}_0}{\omega}\right)^2} \cos(\omega t - \arctan \frac{\dot{x}_0}{\omega x_0}) \tag{7.28}$$

The solution to free undamped vibrations is a harmonic function with frequency (Eq. 7.29):

$$\omega = \sqrt{\frac{k}{m}} \tag{7.29}$$

This frequency is called natural frequency and depends on physical parameters k and m of the system, but not on time or initial conditions. The system will always vibrate at one or more times its natural frequency.

To carry out the vibration analysis of a SDOF system, we can follow these steps:

1. We choose variable x or θ that defines the displacement of the system depending on whether it has linear or rotational motion.
2. We draw a free solid diagram that includes not only forces but also displacement (Fig. 7.8).
3. We apply Newton's Second Law (Eqs. 7.30 and 7.31):

   - To linear motions:

$$\sum_i F_i = m\ddot{x} \qquad (7.30)$$

   - To rotational motions:

$$\sum_i M_i = I\ddot{\theta} \qquad (7.31)$$

   where:

- $\sum_i M_i$ are the moments exerted on the body.
- I is the moment of inertia with respect to the rotation axis.

4. We use static equilibrium equations to eliminate the unknowns that do not depend on displacement or its derivatives.
5. We formulate the motion differential equation.
6. We obtain the value of the natural frequency of the system.

Example 4 The tower-crane in Fig. 7.9 supports a load with a mass of 20 kg. Determine the response of the system to initial conditions of vertical displacement $x_0 = 1$ cm and velocity 1 cm/s both being upward values. The deformation of the cable and the mass of the jib are considered negligible. The elasticity of the jib is known for two different positions of the trolley:

- 8000 N/m
- 16,000 N/m

1. The variable chosen is vertical displacement x of the mass.
2. Figure 7.10b shows the free body diagram with forces acting on it and displacement x.
3. From the dynamic equilibrium equation (Eq. 7.30), we can obtain the motion differential equation of the system (Eq. 7.32):

Fig. 7.9 Tower-crane
supporting a load of 20 kg

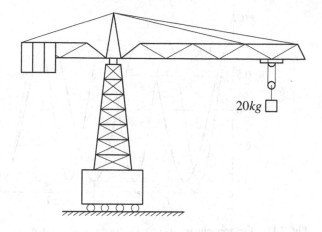

Fig. 7.10 a Concentrated
parameter model of the
system. **b** Free body diagram
of the system

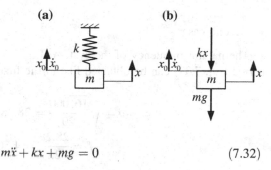

$$m\ddot{x} + kx + mg = 0 \qquad (7.32)$$

4. The natural frequency of the system will depend on the position of the trolley:

- First case:

The natural frequency of the system (Eq. 7.33) can be calculated as:

$$\omega = \sqrt{\frac{8000}{20}} = 20\,\text{rad/s} \qquad (7.33)$$

Its value in cycles per second or hertz is (Eq. 7.34):

$$f = \frac{20}{2\pi} = 3.183\,\text{Hz} \qquad (7.34)$$

And the period (Eq. 7.35) is:

$$T = \frac{1}{3.183} = 0.314\,\text{s} \qquad (7.35)$$

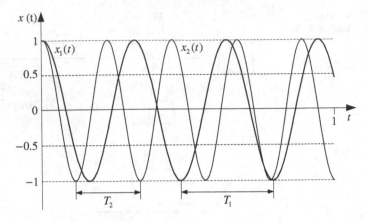

Fig. 7.11 Displacement versus time diagram of each system

- Second case:

The natural frequency of the system (Eq. 7.36) in rad/s and Hz as well as the period (Eq. 7.37) can be obtained like in the first case:

$$\left.\begin{aligned} \omega &= \sqrt{\frac{16,000}{20}} = 28.28\,\text{rad/s} \\ f &= \frac{28.28}{2\pi} = 4.502\,\text{Hz} \end{aligned}\right\} \tag{7.36}$$

$$T = \frac{1}{4.502} = 0.222\,\text{s} \tag{7.37}$$

Figure 7.11 represents the response of the two systems in a graph. As we can see, the response of the system is only affected by its mass and stiffness. This way, in the first case, the frequency of the vibration is 3.183 Hz while in the second case it is 4.502 Hz. Therefore, the stiffer a system undergoing a free undamped vibration is, the higher its vibration frequency will be.

7.3.1.2 Torsional Vibration

In a similar way to systems undergoing a free vibration where the elastic force acts in a straight motion, we can study the case where the elastic action (torque) acts in an angular motion.

For instance, consider a circular disc fixed at the end of a shaft of length L as shown in Fig. 7.12. The other end of the shaft is fixed. The shaft is made of a material with shear modulus G and the polar moment of inertia of its cross-section is J.

Fig. 7.12 Circular disc
fixed at the end of a shaft
of length L

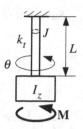

If, by means of an external action **M**, we twist the shaft, the response of the system will make it go back to its equilibrium position due to the recuperating effect of elastic torque $k_t\theta$. The z-axis is directed along the shaft. We can simplify the problem and consider that all elastic actions happen in the shaft while all inertia effects are concentrated in the disc. This way, and by analogy with the rectilinear case, we can write (Eqs. 7.38 and 7.39):

$$M_z - k_t\theta = I_z\ddot{\theta} \tag{7.38}$$

$$I_z\ddot{\theta} - k_t\theta = 0 \tag{7.39}$$

where I_Z is the polar moment of inertia of the disc and $M_z = 0$ (free torsional vibration).

The natural frequency of torsional oscillation (Eq. 7.40) is:

$$\omega = \sqrt{\frac{k_t}{I}} \tag{7.40}$$

The angular displacement of the disc (Eq. 7.41) caused by the perturbation acting on it is:

$$\theta(t) = A\cos\omega t + B\sin\omega t \tag{7.41}$$

Where A and B are integration constants that can be obtained from the initial conditions (Eq. 7.13). Hence, at time $t = 0$ when $\theta_0 = \theta(0)$ and $\dot{\theta}_0 = \dot{\theta}(0)$, we have (Eq. 7.42):

$$\theta(t) = \theta_0 \cos\sqrt{\frac{k_t}{I}}t + \frac{\dot{\theta}_0}{\sqrt{\frac{k_t}{I}}}\sin\sqrt{\frac{k_t}{I}}t \tag{7.42}$$

Remember that the torsional stiffness constant of the shaft, k_t (Eq. 7.43), can be expressed as:

$$k_t = \frac{GJ}{L} \tag{7.43}$$

Substituting Eq. (7.43) in Eq. (7.40) yields:

$$\omega = \sqrt{\frac{GJ}{IL}} = \sqrt{\frac{\pi d^4 G}{32IL}} \tag{7.44}$$

7.3.2 Free Vibrations with Viscous Damping

In real life, mechanical systems have internal resistance, which causes vibration damping and eventually leads to their extinction. This phenomenon can be simulated by a viscous damper that acts with a force equal to $c\dot{x}$.

Remember (Eq. 7.13) that gives us the solution of the characteristic equation (Eq. 7.12). The argument of the radical is zero for the following value of c (Eq. 7.45):

$$\bar{c} = \sqrt{4km} \tag{7.45}$$

This relationship can also be expressed as (Eq. 7.46):

$$\frac{\bar{c}}{2m} = \sqrt{\frac{k}{m}} = \omega \tag{7.46}$$

Thus:

$$\bar{c} = 2m\omega \tag{7.47}$$

We call $\bar{c}$ critical damping (Eq. 7.47).
We also define the concept of the damping ratio, ξ (Eq. 7.48), as:

$$\xi = \frac{c}{\bar{c}} = \frac{c}{2m\omega} = \frac{c}{2m\sqrt{\frac{k}{m}}} = \frac{c}{\sqrt{4km}} \tag{7.48}$$

Plugging this into Eq. (7.13), we obtain (Eq. 7.49):

$$s = -\xi\omega \pm \sqrt{\xi^2\omega^2 - \omega^2} = -\xi\omega \pm \omega\sqrt{\xi^2 - 1} \tag{7.49}$$

Depending on the value of damping ratio ξ, that is, the ratio between c and $\bar{c}$, we can distinguish three different cases.

- Critical damping, $c = \bar{c}$ or $\xi = 1$:

The magnitude of s is (Eq. 7.50):

$$s = -\omega \tag{7.50}$$

Thus, the solution to the problem (Eq. 7.51) has the following form:

$$x(t) = (A_1 + A_2 t)e^{-\omega t} \tag{7.51}$$

This solution does not have an oscillatory character (since it is exponential) and, thus, is of no interest in relation to the dynamic analysis of mechanical systems.

- Overdamping, $c > \bar{c}$, $\xi > 1$:

In this case, the solution (Eq. 7.52) has the following form:

$$x(t) = A_1 e^{-\xi\omega t + \bar{\omega}t} + A_2 e^{-\xi\omega t - \bar{\omega}t} \tag{7.52}$$

Where $\bar{\omega}$ is defined as (Eq. 7.53):

$$\bar{\omega} = \omega\sqrt{\xi^2 - 1} \tag{7.53}$$

The solution can be expressed in terms of a hyperbolic sine and cosine (Eq. 7.54), thus, remaining as:

$$x(t) = e^{-\xi\omega t}(A_1 \cosh \bar{\omega}t + A_2 \sinh \bar{\omega}t) \tag{7.54}$$

Again, this solution is not oscillatory but exponential. Once more, it is of little interest as for the dynamic analysis of mechanisms.

- Underdamping, $c < \bar{c}$, $\xi < 1$:

The value of s has two complex conjugate roots (Eq. 7.55) of the following form:

$$s = -\xi\omega \pm i\omega\sqrt{1 - \xi^2} \tag{7.55}$$

Hence, the solution to the problem (Eq. 7.56) has the following form:

$$x(t) = e^{-\xi\omega t}(A \cos \omega_D t + B \sin \omega_D t) \tag{7.56}$$

Which can also be expressed as (Eq. 7.57):

$$x(t) = Xe^{-\xi\omega t} \cos(\omega_D t - \theta) \tag{7.57}$$

where:

$$\omega_D = \omega\sqrt{1 - \xi^2} \tag{7.58}$$

These expressions yield the solution to the problem, a harmonic function with frequency ω_D (the frequency of the damped vibration) (Eq. 7.58), which has an amplitude that tends to zero. This oscillation frequency is not the natural one, but its value is close, since damping ratio ξ is less than one and becomes even smaller when squared.

Constants A, B (Eq. 7.60), X (Eq. 7.21) and θ (Eq. 7.22) are obtained by imposing the initial conditions (Eq. 7.59):

$$\left. \begin{array}{c} x_0 = x(0) \\ \dot{x}_0 = \dot{x}(0) \end{array} \right\} \tag{7.59}$$

Therefore:

$$\left. \begin{array}{ll} x_0 = x(0) = A & \Rightarrow \quad A = x_0 \\ \dot{x}_0 = \dot{x}(0) = -\xi\omega A + \omega_D B & \Rightarrow \quad B = \frac{\xi\omega x_0 + \dot{x}_0}{\omega_D} \end{array} \right\} \tag{7.60}$$

This yields a solution of the following form:

$$x(t) = e^{-\xi\omega t}\left(x_0 \cos \omega_D t + \frac{\xi\omega x_0 + \dot{x}_0}{\omega_D} \sin \omega_D t\right) \tag{7.61}$$

Otherwise, taking into account (Eq. 7.57) the solution can also be expressed as (Eq. 7.62):

$$x(t) = \sqrt{x_0^2 + \left(\frac{\xi\omega x_0 + \dot{x}_0}{\omega_D}\right)^2} e^{-\xi\omega t} \cos(\omega_D t - \arctan\frac{\xi\omega x_0 + \dot{x}_0}{x_0\omega_D}) \tag{7.62}$$

We can deduce from this explanation that damping has an influence on natural frequency when considering free vibration in systems. Such an influence leads to a reduction in vibration amplitude along time as seen in Fig. 7.13. This is an expected effect, since the damper acts dissipating energy.

Example 5 A precision measuring instrument, shown in Fig. 7.14, has a mass of 1 kg and is supported by a spring and a damper. The spring constant is $k = 3000$ N/m and

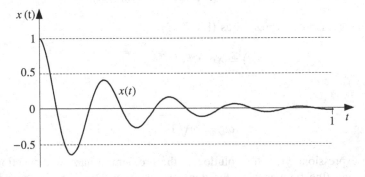

Fig. 7.13 Displacement plot of a free vibration of an underdamped system

Fig. 7.14 Concentrated parameter model of the system

the damping coefficient is $c = 4$ Ns/m. Study the motion of the mass, particularly the vibration amplitude, after 5, 10 and 20 oscillations when it is initially displaced 20 mm of its equilibrium position and then released.

The natural frequency of the system (Eq. 7.63) will be:

$$\omega = \sqrt{\frac{3000}{1}} = 54.7\,\text{rad/s} \tag{7.63}$$

The damping ratio (Eq. 7.64) will be:

$$\xi = \frac{4}{\sqrt{4 \cdot 3000 \cdot 1}} = 0.036 \tag{7.64}$$

Keeping in mind that for $\xi < 1$, the response of the system (Eq. 7.65) is:

$$x(t) = e^{-\xi\omega t}\left(x_0 \cos \omega_D t + \frac{\xi\omega x_0 + \dot{x}_0}{\omega_D}\sin \omega_D t\right) \tag{7.65}$$

We can neglect the second addend in the expression (Eq. 7.65) due to $\dot{x}_0$ being zero and the rest of the values being small. Hence, the response (Eq. 7.66) will be:

$$x(t) = x_0 e^{-\xi\omega t} \cos \omega_D t \tag{7.66}$$

where:

$$\omega_D = 54.7\sqrt{1 - 0.036^2} = 54.66\,\text{rad/s} \tag{7.67}$$

We can see that the natural frequency (Eq. 7.63) and the frequency of the damped vibration (Eq. 7.67) are practically equal. Finally, the solution (Eq. 7.68) remains:

$$x(t) = 0.02e^{-1.97t} \cos 54.66t \tag{7.68}$$

The vibration amplitude (Eq. 7.69) is:

$$X = 0.02e^{-1.97t} \tag{7.69}$$

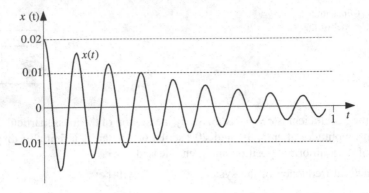

Fig. 7.15 Displacement plot of the vibration

In order to find the time elapsed in 5, 10 and 20 oscillations, we find period of vibration T (Eq. 7.70):

$$T = \frac{2\pi}{\omega_D} = \frac{2\pi}{54.66} = 0.115\,\text{s} \tag{7.70}$$

Hence, after 5 oscillations (Eq. 7.71):

$$t = 5 \cdot 0.115 = 0.575\,\text{s} \tag{7.71}$$

And the amplitude (Eq. 7.72) is:

$$X = 0.00644\,\text{m} = 6.44\,\text{mm} \tag{7.72}$$

After 10 oscillations (Eq. 7.73):

$$X = 0.00207\,\text{m} = 2.07\,\text{mm} \tag{7.73}$$

After 20 oscillations (Eq. 7.74):

$$X = 0.00021\,\text{m} = 0.21\,\text{mm} \tag{7.74}$$

In Fig. 7.15 we can see how the vibration of the system is being damped.

7.4   Forced Vibrations in SDOF Systems

Forced vibrations in mechanical systems are caused by external forces that vary in time. The types of variable loads that we will study in this chapter are harmonic excitations. These forces are interesting to study since they are easy to physically reproduce and can be analytically studied. They also allow us to characterize the dynamic behavior of the system in a very good way.

7.4.1 Forced Vibrations with Harmonic Excitation

Forced vibrations are defined by the general motion equation (Eq. 7.75) already developed in Sect. 7.2.1 of this chapter (Eq. 7.6):

$$m\ddot{x}(t) + c\dot{x}(t) + kx(t) = F(t) \tag{7.75}$$

We assume the excitation force (Eq. 7.76) has the following form:

$$F(t) = f_0 e^{i\omega_e t} = f_0(\cos\omega_e t + i\,\sin\omega_e t) \tag{7.76}$$

The solution to this motion equation has two parts:

- The first one is the general solution to the homogeneous equation, which is the response of the system to a free vibration. Hence, the solution (Eq. 7.77) is the same as the one obtained in Sect. 7.3.2 for underdumping vibrations:

$$x(t) = x_0 e^{-\xi\omega t}\cos(\omega_D t - \theta) \tag{7.77}$$

- The second one is the particular part (Eq. 7.78), which results from the action of an external force. To solve the motion equation, we look for a solution of the following form:

$$x(t) = Ae^{i\omega_e t} \tag{7.78}$$

Where A is a complex constant. By substituting the solution in the differential equation of the forced vibration (Eq. 7.75), we obtain (Eq. 7.79):

$$A(-m\omega_e^2 + c\omega_e + k)e^{i\omega_e t} = f_0 e^{i\omega_e t} \tag{7.79}$$

Hence, constant A is (Eq. 7.80):

$$A = \frac{f_0}{-m\omega_e^2 + c\omega_e + k} = \frac{f_0/k}{1 - \left(\frac{\omega_e}{\omega}\right)^2 + 2i\xi\frac{\omega_e}{\omega}} \tag{7.80}$$

where we know the values of (Eq. 7.81):

$$\left.\begin{array}{l} \omega = \sqrt{\dfrac{k}{m}} \\[2mm] \xi = \dfrac{c}{2m\omega} \\[2mm] \beta = \dfrac{\omega_e}{\omega} \end{array}\right\} \tag{7.81}$$

β is the dimensionless ratio of the excitation force frequency, ω_e, to the natural frequency of the system, ω.

Therefore, the general solution to forced vibrations generated by a harmonic excitation (Eq. 7.82) is:

$$x(t) = Xe^{-\xi\omega t}\cos(\omega_D t - \theta) + \frac{f_0/k}{1 - \beta^2 + 2i\xi\beta}e^{i\omega_e t} \tag{7.82}$$

Based on the nature of harmonic excitation, the solution to the motion equation will change as follows:

- When the excitation is a sine wave (sinusoidal excitation), we take the first addend of the solution and the imaginary part of the second addend.
- In the case of a cosinusoidal excitation, we take the first addend of the solution and the real part of the second addend.

Both terms of the second member of the general solution can be interpreted as follows:

- The first term of the solution represents a transient component of the response that disappears after some time since its amplitude tends to zero.
- The second term of the solution represents a stationary component of the response and it will remain active as long as the force is applied.

By considering the second term (Eq. 7.83), we have:

$$x(t) = \frac{f_0}{k}\frac{1}{1 - \beta^2 + 2i\xi\beta}e^{i\omega_e t} \tag{7.83}$$

We can formulate equation $H(\omega_e)$, called transfer function (Eq. 7.84), which is:

$$H(\omega_e) = \frac{1/k}{1 - \beta^2 + 2i\xi\beta} \tag{7.84}$$

This function is defined so that when we apply an excitation force (Eq. 7.85) equal to:

$$F(t) = f_0 e^{i\omega_e t} \tag{7.85}$$

Then, the response of the system (Eq. 7.86) is:

$$x(t) = f_0 H(\omega_e) e^{i\omega_e t} \tag{7.86}$$

We call ratio f_0/k static displacement, since it is the displacement the system would undergo if the force was statically applied, in other words, if the frequency was zero.

This second term of the solution can also be expressed in magnitude-argument or polar form (Eq. 7.88). To do so, consider (Eq. 7.87):

$$1 - \beta^2 + 2i\xi\beta = \sqrt{(1 - \beta^2)^2 + (2\xi\beta)^2}\, e^{i\varphi} \tag{7.87}$$

$$x(t) = \frac{f_0}{k} \frac{e^{-i\varphi}}{\sqrt{(1 - \beta^2)^2 + (2\xi\beta)^2}} e^{i\omega_e t} = X^{i(\omega_e t - \varphi)} \tag{7.88}$$

where φ (Eq. 7.89) is:

$$\varphi = \arctan \frac{2\xi\beta}{1 - \beta^2} \tag{7.89}$$

And the magnitude X (Eq. 7.90) is:

$$X = \frac{f_0}{k} \frac{1}{\sqrt{(1 - \beta^2)^2 + (2\xi\beta)^2}} \tag{7.90}$$

We introduce the dynamic amplification factor (also known as dynamic magnification factor) (Eq. 7.91), D, as the ratio of the stationary response magnitude, X, to the static displacement, f_0/k

$$D = \frac{X}{f_0/k} = \frac{1}{\sqrt{(1 - \beta^2)^2 + (2\xi\beta)^2}} \tag{7.91}$$

It is interesting to point out that the dynamic amplification factor, D, and the magnitude of the transfer function, $H(\omega_e)$, are related by the stiffness constant, k.

As seen in (Eq. 7.91), we can express the dynamic amplification factor, D, in terms of parameter β, the dimensionless frequency ratio, for the different values of damping ratio ξ. Figure 7.16 shows the relationship between the dynamic amplification ratio and the dimensionless frequency ratio for different values of ξ.

We can observe that for $\xi = 0.1$ (quite usual), the dynamic amplification reaches a value of $D = 5$. When $\xi = 0$, the value of D is infinity. All this occurs when the

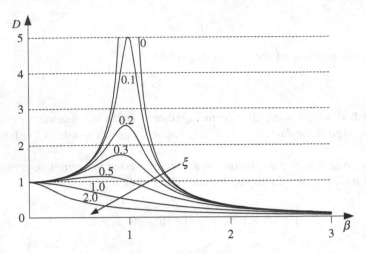

Fig. 7.16 Plot of the relationship between dynamic amplification factor D and dimensionless frequency ratio β for different values of damping ratio ξ

values of dimensionless frequency ratio β are close to 1, that is, when the value of excitation frequency ω_e is close to the natural frequency of the system. When this happens we say that the system is at resonance.

The excitation frequency is directly related to the machine speed. The speed at which the machine is at resonance is called critical speed. This speed has to be avoided since material stress increases and can lead to considerable damage or even the breakage of the part undergoing vibration.

However, we can work at speeds above the critical one ($\beta > 1.5$) without any danger. In order to do so, it is necessary to get past the critical speed as quickly as possible. As speed exceeds resonance, the amplitude reduces, its value tending to zero for higher speeds.

It is interesting to find the maximum value of the dynamic amplification factor, D. To do so, we have to equate the first derivative of D with respect to β to zero (Eq. 7.92):

$$\frac{dD}{d\beta} = \frac{-1}{2} \frac{8\xi^2\beta - 4\beta(1 - \beta^2)}{\sqrt{(1 - \beta^2)^2 + (2\xi\beta)^2}} = 0 \tag{7.92}$$

The solution to this equation yields the maximum dynamic amplification value (Eq. 7.93):

$$D_{max} = \frac{1}{2\xi\sqrt{1 - \xi^2}} \tag{7.93}$$

For small values of damping ratio ξ, we can obtain a good approximation by considering $\sqrt{1 - \xi^2}$ slightly lower than 1. Hence, the maximum value of the dynamic amplification factor results (Eq. 7.94):

$$D_{max} = \frac{1}{2\xi} \qquad (7.94)$$

Obtaining damping ratio experimentally is a question of great interest. Different methods can be applied to this issue, but none of them are simple. The relative logarithmic decay method or methods based on the dynamic amplification factor are some of the most commonly used.

7.4.2 Harmonic Excitations in Machines

Two different cases are usually studied in Machine Theory:

- An unbalanced machine that transmits vibrations to its support.
- A machine that undergoes a vibration due to a harmonic disturbance of the support.

7.4.2.1 Unbalanced Rotation

Unbalanced machine elements with turning motion are one of the main vibration sources.

Consider a mass-spring-damper system restricted to moving only in a vertical direction and which is excited by a machine part that rotates in an unbalanced state. This is represented by mass m in Fig. 7.17, which has an offset, r, and rotates at angular velocity ω_e. The excitation force (Eq. 7.95) will be:

$$F(t) = mr\omega_e^2 \sin \omega_e t \qquad (7.95)$$

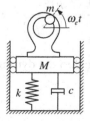

Fig. 7.17 Mass-spring-damper system restricted to moving only in a vertical direction

We can write this equation using the imaginary part of complex number $e^{i\omega_e t}$ (Eq. 7.96) instead of $\sin \omega_e t$:

$$F(t) = mr\omega_e^2 Im(e^{i\omega_e t}) \tag{7.96}$$

The response to the harmonic excitation force (Eq. 7.97) is the imaginary part of the product of complex force $F(t)$ and transfer function $H(\omega_e)$:

$$x(t) = mr\omega_e^2 Im(H(\omega_e)^{i\omega_e t}) \tag{7.97}$$

The evaluation of this expression yields (Eq. 7.98):

$$
\begin{aligned}
x(t) &= mr\omega_e^2 \frac{1/k}{\sqrt{(1-\beta^2)^2 + (2\xi\beta)^2}} Im(e^{i(\omega_e t - \varphi)}) \\
&= \frac{m}{M} r\beta^2 \frac{1}{\sqrt{(1-\beta^2)^2 + (2\xi\beta)^2}} Im(e^{i(\omega_e t - \varphi)})
\end{aligned}
\tag{7.98}
$$

To have a notion of vibration transmissibility, we have to evaluate the magnitude of the exciting force as well as the force transmitted through the spring (Eq. 7.99) and damper (Eq. 7.100) to the support. This force can be obtained by adding up the forces in the spring and damper:

$$F_k = kX \tag{7.99}$$

$$F_c = c\omega_e X = 2k\xi\beta X \tag{7.100}$$

Hence, force transmitted to the support of the mechanism (Eq. 7.101) is:

$$F_s = \sqrt{F_k^2 + F_c^2} = kX\sqrt{1 + (2\xi\beta)^2} \tag{7.101}$$

We define the transmissibility ratio or just the transmissibility, T_r (Eq. 7.102), as the ratio of the magnitude of the force transmitted to the support, F_s, to the magnitude of the exciting force, $|F(t)|$:

$$T_r = \frac{F_s}{|F(t)|} = \frac{\sqrt{1 + (2\xi\beta)^2}}{\sqrt{(1-\beta^2)^2 + (2\xi\beta)^2}} = D\sqrt{1 + (2\xi\beta)^2} \tag{7.102}$$

We can see that its value is equal to the dynamic amplification factor corrected by term $\sqrt{1 + (2\xi\beta)^2}$.

7.4.2.2   Support in Motion

In many cases, the vibration of a system can be excited by the motion of the support.

In the system shown in Fig. 7.18, the support undergoes a harmonic motion with amplitude X_0.

Let $x(t)$ be the absolute displacement of the mass, m, and let $x_r(t)$ be the relative displacement of the mass with respect to the support (Eq. 7.103), which has to be equal to the difference between the absolute displacements of the mass and the support.

$$x_r(t) = x(t) - x_0(t) \qquad (7.103)$$

After establishing the force equilibrium of mass m, we obtain the differential equation (Eq. 7.104) (Fig. 7.19):

$$m\ddot{x}(t) + c(\dot{x}(t) - \dot{x}_0(t)) + k(x(t) - x_0(t)) = 0 \qquad (7.104)$$

We subtract a quantity of $m\ddot{x}_0(t)$ from both terms (Eq. 7.105):

$$m\ddot{x}(t) - \ddot{x}_0(t)) + c(\dot{x}(t) - \dot{x}_0(t)) + k(x(t) - x_0(t)) = -m\ddot{x}_0(t) \qquad (7.105)$$

Therefore, the differential equation of the basic discrete system is fulfilled when applied to the relative motion of mass m with respect to the support (Eq. 7.106). The exciting forces are the ones generated by the motion of the support.

$$m\ddot{x}_r(t) + c\dot{x}_r + kx_r(t) = -m\ddot{x}_0(t) \qquad (7.106)$$

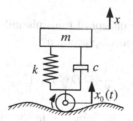

Fig. 7.18 Mass-spring-damper system undergoes vibration due to harmonic disturbance of the support

Fig. 7.19 Free body diagram of the mass-spring-damper system

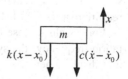

Consider that the support moves with sinusoidal motion (Eq. 7.107):

$$\left.\begin{array}{l} x_0(t) = X_0 \sin \omega_e t = X_0 \mathrm{Im}(e^{i\omega_e t}) \\ \dot{x}_0(t) = X_0 \omega_e \cos \omega_e t = X_0 \omega_e Re(e^{i\omega_e t}) \\ \ddot{x}_0(t) = -X_0 \omega_e^2 \sin \omega_e t = -X_0 \omega_e^2 Im(e^{i\omega_e t}) \end{array}\right\} \qquad (7.107)$$

The excitation force (Eq. 7.108) is the inertia force mentioned above, which can be defined as:

$$F(t) = mX_0 \omega_e^2 \mathrm{Im}(e^{i\omega_e t}) \qquad (7.108)$$

The response of the system (Eq. 7.109) is the product of the imaginary part of the complex force and the transfer function:

$$x_r(t) = mX_0 \omega_e^2 \mathrm{Im}(H(\omega_e)e^{i\omega_e t}) \qquad (7.109)$$

Then, absolute motion (Eq. 7.110) is the sum of support motion and relative motion:

$$\begin{aligned} x(t) &= X_0 \mathrm{Im}(e^{i\omega_e t}) + mX_0 \omega_e^2 \mathrm{Im}(H(\omega_e)e^{i\omega_e t}) \\ &= X_0 \mathrm{Im}((1 + m\omega_e^2 H(\omega_e))e^{i\omega_e t}) \end{aligned} \qquad (7.110)$$

The magnitude of the absolute displacement (Eq. 7.111) is:

$$\begin{aligned} X &= X_0 \left|1 + m\omega_e^2 H(\omega_e)\right| = X_0 \left|1 + \frac{\beta^2}{1 - \beta^2 + 2i\xi\beta}\right| \\ &= X_0 \left|\frac{1 + 2i\xi\beta}{1 - \beta^2 + 2i\xi\beta}\right| = X_0 D\sqrt{1 + (2\xi\beta)^2} \end{aligned} \qquad (7.111)$$

In this case, we define the concept of transmissibility (Eq. 7.112) as the ratio of mass m displacement to support displacement:

$$T_r = \frac{X}{X_0} = D\sqrt{1 + (2\xi\beta)^2} \qquad (7.112)$$

The expression obtained is the same as in the previous section. Figure 7.20 shows the variation of transmissibility T_r against β for different values of damping ratio ξ. We can see that, in order to have a low transmissibility ratio, the value of the natural frequency of the system should be less than half of the excitation frequency value, ($\beta > 2$). Therefore, parameters k and m will be chosen so that this requirement is fulfilled.

For β values close to one (natural frequency of the system similar to the excitation frequency) the system starts vibrating at resonance, which causes high stress and displacements.

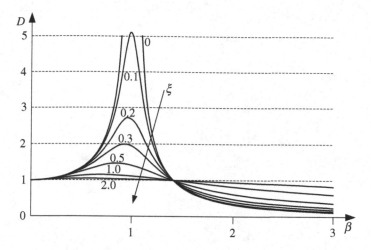

Fig. 7.20 Plot of the relationship between T_r and β for different values of damping ratio ξ

Fig. 7.21 Example of a
vibration isolation solution

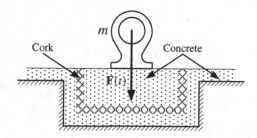

When a system has to work at its resonance speed, it is convenient to use a high damping coefficient in order to reduce the transmissibility ratio. On the other hand, when the frequency of the external force is higher than the natural frequency of the system, low damping is used.

7.4.3   Vibration Isolation

From the results obtained in the previous section, we can study each particular vibration isolation problem and find those values of k and c that minimize transmissibility ratio T_r.

The general approach to this kind of problem is to reduce the natural frequency of the system. In order to do so, we increase its mass. An example is shown in Fig. 7.21.

Chapter 8
Gears

Abstract Before studying how gear transmission works, we need to understand how motion is transmitted between a pair of links with curved surfaces in direct contact. These contact surfaces can be of any shape. Motion between them can be pure rolling, pure sliding or a combination of both. In this chapter we will learn about the kinematic requirements for a curve to be used to manufacture a gear and the commonly-used gear tooth shapes. We will study involute gears and their kinematics. Finally, we see a classification of toothed wheels and their fundamental concepts.

8.1 Introduction

Before studying how gear transmission works, we need to understand how motion is transmitted between a pair of links with curved surfaces in direct contact. These contact surfaces can be of any shape. Motion between them can be pure rolling, pure sliding or a combination of both.

In this chapter we will learn about the kinematic requirements for a curve to be used to manufacture a gear and the commonly-used gear tooth shapes. We will study involute gears and their kinematics. Finally, we see a classification of toothed wheels and their fundamental concepts.

8.1.1 Characteristics of Motion Transmitted by Curves in Contact

Consider Fig. 8.1, where all points of the motor link, 2, turn around O_2 and all points of the follower, link 3, turn around O_3. The velocity of any point of links 2 and 3 is perpendicular to its turning radius.

© Springer International Publishing Switzerland 2016

A. Simón Mata et al., *Fundamentals of Machine Theory and Mechanisms*,
Mechanisms and Machine Science 40, DOI 10.1007/978-3-319-31970-4_8

Fig. 8.1 Transmission
between two surfaces in direct
contact with planar movement

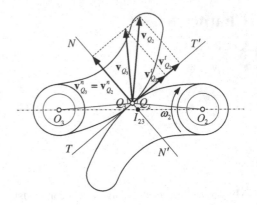

The contact point coincides with points Q_2 and Q_3 of links 2 and 3 respectively.
Motion can be transmitted by direct contact as long as there is a force perpendicular
to the surfaces in contact.

Whatever the velocity of points Q_2 and Q_3 is, their components in the common
normal direction, v_{Q2}^n and v_{Q3}^n, have to be the same so that the links in contact do not
separate or enter each other. Their tangential components are located at tangent line
TT'. The difference between these components yields the sliding velocity.

When tangential components are equal, the sliding velocity is zero, so both
components move with pure rotation. When this occurs, contact point Q coincides
with instantaneous center I_{23}, which is located on the line that joins both turning
centers (Fig. 8.1).

8.1.2 Velocity Relationship Between Two Curves in Contact

In Fig. 8.1, assume that body 2 moves with angular velocity ω_2. The magnitude of
the linear velocity of point Q_2 is given by Eq. (8.1):

$$v_{Q_2} = \overline{O_2 Q_2}\,\omega_2 \qquad (8.1)$$

The angular velocity of body 3 (Eq. 8.2) is:

$$\omega_3 = \frac{v_{Q_3}}{\overline{O_3 Q_3}} \qquad (8.2)$$

Therefore, the velocity ratio (Eq. 8.3) is:

$$\frac{\omega_2}{\omega_3} = \frac{v_{Q_2}}{\overline{O_2 Q_2}}\frac{\overline{O_3 Q_3}}{v_{Q_3}} \qquad (8.3)$$

According to Kennedy's theorem, the instantaneous rotation center of bodies 2 and 3 (I_{23}) has to be located on the line that passes through centers O_2 and O_3. It also has to be located on common normal NN' so that the normal relative velocity is zero. In other words, the normal velocity components of both points in contact have to be equal.

Hence, we can express velocity $v_{I_{23}}$ (Eq. 8.4) in terms of link 2 or link 3 as:

$$v_{I_{23}} = \overline{O_2 I_{23}} \omega_2 = \overline{O_3 I_{23}} \omega_3 \tag{8.4}$$

Then, the velocity ratio (Eq. 8.5) is:

$$\frac{\omega_2}{\omega_3} = \frac{\overline{O_3 I_{23}}}{\overline{O_2 I_{23}}} \tag{8.5}$$

Therefore, angular velocities are inversely proportional to the segments in which common normal NN' at contact point Q divides the line that joins centers O_2 and O_3.

Moreover, in order for the angular velocity ratio to be constant, common normal NN' has to divide the line of centers $O_2 O_3$ in segments with a constant proportion. Hence, if the position of centers O_2 and O_3 is fixed, point I_{23} has to be fixed as well.

8.1.3 Rolling Wheels

The easiest way to transmit forces and motion consists of two wheels with a circular cross section set in contact, so that the friction force between them transmits turning motion from one shaft to another.

In this case, the angular velocity ratio remains constant as long as the force transmitted does not exceed the friction force. The contact point is located on the line that joins the centers and coincides with instantaneous center I_{23}. The system works in pure rolling without slipping and the linear velocities of the wheels at the point of contact are the same.

The condition (Eq. 8.6) that has to be met is:

$$F = F_{Tr} \leq \mu N \tag{8.6}$$

where μ is the coefficient of friction between the surfaces in contact and N is the normal force between both surfaces at the point of contact. Figure 8.2 shows two rolling wheels that transmit force F where the normal force is due to the action of a spring that acts on the center of wheel 3.

Therefore, power transmission (Eq. 8.7) in this type of system is limited by friction resistance:

Fig. 8.2 Rolling wheels

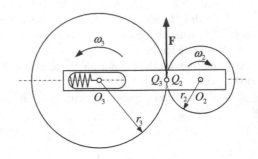

$$P_{Tr} = M_2\omega_2 = M_3\omega_3 \tag{8.7}$$

where the torque in body 2 and 3 is (Eq. 8.8):

$$\left.\begin{array}{l} M_2 = r_2 F \\ M_3 = r_3 F \end{array}\right\} \tag{8.8}$$

The angular velocity ratio (Eq. 8.9) when $F \leq \mu N$ is:

$$\frac{\omega_2}{\omega_3} = \frac{M_2}{M_3} = \frac{r_3}{r_2} = \text{constant} \tag{8.9}$$

8.2   Toothed Wheels (Gears)

We can consider toothed wheels as rolling wheels to which we have added some "teeth" in order to allow power transmission with forces higher than the friction force. The tooth profile has to be such that motion transmitted is equal to the one in rolling wheels. Kinematics of rolling wheels and toothed wheels are, therefore, the same. However, power transmission is no longer restricted by friction when talking about toothed wheels.

8.3   Condition for Constant Velocity Ratio. Fundamental Law of Gearing

In order for a toothed wheel to transmit the same motion as its equivalent friction wheel, the profile of its teeth has to fulfill the following kinematic condition: both wheels have to move with a constant angular velocity ratio at every instant.

Figure 8.3 shows the contact between a pair of geared teeth. The circles that would define two rolling wheels with the same motion as the actual gears are called pitch circles. Such wheels are tangent at point P (called pitch point) located on the

Fig. 8.3 Two gearing tooth profiles

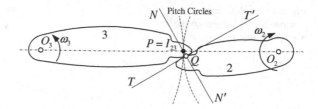

line that joins both centers. In order for the tooth profile to comply with the constant relative velocity condition, the common normal NN' at contact point Q has to divide the line of centers O_2O_3 following a constant ratio. As the distance between centers does not change, point P has to be a fixed point on the line of centers. This condition is known as fundamental law of gearing and the two tooth profiles which satisfy this condition are called conjugate profiles.

Since the pitch point is the tangent point of the pitch circles, the velocity ratio (Eq. 8.10) is the same as the one for two rolling wheels:

$$\frac{\omega_2}{\omega_3} = \frac{\overline{O_3P}}{\overline{O_2P}} = \text{constant} \qquad (8.10)$$

8.4   Involute Teeth

Of all possible configurations for the tooth profile, the involute and cycloid ones are the most commonly used. Of these two, the involute is the one used in almost every case except for clock mechanisms and measuring instruments such as micrometers in which the cycloid form is used. This type of tooth profile has a more precise functioning but it demands higher precision in manufacturing and assembling as well.

The involute profile has several advantages of which the two most important ones are the following:

- It is easy to manufacture with simple tools.
- It allows high manufacturing tolerance since the transmission relationship remains constant even when the distance between centers is modified.

The involute profile is the curve generated by a point of a straight line that rolls on a circle without slipping (Fig. 8.4). The circle is known as base circle and it is the locus of all the curvature centers of the involute curve (0, 1, 2, 3, ...). The straight line is perpendicular to the generated curve at every instant while it is tangent to the base circle.

Two methods of generating an involute profile are:

- The involute can be generated by a point on a cord that is unwrapped from a fixed circumference (the base circle). This is shown in Fig. 8.5.

Fig. 8.4 Involute curve
generated by a tangent to the
base circle that rolls over it

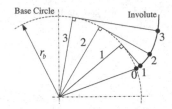

Fig. 8.5 Involute generated
by a point on a cord that is
unwrapped from a base circle

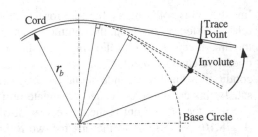

Fig. 8.6 Involute generated
on a rotating paper by a point
moving on a straight path

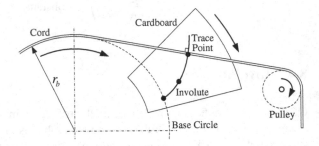

- The involute can also be traced on a piece of cardboard that rotates jointly with the base circle. The tracer point moves along a straight path on a cord that is maintained taut by means of a pulley. This method is shown in Fig. 8.6.

In the method described by Fig. 8.6 we can substitute the pulley by another circle that turns in opposite direction to the base circle. The cord is maintained taut and a piece of cardboard is attached to each circle. The point of the cord that generates the involute profile on the first wheel will also generate another involute curve on the cardboard attached to the second wheel. These two profiles are known as conjugate profiles. The normal to the profiles at the point generated at every instant of time is defined by the direction of the cord, which is tangent to both base circumferences. This direction is maintained invariable and, thus, it intersects the line that joins the centers of both wheels always at the same point (Fig. 8.7). Hence, we have just demonstrated that two toothed wheels with an involute tooth profile comply with the law of gearing.

Fig. 8.7 Involute profiles of
a pair of mating teeth in two
different positions

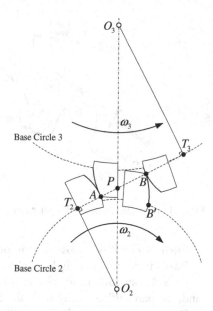

The involute curve has the following properties:

- The normal to the involute is tangent to the base circle at every instant.
- As shown in Fig. 8.7, the length of the normal to the tooth profile at point B,
  $\overline{T_2B}$, is equal to the length of arch $\overparen{T_2B'}$. This is fulfilled for every point in an involute profile.
- The distance between two points of two involute curves traced with the same base circle is always the same when it is measured along the common normal.
- Two involute profiles are conjugate profiles independently from the diameters of their base circles.
- Two involute profiles remain conjugate even when the distance between the centers varies.

8.5   Definitions and Nomenclature

In order to make the study of gear systems easier, we define a set of normalized parameters. Definitions referring to toothed wheels are expressed in relative to their cross-section (Fig. 8.8). Here we will see the terminology used for spur gears (with a cylindrical pitch surface) but most of the definitions are common to all types of gears.

Pitch surface: it is a cylinder whose diameter is the pitch diameter. This cylinder defines a rolling wheel that is kinematically equivalent to the toothed wheel.

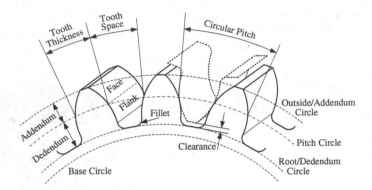

Fig. 8.8 Terminology used for spur gears

Pitch circle: it is a cross-section of the pitch surface.

Base circle: it is the circle used to generate the involute tooth profile.

Outside or Addendum circle: it is the circle that limits teeth size on their top land, the part furthest away from the center of the base circle.

Root or dedendum circle: it is the circle that limits teeth size on their closest side to the center of the pitch circle.

Addendum: it is the radial distance from the pitch circle to the addendum circle.

Dedendum: it is the radial distance from the pitch circle to the dedendum circle.

Whole depth: it is the distance from the top of the tooth to the root of the tooth, that is, the sum of the addendum and dedendum.

Clearance: it is the distance from the addendum circle of a wheel to the root circle of the wheel it is in mesh with.

Face of a tooth: it is the surface of the tooth above the pitch surface.

Flank of a tooth: it is the surface of the tooth below the pitch surface.

Tooth thickness: it is the width of the tooth measured along the pitch circle.

Tooth space: it is the space between two adjacent teeth measured along the pitch circle.

Backlash: it is the difference between the tooth thickness of one gear and the tooth space of the mating gear.

Circular pitch: it is the sum of tooth thickness and tooth space measured along the pitch circle. Its value (Eq. 8.11) can be computed as the ratio of the perimeter of the pitch circle (πD) to the number of teeth in the wheel (Z).

$$p_c = \frac{\pi D}{Z} \tag{8.11}$$

Module: it is the ratio of the pitch circle diameter to the number of teeth in the wheel (Eq. 8.12). It is normally expressed in mm:

Fig. 8.9 Portions of a pair of involute gear in mesh with a pair of mating teeth shown in two different positions

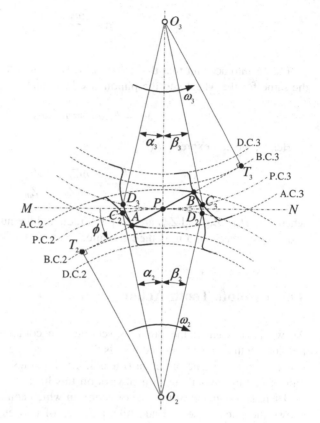

$$m = \frac{p_c}{\pi} = \frac{D}{Z} \tag{8.12}$$

The following basic concepts are defined for two meshed gears (Fig. 8.9):

Pinion: it is the wheel with the smaller number of teeth in a gear train.

Pitch point P: it is the point of tangency between the two pitch circles. It is the contact point between the two kinematically equivalent rolling wheels.

Line of action or pressure line: it is the normal to two gear teeth at the point of contact. It is tangent to the base circles. Figure 8.9 shows a segment of the action line between tangential points T_2 and T_3.

Common tangent: it is the line tangent to both pitch circles at the pitch point. Hence, it is normal to the line of centers, MN in Fig. 8.9.

Pressure angle ϕ: it is the angle formed by the line of action and the common tangent to the pitch circles at the pitch point.

Velocity ratio of engaged gears μ: it is the ratio of the velocity of the input or driver wheel, ω_{in}, to the velocity of the output or driven one, ω_{out}, (Eq. 8.13).

$$\mu = \frac{\omega_{in}}{\omega_{out}} \tag{8.13}$$

Taking into account that the linear velocity of the pitch point, v_P (Eq. 8.14), is the same for the wheel and the pinion, we have that:

$$v_P = R_{in}\omega_{in} = R_{out}\omega_{out} \tag{8.14}$$

Hence, we can express μ (Eq. 8.15) as:

$$\mu = \frac{R_{out}}{R_{in}} = \frac{mZ_{out}}{mZ_{in}} = \frac{Z_{out}}{Z_{in}} \tag{8.15}$$

where R_{in}, R_{out}, Z_{in}, and Z_{out} are the pitch radii and the number of teeth in the input and output wheels and m is their module.

8.6 Involute Tooth Action

As we have seen in the previous section, the common normal to a couple of involute teeth at the point of contact is the common tangent to both base circles and it is known as the line of action (Figs. 8.7, 8.8 and 8.9). The point of contact of a pair of mating involute teeth is always on this line.

There is a segment of the line of action in which contact takes place. Figure 8.9 shows the pitch, base and addendum circles of two engaged gears. Considering wheel 2 to be the input wheel, contact starts at point A, the intersection point of the line of action and the addendum circle of the driven wheel (A.C.3). During the motion of the gears, the point of contact moves over the line of action, passes through point P (pitch point) and contact finishes when it reaches point B, the intersection of the line of action and the addendum circle of the drive wheel (A.C.2).

This way, point A defines the beginning of contact and point B the end of contact. Line APB defines the path of the point of contact and its length is the length of the path of contact.

The tooth profile of wheel 2 at the beginning of engagement intersects the pitch circle at point C_2. Arch $\overset{\frown}{C_2P}$ is known as arc of approach, since it is defined by the motion of the tooth towards the center line.

Point D_2 is the intersection of the profile of the same tooth and the pitch circle at the instant when contact ends. Arc $\overset{\frown}{PD_2}$ is known as arc of recess since it is defined by the motion of the tooth away from the center line.

Angle α_2, angle of approach, is the angle formed by line O_2C_2 and the center line. In other words, the angle subtended by the arc of approach.

Angle β_2, angle of recess, is the angle formed by line O_2D_2 and the center line. That is to say, the angle subtended by the arc of recess.

The sum of the arcs of approach and recess is known as the arc of action ($\overgroup{C_2D_2}$). Thus, the angle of action is equal to the sum of the latter two.

The arcs of approach and recess of wheel 3 are the same as the ones of wheel 2, since their pitch circles represent two rolling wheels that roll without slipping. However, angles α_3 and β_3 are different to α_2 and β_2, since the radii of the two pitch circles are different.

8.7 Contact Ratio

The contact ratio, R_C, is defined as the ratio of the length of the arc of contact to the circular pitch. It can be considered the average number of teeth in contact during a gearing cycle. Its value always has to be greater than one, since before contact between a pair of mating teeth finishes, another pair has to start so that there is always a pair of teeth in contact. A contact ratio of 1.4 indicates that during 60 % of the time a pair of teeth is in contact and during 40 % of the time two pairs of teeth are in contact. A high contact ratio value usually implies a more uniform performance of geared wheels as well as an increase in load transmission capacity, since the same amount of load will be transmitted by more than one pair of teeth. However, in order to achieve a high contact ratio, tooth profiles have to be manufactured with great precision.

Figures 8.10 and 8.11 show two wheels engaged with two pairs of teeth in contact at points A, E and F, B respectively. In Fig. 8.10 one of the pairs is at the start-of-contact position while in Fig. 8.11 the same pair is at the end-of-contact position. Between both positions, the contact point between the teeth moves along the line of action. As explained previously in this chapter, initial and end contact points are located at the intersection of the addendum circles of the wheel and the

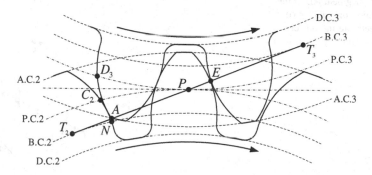

Fig. 8.10 Two wheels gearing with two pairs of teeth in contact at points A and E

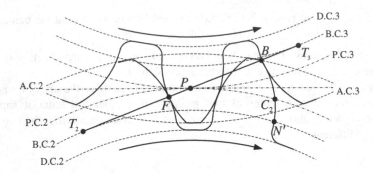

Fig. 8.11 Two wheels engaged with two pairs of teeth in contact at points F and B

pinion with the line of action respectively. Contact starts at point A (Fig. 8.10) and ends at point B (Fig. 8.11).

Points T_2 and T_3 are tangent points of the base circles of each wheel and the action line.

Contact ratio R_C has already been defined as the quotient of the arc of action and the circular pitch (Eq. 8.16).

We can also define the contact ratio by using parameters measured on the base circle. In this case, the contact ratio is the quotient between the arc of action measured on base circle $\overset{\frown}{NN'}$ and base pitch p_b. In Figs. 8.10 and 8.11 the length of arc of action $\overset{\frown}{NN'}$ coincides with the length of segment $\overline{AB}$, since the distance the contact point has to travel along the line of action is equal to the distance of the curve measured on the base circle as long as the tooth profile is an involute generated on the base circle.

Figure 8.12 shows segments $\overline{T_2A}$, $\overline{T_3B}$, $\overline{T_2P}$ and $\overline{T_3P}$ as well as the radii of the addendum (r_{a2} and r_{a3}), base (r_{b2} and r_{b3}) and pitch (r_2 and r_3) circles. The contact ratio can be calculated by means of these segments and radii as follows:

Fig. 8.12 Segment AB of the line of action on which two mating teeth are in contact during a gearing cycle

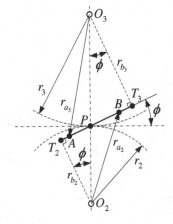

$$R_C38; = \frac{\overparen{CC'}}{p_c} = \frac{\overline{AB}}{p_b} = \frac{\overline{AP} + \overline{PB}}{p_b} = \frac{(\overline{T_3 A} - \overline{T_3 P}) + (\overline{T_2 B} - \overline{T_2 P})}{\frac{2\pi r_b}{Z}}$$

$$38; = \frac{(\sqrt{r_{a3}^2 - r_{b3}^2} - r_3 \sin \phi) + (\sqrt{r_{a2}^2 - r_{b2}^2} - r_2 \sin \phi)}{\pi \frac{2r}{Z} \cos \phi} \qquad (8.16)$$

$$38; = \frac{\sqrt{r_{a3}^2 - r_{b3}^2} + \sqrt{r_{a2}^2 - r_{b2}^2} - (r_2 + r_3) \sin \phi}{\pi m \cos \phi}$$

where:

- T_2, T_3: are the tangent points of the line of action to the base circle.
- A, B: are the intersection points of the line of action and the addendum circle of the wheels engaged.
- P: is the Pitch point.
- r_{a2}, r_{a3}: are the radii of the addendum circles.
- r_{b2}, r_{b3}: are the radii of the base circles.
- ϕ: is the pressure angle. Its cosine (Eq. 8.17) can be calculated as:

$$\cos \phi = \frac{r_{b3}}{r_3} = \frac{r_{b2}}{r_2} \qquad (8.17)$$

8.8 Relationship Between Velocity Ratio and Base Circles

The standard center distance of two mating gears can easily be calculated by summing up the pitch radii of both gears. If the gears are assembled with a distance between centers that is different to this standard center distance, pressure angle ϕ will change.

Therefore, the pressure angle is not a property of a single gear as it depends on how the wheel is assembled. The same happens with the pitch radii. However, involute tooth profiles of a gear do not change and, therefore, their base circles are invariant. Thus, the base circle is a property of a single toothed wheel and is fixed once the gear has been manufactured.

Consider the case shown in Fig. 8.13a, in which, originally, point P was tangent point of standard pitch circles P.C.2 and P.C.3. If center O_3 of wheel 3 moves to a new point O_3' away from O_2, (Fig. 8.13b), the direction of the common tangent to both base circles, $T_2 T_3$, varies to $T_2' T_3'$ as it makes an anticlockwise rotation in order to maintain tangency and cuts through the center line at a new point P'. Thus, pressure angle ϕ and the pitch circle radii have been increased and point P' is the pitch point of the new assembly.

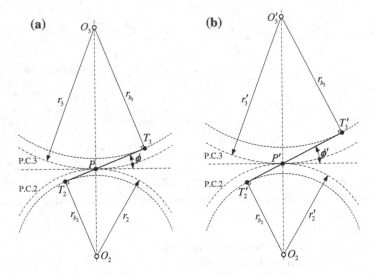

Fig. 8.13 a Two mating gears. **b** The same gears with a new P' pitch point after moving center O_3 away from O_2

It can be verified (Eq. 8.18) from Fig. 8.13 that:

$$\left.\begin{array}{c} r_2 = \dfrac{r_{b2}}{\cos\phi} \\[2mm] r_3 = \dfrac{r_{b3}}{\cos\phi} \end{array}\right\}$$

$$\left.\begin{array}{c} r_2' = \dfrac{r_{b2}}{\cos\phi'} \\[2mm] r_3' = \dfrac{r_{b3}}{\cos\phi'} \end{array}\right\} \tag{8.18}$$

Using the relationships in Eq. (8.18), it can be proved that even though the pitch radii have changed, their ratio remains the same (Eq. 8.19) as the ratio between the base radii, thus invariable:

$$\frac{r_3}{r_2} = \frac{r_3'}{r_2'} = \frac{r_{b3}}{r_{b2}} = \text{constant} \tag{8.19}$$

Therefore, the angular velocity ratio will also be constant (Eq. 8.20) for small variations of the distance between centers:

$$\mu = \frac{\omega_2}{\omega_3} = \frac{r_3}{r_2} = \frac{r_{b3}}{r_{b2}} = \text{constant} \tag{8.20}$$

In other words, the ratio between the angular velocities is inversely proportional to the ratio of the base circle radii. Since the base circles do not change when the centers move, the relationship between the angular velocities remains invariable and the pitch circles have to change their dimensions. However, their radii maintain a constant ratio.

8.9   Interference in Involute Gears

Provided the gears are working correctly, the gearing curve should be fully comprehended between points of tangency T_2 and T_3 of the line of action and the base circles.

Figure 8.14 shows a mating pair of teeth starting and ending the gear cycle. The addendum circle of wheel 3 does not cut segment $\overline{T_2T_3}$ of the line of action, while the addendum circle of pinion 2 cuts the line of action in B.

So, initial contact between both teeth at point A happens outside segment $\overline{T_2T_3}$ of the pressure line. This means that initial contact occurs at a point of the pinion tooth profile which is below the base circle where the profile is not involute. As a result, the fundamental law of gearing will not be complied and the motion transmission will not have a constant angular velocity ratio. This phenomenon is called interference.

Interference will occur when contact between a pair of tooth profiles does not take place between tangent points T_2 and T_3. This happens when the number of teeth is low.

Once they have been performed, the interference from the wheels could be eliminated by following any of the following procedures:

- Contact out of the tangent points can be avoided by making the addendum circle of the wheel smaller.

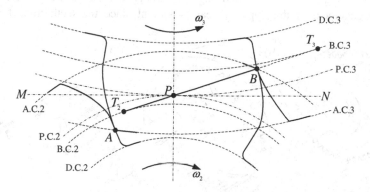

Fig. 8.14 A pair of mating teeth at the beginning and at the end of the gearing cycle with interference at the first point of contact (A)

- Machining the tooth profile and removing part of the pinion tooth flank.
- By increasing the pressure angle, the tangent point can be displaced until it reaches the addendum circle. This way, contact will not take place outside the segment defined by the tangency points of the line of action to the base circle and, therefore, the interference will have disappeared. In order to do so, the distance between centers O_2 and O_3 has to be increased.

Usually, before machining a couple of geared wheels, a theoretical study of the possible interferences has been carried out. When malfunctioning is foreseen, the pinion teeth are cut with a hob further away from the center of rotation and the distance between centers is increased. If we do not want to change the distance between centers, the wheel can be cut closer to the center of rotation.

This way, both the beginning and the end of contact are delayed, which makes the tooth profile involute at the initial contact point.

8.10   Gear Classification

Depending on their surface pitch, gears can be classified as.

8.10.1   Cylindrical Gears

Their teeth are machined on a cylindrical surface. Depending on how the teeth are placed on the wheel, these gears are:

- Parallel gears.
  They transmit power between parallel shafts. This category is subdivided depending on the tooth type:

  - Spur gears:
    The tooth is straight-cut and aligned parallel to the axis (Fig. 8.15). Tooth engagement in this type of gear is violent since the teeth impact at the

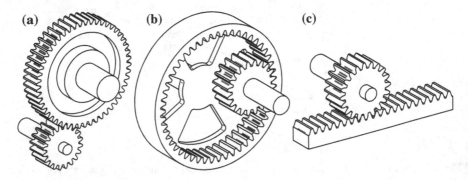

Fig. 8.15 a External spur gears. **b** Internal spur gears. **c** Rack and pinion spur gears

Fig. 8.16 a Parallel helical
gears. **b** Double helical gears

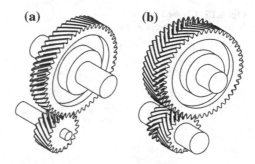

beginning of contact. The main advantage of spur gears is that they are
simple and less expensive to manufacture. Their main limitations are that
they are noisy and their maximum speed is lower than other gears.
– Parallel helical gears
– The tooth edge is inclined to the axis of the pitch cylinder. If they are
   mounted on parallel shafts, the helix angles of the two wheels have to be
   equal and symmetrical (Fig. 8.16). Compared to spur gears, they are less
   noisy since engagement is produced gradually. Their contact ratio is higher,
   which enables them to transmit bigger torque and rotate at higher speed. The
   main disadvantage of these gears is the action of an axial load on the shaft.
   Gears have to be affixed on the axial direction, which makes support design
   more expensive. Also, parallel helical gears are less efficient and more
   complicated to machine and, therefore, more expensive.

• Crossed axes gears.
   They can be classified as follows:

– Worm gears:
   When the inclination of the tooth is close to 90°, the pinion becomes a worm
   screw that could even have a single tooth (Fig. 8.17). This type of gear is
   used for high speed ratios. An interesting characteristic of this type of
   gearing is the possibility to make transmission irreversible. When the
   inclination of the helix is greater than 85°, the wheel cannot drive the screw.

Fig. 8.17 Worm screw and
worm wheel

Fig. 8.18 Crossed helical
gears with a 45° helix angle

- Helical gears. The angle formed by the shafts is equal to the sum of both
  helix angles, which can take any value. Efficiency is low in this type of
  gearing due to high friction rates. This type of gear is not commonly used.
  Figure 8.18 shows a pair of crossed helical gears. The helix angle of both
  gears is 45° and they transmit power between two shafts whose axes cross
  with an angle of 90°.

8.10.2 Bevel Gears

Their pitch surfaces are portions of rolling cones. They are used to transmit power
between intersecting shafts. The two pitch cones must have a common apex that
coincides with the intersection point of the axes of both shafts.

Depending on the shape of the teeth, these can be:

- Straight teeth (Fig. 8.19):
  They are less expensive to manufacture and there are many different sizes
  available. They become noisy at high speeds.

Fig. 8.19 Bevel gears with
straight teeth

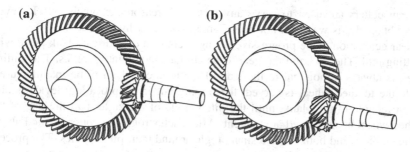

Fig. 8.20 a Helical bevel gears with shafts whose axes intersect. **b** Hypoid gears with crossing shafts

- Spiral bevel (Fig. 8.20a):
  This type of gear has the same advantages and disadvantages as cylindrical helical gears with parallel teeth when compared with gears with straight teeth.

8.10.3 Hypoid Gears

The pitch surface is a portion of rolling hyperboloid (Fig. 8.20b). They transmit motion between nonparallel, nonintersecting shafts. They are only manufactured with spiral teeth. Performance is smooth and silent. They have stronger teeth than the equivalent spiral bevel gear but, at the same time, they are more expensive to manufacture.

8.11   Manufacturing of Toothed Wheels

Toothed wheels can be manufactured by means of a wide range of processes. Some of them are scarcely employed, such as casting (sand casting, lost wax casting, pressure casting). Usually a workpiece that has been smoothed by electrical discharge machining (EDM) is used and shaped by a CNC milling machine or even by extrusion. Toothed wheels made from polymers, aluminum, magnesium and materials of this kind, can be extruded and cut to fit any width needed. Thin toothed wheels can be cut from a metal sheet.

Form milling and gear generation are the two mainly used processes to manufacture toothed wheels.

The cutter used in form milling has the same form as the space between teeth. Hence, technically, a different cutter would be needed for each gear module and number of teeth. The change of the profile is not significant in many cases and normally, for a given module, and depending on the number of teeth, only eight different cutters are used to carve any gear with reasonable precision. However, gears obtained this way will not be suitable to work at high speeds.

Gear generation is the progressive engagement of a cylindrical blank piece with a cutting tool. The tool and the gear to be cut have conjugate profiles. The cutting action is always orthogonal to the side of the toothed wheel to be manufactured. While the toothed wheel is carved, the milling cutter reciprocating device is also moved relative to the blank, until the pitch circle of the cutter and the one of the toothed wheel being carved are tangent. After each stroke, the cutter is lifted above the blank piece and both rotate a small angle around their pitch circles. The process continues until all the teeth are carved.

The cutter can be shaped as a rack. In this case, the pitch circle of the cutting tool is a straight line tangent to the pitch circle of the wheel being carved. The teeth on the gear are carved by a reciprocating motion of the teeth in the rack as shown in Fig. 8.21.

Gear hobbing is a tooth generation process similar to tooth carving by a rack. However, in this case, the teeth of the cutting tool have the same shape as the teeth in the wheel being carved and it extends along a helical curve. The cutting tool looks like a worm screw with the teeth carved on it. The action of the cutting tool is shown in Fig. 8.22.

The teeth of the gear hob are aligned with the ones of the toothed wheel in the machine. The cutting tool and the workpiece are rotating continuously with the

Fig. 8.21 Gear generation using a cutter with rack tooth profiles

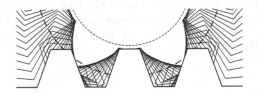

Fig. 8.22 a Lateral view of the machining of a toothed wheel by gear hobbing. **b** Front view

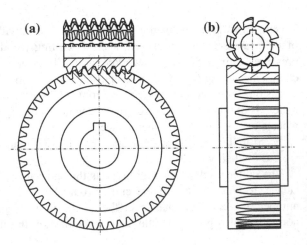

proper angular velocity. The cutting tool moves over the width of the blank piece until the tooth is completed. Among machining processes, carving is the most popular since it allows a high production rate as well as using the same tool either for straight or helicoidal teeth.

Toothed wheels that have to support high loads with respect to the size of their teeth are generally made of resistant materials such as steel. When production is low, they can be manufactured by EDM, CNC or gear hobbing. However, when large numbers of units need to be manufactured, gears are normally generated by a rack. High production of toothed wheels requires the following steps:

- Manufacturing of the base cylinder.
- Tooth generation.
- Tooth refinement.
- Heat treating.
- Machining, swarf removal and cleaning.
- Final coating.

Complete manufacturing of the piece includes the creation of the hub as well as the keyway. Tooth generation includes tooth manufacturing by any of the processes described above. Gear tooth refinement includes the correction of small errors in the pitch and profile, burnishing (by engaging the wheel with another one that has been tempered and burnished) and honing. These processes are necessary to eliminate manufacturing scratches and to improve the accuracy of the gear manufactured. By engaging gear pairs, higher quality is achieved as well as noise reduction. Heat treatment includes superficial tempering, a process necessary in toothed wheels in order to improve their resistance to pressure overload and lack of lubrication. Swarf removal and final cleaning of the piece are essential tasks in the manufacturing of toothed wheels regardless of the process employed. Final coating includes processes such as aluminum anodizing or electrolytic deposition of diamond particles but it may also include only lubrication or painting. The aim of this step is to improve the resistance of the piece to corrosion, reduce friction, provide a coating or just improve the appearance of the part.

8.12   Gear Standardization

A couple of gears need to have teeth of the same size in order to engage correctly. Since the circular pitch and module are unambiguously related with the size of the teeth, these measurements need to be the same for both toothed wheels.

Toothed wheels manufactured with standardized dimensions are interchangeable, that is, any two toothed wheels with the same parameters can be meshed together. The use of interchangeable gears is advisable due to market availability of both gears and milling tools. This means that damaged gears can easily be replaced. Standardization is more a matter of convenience than a technological necessity.

In order for a couple of gears to be interchangeable, it is necessary that the following parameters of both toothed wheels are the same:

- Module.
- Circular pitch.
- Tooth thickness (equal to half of the circular pitch).
- Pressure angle.
- Addendum and dedendum.

The following relationships are established between module, addendum, dedendum, tooth thickness and pressure angle in order to achieve interchangeability as long as both toothed wheels have the same module and for any number of teeth. ISO standard uses a linear rack with $\phi = 20°$ and defines two different types of teeth, where module m is expressed in millimeters:

- Regular teeth (full-depth system):

  - Addendum = m
  - Dedendum = $1.25m$
  - Tooth thickness = $0.5m\pi$
  - Clearance = $0.25m$

  With these values, the tooth whole depth is $2.25m$ while the working depth is $2m$.
- Short teeth (stub-tooth system):

  - Addendum = $0.75m$
  - Dedendum = m
  - Tooth thickness = $0.5m\pi$
  - Clearance = $0.25m$

8.13 Helical Gears

Helical gears are generated the same way as spur gears by carving the teeth profile on a cylindrical surface. However, while the latter have their teeth parallel to the shaft, helical teeth are at an angle to the shaft (Fig. 8.23a). Their teeth are carved following a helix curve while maintaining the involute curve for the contact profile.

This difference gives this configuration a series of advantages with respect to the spur gears.

- Contact is progressive. It starts at the edge of one tooth and increases gradually by following successive lines of contact (Fig. 8.23b). It finishes at a point at the other end of the tooth. In straight teeth, contact starts suddenly across the entire face of the teeth causing more stress and noise.

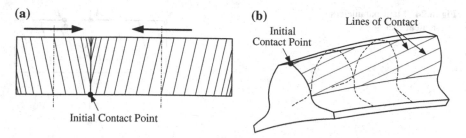

Fig. 8.23 a A pair of parallel helical gears in mesh with the initial contact point between a pair of teeth. **b** Initial contact point of a tooth and successive lines of contact

- Due to its progressive contact, process performance is smoother and produces less noise. Dynamical loads are also reduced.
- This type of gears can work at higher speeds than spur gears.
- The contact ratio is also higher.
- The transmitted load is spread along an angled line making bending loads smaller.
- Tooth wear is reduced and they can transmit more power than spur gears with teeth of the same size.
  However, there are also some disadvantages when compared with spur gears.
- Load is transmitted perpendicularly to the tooth profile which generates an axial component of the load that has to be taken into consideration when designing the shaft supports.
- The teeth carving process is more complicated and should be carried out by generation.
- Higher manufacturing costs.

The general balance is, however, clearly favorable to helical gears, especially for high speed and high power applications.

8.13.1 Helical Gear Parameters

The shape of the teeth of helical gears is an involute helicoid. Each section of the tooth is an involute but any point on it describes a helix, not a straight line, on the base cylinder.

Figure 8.24 shows a section of a helical rack and its main parameters:

- Ψ is the helix angle.
- AB is the traverse circular pitch or just the circular pitch, p_c, measured in the plane of rotation.
- AC is the normal circular pitch, p_n, measured over a plane perpendicular to the tooth.

Fig. 8.24 Main parameters
in a helical rack

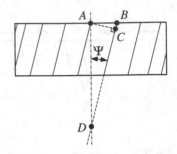

In Fig. 8.24 we can see that these two parameters are related (Eq. 8.21) by:

$$p_n = p_c \tan \Psi \qquad (8.21)$$

- AD is the axial pitch, p_x, that is, the pitch measured in the direction of the axle. Its value (8.22) is:

$$p_x = \frac{p_c}{\tan \Psi} \qquad (8.22)$$

- The normal module, m_n, the module measured on a plane orthogonal to the tooth. This magnitude verifies (Eq. 8.23):

$$p_n = \pi m_n \qquad (8.23)$$

- The circular module, m_c, the module measured along the pitch circle on a traverse plane. It verifies (Eq. 8.24):

$$p_c = \pi m_c \qquad (8.24)$$

Both modules are related as Eqs. (8.25) and (8.26):

$$\left. \begin{array}{l} m_n = \frac{p_n}{\pi} = \frac{p_c}{\pi} \cos \Psi \\ m_c = \frac{p_c}{\pi} \end{array} \right\} \Rightarrow \frac{m_n}{m_c} = \cos \Psi \qquad (8.25)$$

Therefore:

$$m_n = m_c \cos \Psi \qquad (8.26)$$

The same way we can define the normal pressure angle, ϕ_n, in the plane orthogonal to the tooth, and the circular pressure angle, ϕ_c, in the front plane (a plane perpendicular to the shaft).

The relationship between both angles is given by Eq. (8.27):

$$\frac{\tan \phi_n}{\tan \phi_c} = \cos \Psi \tag{8.27}$$

Other relationships defined for gears with straight teeth can also be applied to the normal plane of a helical gear.

8.13.2 Correlation with Spur Gears

Figure 8.25 shows that if we cut the pitch cylinder of a helical gear with a plane normal to the teeth, we get an ellipse with a minor semi-axis equal to $D/2$, D being the diameter of the pitch cylinder.

The equivalent diameter, D_{eq}, of a helical gear is the pitch diameter of the equivalent spur gear. Its value is the radius of curvature of the ellipse at the pitch point. To find it, we start from the ellipse equation (Eq. 8.28):

$$\frac{x^2}{a^2} + \frac{y^2}{b^2} = 1 \tag{8.28}$$

Taking into account (Eq. 8.29) that defines the radius of curvature:

$$\rho = \frac{\sqrt{\left(1 + \left(\frac{dy}{dx}\right)^2\right)^3}}{\frac{d^2y}{dx^2}} \tag{8.29}$$

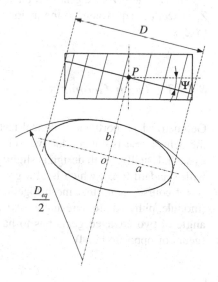

Fig. 8.25 Ellipse obtained as the intersection of the pitch cylinder of a helical gear with a plane normal to the teeth

Combining Eqs. (8.28) and (8.29), we obtain, for $x = 0$, the radius of curvature at the pitch point (Eq. 8.30):

$$\rho = \frac{a^2}{b} \tag{8.30}$$

Considering Eqs. (8.31) and (8.32)

$$\left.\begin{array}{l} a = \frac{D}{2\cos\Psi} \\[2mm] b = \frac{D}{2} \end{array}\right\} \tag{8.31}$$

$$\rho = \frac{D_{eq}}{2} \tag{8.32}$$

We finally obtain (Eq. 8.33):

$$D_{eq} = \frac{D}{\cos^2\Psi} \tag{8.33}$$

where D_{eq} is the diameter of an equivalent toothed wheel with straight teeth. We can define the equivalent number of teeth (Eq. 8.34) as:

$$Z_{eq} = \frac{D_{eq}}{m_n} \tag{8.34}$$

Substituting Eqs. (8.26) and (8.33) in Eq. (8.34), we obtain (Eq. 8.35):

$$Z_{eq} = \frac{D}{\cos^2\Psi}\frac{1}{m\cos\Psi} = \frac{Z}{\cos^3\Psi} \tag{8.35}$$

That is, the involute gear-tooth action in a helical toothed wheel with a number, Z, of teeth is equivalent to the action in a spur gear with a number of teeth equal to Z_{eq}.

8.13.3 Teeth Geometry

Geometrical proportions of helical teeth are not standardized since the concept of the helical gear is not conductive to interchangeable parts and are rarely employed as such. Usually, tooth design is slightly modified so that it is possible to carve them with standard tools, which is of a great interest.

Of course, gears that mesh together need to have the same normal parameters (module, pith and pressure angle). Apart from that, in helical parallel gears the helix angle of two engaged gears has to have the same value but an opposite direction (gears of opposite hand).

Values commonly used for the helix angle and the normal pressure angle are:

- $30° \leq \Psi \leq 15°$
- $\phi_n = 20°$

Tooth height is generally equal to the ones used in gears with straight teeth, but in this case parameters are related with the normal module:

- Addendum = m_n
- Dedendum = $1.25m_n$
- Tooth thickness = $0.5m_n\pi$
- Clearance = $0.25m_n$

With these values, the tooth working depth is $2m_n$ while its whole depth is $2.25m_n$.

8.14 Bevel Gears

These types of gears have their teeth carved on the surface of a circular cone. This type of gear is used to transmit motion between crossing axles. There is one condition to be met so that the kinematic performance of the gears is correct and there is no slip: the vertexes of both cones have to be located at the same point (point O in Fig. 8.26). The most common angle between shafts is 90° but they can work at any angle as long as the previous condition is fulfilled met.

The parameters that need to be considered for their design are (Fig. 8.26):

- Σ: the angle between shafts.
- φ_2: the cone pitch angle of the pinion.
- φ_3: the cone pitch angle of the wheel.
- φ_o: the outside angle.
- φ_r: the root angle.
- α: the addendum angle.
- δ: the dedendum angle.
- D: the pitch diameter.
- D_o: the outside diameter.
- D_r: the root diameter.

Teeth can be straight, helical or spiral. The first one has straight teeth that are parallel to the generatrix of the pitch cone. They are easier to design and manufacture than the other two types. However, they also produce more noise when working at high speeds. The tooth edge of helical bevel gears is inclined to the generatrix of the pitch cone. So, they are carved along a helical curve on the cone surface. They have similar advantages to parallel helical gears versus spur gears. The teeth in spiral bevel gears are spiral curves. However, generally, circular arcs

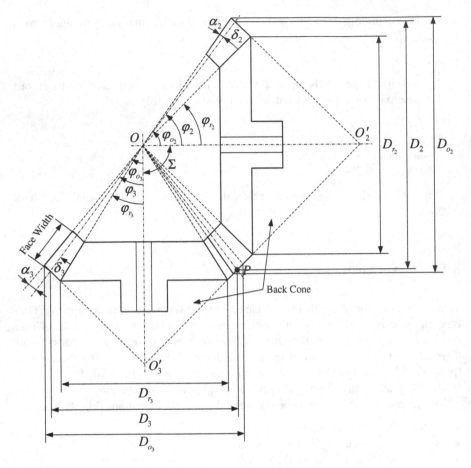

Fig. 8.26 Main parameters in bevel gears

are manufactured which are an approximation to the spiral curves. They are smoother and quieter than the previous types and are generally used to work at high speeds.

8.14.1   Design of a Bevel Gear System

When we study two cones in contact like the ones in Fig. 8.26, the following parameters need to be defined:

- The pitch angles of both surfaces are φ_2 and φ_3.
- The pitch diameters, which are measured on the bigger edge, are D_2 and D_3.

- Like in cylindrical gears, the smaller gear is called pinion and the bigger one is called wheel. When the angle between the shafts is 90°, the second one is known as the crown wheel.

Generally, when projecting a system of bevel gears, angle Σ between the shafts and the velocity ratio are known. With these values, the pitch diameters and the number of teeth needed, Z_2 and Z_3, can be obtained.

The velocity ratio (Eq. 8.36) is:

$$\frac{\omega_3}{\omega_2} = \frac{D_2}{D_3} = \frac{Z_2}{Z_3} \tag{8.36}$$

Cone distance $\overline{OP}$ (Eqs. 8.37 and 8.38) can be obtained as (Fig. 8.26):

$$\overline{OP} = \frac{D_2}{2\sin\varphi_2} \tag{8.37}$$

or

$$\overline{OP} = \frac{D_3}{2\sin\varphi_3} \tag{8.38}$$

And since both expressions are equal:

$$\sin\varphi_2 = \frac{D_2}{D_3}\sin\varphi_3 = \frac{D_2}{D_3}\sin(\Sigma - \varphi_2) = \frac{D_2}{D_3}(\sin\Sigma\cos\varphi_2 - \cos\Sigma\sin\varphi_2) \tag{8.39}$$

By dividing Eq. (8.39) by $\cos\varphi_2$, we obtain (Eq. 8.40):

$$\tan\varphi_2 = \frac{D_2}{D_3}(\sin\Sigma - \cos\Sigma\tan\varphi_2) \tag{8.40}$$

Thus, the pitch angle φ_2 (Eq. 8.41) will be:

$$\tan\varphi_2 = \frac{\frac{D_2}{D_3}\sin\Sigma}{1 + \frac{D_2}{D_3}\cos\Sigma} = \frac{\sin\Sigma}{\frac{Z_3}{Z_2} + \cos\Sigma} \tag{8.41}$$

Similarly, the pitch angle φ_3 (Eq. 8.42) will be:

$$\tan\varphi_3 = \frac{\sin\Sigma}{\frac{Z_2}{Z_3} + \cos\Sigma} \tag{8.42}$$

When the angle formed by both shafts is $\Sigma = 90°$, it yields (Eqs. 8.43 and 8.44):

$$\left.\begin{array}{l}\cos \Sigma = 0 \\ \sin \Sigma = 1\end{array}\right\} \Rightarrow \tan \varphi_2 = \dfrac{1}{\dfrac{Z_3}{Z_2}+0} = \dfrac{Z_2}{Z_3} \qquad (8.43)$$

And thus,

$$\tan \varphi_3 = \dfrac{Z_3}{Z_2} \qquad (8.44)$$

8.14.2 Bevel Gears with Straight Teeth

In Sect. 8.4 we explained the way an involute profile is traced by a point on a cord as it unwinds from the base circle (Fig. 8.5). In that example the profile was generated on the rotation plane of the base circle. To be able to generate the involute surface of a tooth face instead of just a profile, we need to unroll a plane over the base cylinder.

In the case of bevel gears, the tooth profile is generated by unrolling a plane over a cone. This way every point on the plane generates an involute profile. The problem is that such profile is not generated over a cone but over a sphere, since, as it can be seen in Fig. 8.27, when the plane tangent to the base cone unrolls, distance $\overline{OP}$ stays fixed.

Actually, the teeth are not traced over a sphere but over the back cone. This is called Tredgold approximation.

8.14.3 Equivalent Spur Gear

The pith point of a pair of bevel gears is on the back cone (Fig. 8.26). We can consider that each bevel gear forms a portion of a spur gear on the back cone. This is called equivalent spur gear and its radius $D_{eq}/2$ is equal to the back cone radius.

Fig. 8.27 Generation of the involute profile with a tangent plane rolling over the base cone and back cone with its generatrix perpendicular to the pitch cone one

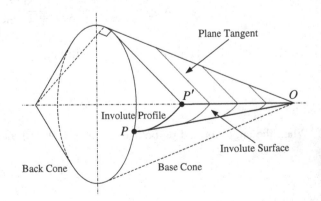

Fig. 8.28 Pitch cone and
back cone of a bevel gear

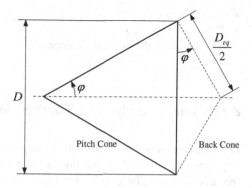

From Fig. 8.28 we can deduce (Eq. 8.45):

$$\cos \varphi_2 = \frac{D_2/2}{D_{2eq}/2} \Rightarrow D_{2eq} = \frac{D_2}{\cos \varphi_2} \tag{8.45}$$

And similarly, (Eq. 8.46):

$$D_{3eq} = \frac{D_3}{\cos \varphi_3} \tag{8.46}$$

Hence, the equivalent number of teeth (Eq. 8.47) is:

$$Z_{eq} = \frac{D_{eq}}{m} = \frac{D}{m \cos \phi} = \frac{Z}{\cos \phi} \tag{8.47}$$

Its value will usually be a decimal number (*m* being the module of the larger part
of the tooth).

8.14.4 Teeth Geometry in Bevel Gears with Straight Teeth

The pressure angle used in the majority of bevel gears with straight teeth is 20°.
Bevel gears are not interchangeable. Each wheel is designed for a determined
pinion and teeth are usually cut with an unequal addendum in order to avoid
interference and increase the contact radius.

The American Gear Manufacturers Association (AGMA) standardized this
modification with the following values:

- Velocity ratio: Z_2/Z_3
- Working depth: $2m$
- Whole depth: $2.188m$

- Angle between axles: 90°
- Pressure angle: 20°
  The value of the addendum depends on the tooth size, which is also tabulated.

8.15   Force Analysis in Toothed Wheels

Loads acting on gears when transmitting power are transmitted to the shafts. It is important to fully understand directions and values of the forces that appear so that the driveshaft and bearing design can be carried out properly.

As we did in Chap. 4, we will use nomenclature F_{ij} to express the force exerted by gear i on gear j or the force transmitted by gear i to shaft j. In order to differentiate them, gears will be named with letters while numbers will be used for shafts.

Next we will carry out a dynamic analysis in spur, helical and bevel gears.

8.15.1   Forces in Spur Gears

Figure 8.29 shows the pitch circles of a pair of mating gears in mesh. Pinion 2 is mounted on axle O_2 that rotates clockwise at angular velocity ω_2 and wheel 3 is installed on axle O_3 that rotates at velocity ω_3 counterclockwise. The value of ω_3 can be computed from the velocity ratio. The pinion is the drive gear.

Forces will the applied to the contact point of both gears. As explained before in this chapter, the contact point moves along the line of action during the gear cycle. This means that force transmission between teeth in contact happens along the line of action. In this situation, the tooth of pinion 2 that is in contact with wheel 3 pushes it with a force equal to F_{23}, while the resistant force exerted by the wheel on the pinion is F_{32} with the same value and opposite direction.

Figure 8.30 shows the forces that are acting on each gear.

Fig. 8.29 Pitch circles and the line of action of a pair of meshing gears

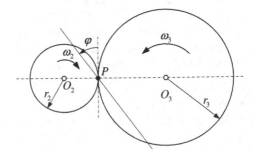

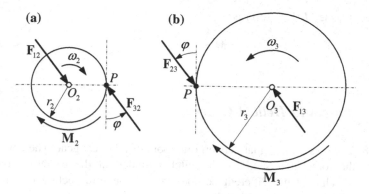

Fig. 8.30 Forces acting on the pinion (**a**) and on the wheel (**b**) and reactions exerted by the shafts

Fig. 8.31 Forces acting on
the pinion

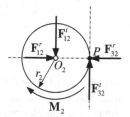

We will focus on the pinion (Fig. 8.30a) and study the system of forces and torques acting on it. To do so, we break the forces into their radial and tangential components (Fig. 8.31). The torque transmitted is produced by tangential loads, while the radial component generates a load on the axle that does not transmit any power. We call the tangential component transmitted force, its magnitude (Eq. 8.48) being:

$$\mathbf{F}'_{32} = \mathbf{F}_{32} \cos \phi \qquad (8.48)$$

The value of the torque transmitted (Eq. 8.49) is:

$$M_2 = r_2 F'_{32} \qquad (8.49)$$

where r_2 is the pitch radius of the pinion.

It is interesting to find the relationship between the torque and the power (Eq. 8.50). To do so, we need to remember that:

$$P = vF^t = r\omega F^t \qquad (8.50)$$

Hence, torque M (Eq. 8.51) is:

$$M = rF^t = \frac{P}{\omega} \tag{8.51}$$

8.15.2 Forces in Helical Gears

Figure 8.32 shows force F that acts on one tooth of a helical gear. The application point of the force is located on the pitch cylinder, at the center of the face width. The value of the different components of the force acting on the tooth (Eqs. 8.52–8.54) can be extracted from Fig. 8.32:

$$F^r = F \sin \phi_n \tag{8.52}$$

$$F^t = F \cos \phi_n \cos \Psi \tag{8.53}$$

$$F^a = F \cos \phi_n \sin \Psi \tag{8.54}$$

where F^r is the radial component, F^t is the tangential component or transmitted load and F^a is the axial component or thrust load. Also Ψ is the helix angle, ϕ_n is the pressure angle measured on the normal plane and ϕ_t is the same angle measured on the tangential plane. Their relationship is (Eq. 8.55):

$$\frac{\tan \phi_n}{\tan \phi_t} = \cos \Psi \tag{8.55}$$

The three components can be determined once F^t is known, which can be worked out with Eq. (8.51) knowing the power that has to be transmitted, the angular velocity and the pitch radius.

Fig. 8.32 Forces on a helical gear

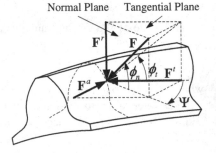

The main difference with respect to spur gears is the action of a force in the direction of the axle that has to be taken into account during the design process.

8.15.3 Forces on Bevel Gears

In order to determine the loads on a driveshaft when power is transmitted by bevel gears, we consider that the resultant of the transmitted forces acts on the middle point of the face width. This does not happen this way since the real resultant force would be located closer to the larger edge of the tooth. Assuming this hypothesis, the magnitude of the transmitted load (tangential component) (Eq. 8.56) can be obtained as:

$$F^t = \frac{M}{r} \tag{8.56}$$

where:

- M is the transmitted torque.
- r is the pitch radius at the center of the face width of the tooth.

Figure 8.33 shows how force $\mathbf{F}$ has three components (tangential, radial and axial) which are perpendicular to each other like in helical gears. Knowing the tangential (Eq. 8.56) component, the radial (Eq. 8.57) and axial (Eq. 8.58) components can easily be calculated (Fig. 8.33):

$$F^r = F^t \tan \phi \cos \gamma \tag{8.57}$$

$$F^a = F^t \tan \phi \sin \gamma \tag{8.58}$$

Fig. 8.33 Resulting force **F** acting on the center of the face width of a tooth and its tangential, radial and axial components

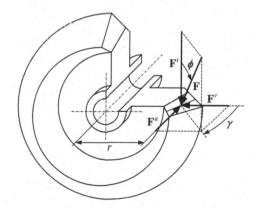

Chapter 9
Gear Trains

Abstract We call a set of two or more gears meshed together a gear train. A gear train is, therefore, a kinematic chain in which motion is transmitted by gears. Some of the reasons to use gear trains to transmit motion are: (a) Space restrictions. When motion has to be transmitted between two axles that are far apart. In that case, two large wheels would be needed. A gear train with several small gears could be used, occupying much less space. (b) The transmission ratio is too high or low. To achieve such a transmission ratio with only two wheels it would be necessary to use one toothed wheel with a much larger pitch diameter than the other. This solution has some problems such as interference and more energy losses than a solution with more than two wheels. (c) Opposite output and input angular velocities are required. (d) Multiple outputs are needed for a single input.

9.1 Classification of Gear Trains

Gear trains are normally classified depending on the movement of the shafts relative to the frame:

- Ordinary gear train. All axles are fixed in space and their velocity with respect to the frame is null. We find two different trains depending on the number of wheels per shaft:

  - Simple gear train: There is only one wheel per shaft.
  - Compound gear train: There is more than one wheel in at least one of the shafts of the train.

- Epicyclic gear train. At least one of the axles moves with respect to the frame.

© Springer International Publishing Switzerland 2016

A. Simón Mata et al., *Fundamentals of Machine Theory and Mechanisms*,

Mechanisms and Machine Science 40, DOI 10.1007/978-3-319-31970-4_9

9.2   Ordinary Trains

Ordinary trains are used most in machines. In this type of train all gears rotate about
their center but none of the gear centers move relative to the frame link. In this
section we will learn how to calculate their velocity ratio and train value.

We call the inverse value of velocity ratio μ gear transmission ratio or just train
value i. Normally, it is more practical to use this parameter since, when multiplied
by the input speed (the known value), we directly obtain the output speed (the
unknown value) (Eq. 9.1).

$$\omega_{out} = i\omega_{in} \tag{9.1}$$

9.2.1   Simple Trains

As explained in the previous section, in simple trains there is only one toothed
wheel per shaft. The velocity ratio μ (Eq. 9.2), that is, the ratio of the angular
velocity of the input or driver gear, ω_{in}, to the one of the output or driven wheel,
ω_{out}, can be written as:

$$\mu = \frac{\omega_{in}}{\omega_{out}} = \frac{\omega_1}{\omega_2}\frac{\omega_2}{\omega_3}\frac{\omega_3}{\omega_4}\ldots\frac{\omega_{n-2}}{\omega_{n-1}}\frac{\omega_{n-1}}{\omega_n} \tag{9.2}$$

where ω_1 is the speed of the driver wheel and ω_n is the speed of the output gear. So,
in this chapter we will not use number 1 to name the frame link like in the previous
chapters. Number 1 will be used to name the first gear of the train.

Direction of rotation can be obtained by analyzing the connection scheme of the
train. It is better to obtain the direction of rotation by analyzing the connection
scheme of the train in order to avoid possible computing mistakes that sometimes
arise when using a sign criterion.

Since each pair of toothed wheels that mesh together has the same module
(Eq. 9.3), the following relationship is verified:

$$m_i = m_j \tag{9.3}$$

Writing the modules in terms of the pitch diameters and the number of teeth
(Eq. 9.4):

$$\frac{D_i}{Z_i} = \frac{D_j}{Z_j} \tag{9.4}$$

So, the velocity ratio of a pair of gears in mesh (Eq. 9.5) can be written in terms of the number of teeth:

$$\frac{\omega_i}{\omega_j} = \frac{R_j}{R_i} = \frac{D_j}{D_i} = \frac{Z_j}{Z_i} \tag{9.5}$$

Hence, the velocity ratio can be calculated as:

$$\mu = \frac{\omega_{in}}{\omega_{out}} = \frac{R_2\,R_3\,R_4}{R_1\,R_2\,R_3} \cdots \frac{R_{n-1}}{R_{n-2}} \frac{R_n}{R_{n-1}} = \frac{R_n}{R_1} = \frac{Z_n}{Z_1} \tag{9.6}$$

It can be seen that the velocity ratio does not depend on the number of teeth of the intermediate wheels and it is only affected by the number of teeth of the input and output gears (Eq. 9.6).

The train value Eq. (9.7) will be:

$$i = \frac{\omega_{out}}{\omega_{in}} = \frac{Z_1}{Z_n} \tag{9.7}$$

So, the train value is equal to the ratio of the number of teeth of the input wheel to the number of teeth of the output gear. The velocity ratio and the train value have a minus sign when the input and output rotations have opposite directions (Fig. 9.1).

Intermediate wheels are called idle gears since they have no effect on the velocity ratio. They only have an impact on the turning direction. Idle gears allow us to use gears with a small diameter independently of the distance between the input and output shafts (Fig. 9.2).

Fig. 9.1 Simple gear train with velocity ratio $\mu = -Z_4/Z_1$ and train value $i = -Z_1/Z_4$

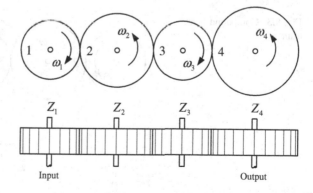

Fig. 9.2 From *left* to *right*, driving gear, idle gears and driven gear in a simple gear train

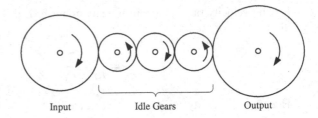

Input   Idle Gears   Output

9.2.2 Compound Trains

Compound trains have more than one gear in, at least, one of the shafts. Figure 9.3 shows an example of this. The velocity ratio of a compound gear train can be obtained by analyzing the velocity ration of each pair of meshed gears (Eq. 9.8).

The velocity ratio of the train shown in Fig. 9.3 can be calculated as:

$$\mu = \frac{\omega_{in}}{\omega_{out}} = \frac{\omega_1}{\omega_5} = -\frac{Z_2 Z_3 Z_5}{Z_1 Z_2 Z_4} = -\frac{Z_3 Z_5}{Z_1 Z_4} \tag{9.8}$$

Thus, the velocity ratio is equal to the quotient of the product of the number of teeth of the driven wheels and the product of the number of teeth of the driver ones (Eq. 9.9). It is affected by a minus sign whenever the direction of rotation of the output is contrary to the one of the input:

$$\mu = \frac{\text{product of number of teeth on driven gears}}{\text{product of number of teeth on driver gears}} \tag{9.9}$$

The number of teeth of gear 2 does not appear in final (Eq. 9.8) as it is a driver and driven wheel at the same time. In Fig. 9.3 we can see that gear 2 is driven by

Fig. 9.3 Compound gear train

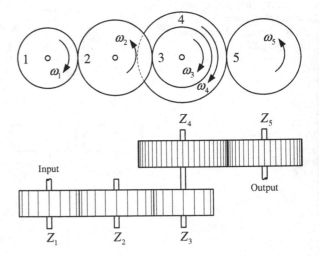

Fig. 9.4 Reverted gear train

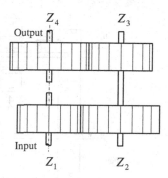

gear 1 and it drives gear 3. So, gear 2 is an idle gear and has effect on the turning direction but not on the velocity ratio.

The train value (Eq. 9.10) of the train shown in Fig. 9.3 will be the inverse of Eq. (9.8). Again, the minus sign shows that the direction of the output rotation is opposite to the input one:

$$i = \frac{\omega_{out}}{\omega_{in}} = \frac{\omega_5}{\omega_1} = -\frac{Z_1 Z_4}{Z_3 Z_5} \tag{9.10}$$

In general we can calculate the train value (Eq. 9.11) as:

$$i = \frac{\text{product of number of teeth on driver gears}}{\text{product of number of teeth on driven gears}} \tag{9.11}$$

When the input and output gears are coaxial (Fig. 9.4), the train receives the name of reverted gear train, the velocity ratio being computed the same way as explained above.

In the case of the train in Fig. 9.4, the driver wheels arc 1 and 3 while the driven wheels are 2 and 4. The train value (Eq. 9.12) can be computed as:

$$i = \frac{\omega_{out}}{\omega_{in}} = \frac{\omega_4}{\omega_1} = \frac{Z_1 Z_3}{Z_2 Z_4} \tag{9.12}$$

9.2.3   Gearboxes

One of the main applications of gear trains in machines is to transmit torque to the output shaft with multiple velocity ratios. This is achieved by changing the pairs of engaged wheels.

The main application is, doubtlessly, gearboxes in motor vehicles. We will study how a three-speed plus reverse manual gearbox works. A gearbox with more speeds will operate the same way with the main difference being the number of gears in the train.

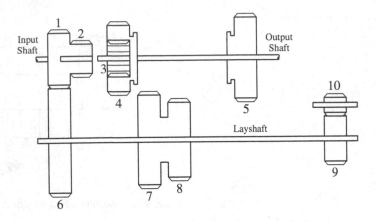

Fig. 9.5 Three-speed plus reverse manual gear box

Figure 9.5 shows a reverted gear train with three different shafts:

- The input shaft is connected to the engine crankshaft generally through the clutch.
- The layshaft or countershaft is the intermediate shaft that mounts several gears all of which turn jointly with it with opposite direction to the input shaft. All the gears on this shaft are idle gears.
- The output shaft transmits motion from the gearbox to the driver wheels via one or more differentials (see Sect. 9.3.3).

Gears 1 and 6 are always engaged (Fig. 9.5). This is what we call a "constant mesh" configuration. Therefore, whenever the input shaft rotates, so does the countershaft, even when no gear of the output shaft is meshed (that is, when it is in neutral position).

In order to transmit motion to the vehicle wheels, we need to manipulate the gear lever and engage the desired speed. Depending on the couple of gears meshed we will be able to reach more or less speed and at the same time more or less torque. To do so, the gear lever slides gears 4 or 5 through the output shaft, which is grooved, obtaining the following different combinations:

- Gears 1–6–8–5. First speed, the torque is maximum and the velocity is minimum.
- Gears 1–6–7–4. Second speed, torque and velocity are intermediate.
- Gears 2–3. Third speed or sometimes called, direct speed, since it connects the input and output shafts through clutch teeth. Thus, the engine output torque and turning speed are transmitted with a conversion ratio of 1:1. The torque transmission by engagement of interlocking teeth is not used in modern gearboxes.
- Gears 1–6–9–10–5. Reverse speed. Turning direction of the output shaft is reversed by idle gear 10.

There are other elements in manual gear boxes that have not been mentioned such as synchronizers. These elements are used to make the countershaft and output shaft turning speeds equal in order to facilitate the meshing operation of a couple of gears.

9.3   Planetary or Epicyclic Trains

In epicyclic gear trains at least one of the shafts changes its position with respect to the frame turning about another shaft. By using this type of gear train high speed ratios can be achieved with a few toothed wheels.

Epicyclic trains have two degrees of freedom. In other words, we need to know the motion of two different bodies in the train in order to be able to define the motion of the rest. In the train shown in Fig. 9.6, we can distinguish:

- A wheel called sun (gear 1). It can be fixed the same way as the one in Fig. 9.6 or move with known parameters.
- One or more gears that rotate around the sun. These gears are called planets or satellites (gears 3).
- A carrier arm for the planets that rotates around the sun (arm 2).

Two different methods are used to determine the speed ratio and turning direction of all gears and arms in a planetary gear train. These are the tabular method and the algebraic method (Willis formula) and they will be described in detail as follows.

9.3.1   Tabular Method

The main hypothesis of the tabular method is that the angular velocity of any gear is equal to the angular velocity of the arm plus the relative motion of the gear with respect to the arm.

A simple planetary gear train will be used to simplify the understanding of this method. Clockwise turns will be considered positive and counterclockwise negative.

Fig. 9.6 Planetary gear train

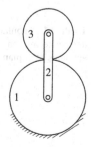

Fig. 9.7 Simple planetary
gear train

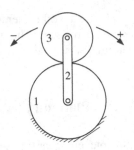

Gear 1 is fixed and meshed with gear 3. Both gears are held by carrier arm 2
which can rotate around the center of 1 (Fig. 9.7).

The application of the method starts by creating a table with as many columns as
the number of gears in the train plus the carrier arm and three rows. The real motion
of the train can be obtained as the superposition of two relative motions as follows:

- A—All gears are locked and move jointly with the arm.
- B—The arm is locked and motion of the gears can be studied as an ordinary
  train.
- C—The previous relative motions are added up to obtain the real motion of the
  gears and arm in the planetary train.

Hence, in the example shown in Fig. 9.7, we fill row A with the results obtained
when considering the system joined to the carrier arm as it rotates one turn in
positive direction (Table 9.1).

Row B shows the results of considering the arm fixed and making the sun gear
rotate one turn in opposite direction, that is, a negative turn. This way its angular
speed in section C is zero. Since the arm is considered fixed, if gear 1 rotates 1 turn
in negative direction, gear 3 will rotate Z_1/Z_3 in positive direction.

Row C is the sum of the rotations performed in rows A and B, the result being
the real motion of each of the wheels as well as the planetary arm.

The speed ratio between arm 2 and wheel 3 (Eq. 9.13) of the planetary train is:

$$\frac{\omega_2}{\omega_3} = \frac{1}{1 + \dfrac{Z_1}{Z_3}} \qquad (9.13)$$

Table 9.1 Tabular method to
determine motion of the
elements of a planetary gear
train

	Gear 1	Arm 2	Gear 3
A	1	1	1
B	−1	0	$\dfrac{Z_1}{Z_3}$
C	0	1	$1 + \dfrac{Z_1}{Z_3}$

9.3.2   Formula or Algebraic Method

We define the apparent velocity ratio, μ_a, as the velocity ratio between the gears of the planetary train with respect to the arm. In other words, the velocity ratio of an ordinary gear train resulting from locking the arm in the planetary train (Eq. 9.14).

$$\mu_a = \frac{\text{product of number of teeth on driven gears}}{\text{product of number of teeth on driver gears}} \tag{9.14}$$

Its value is negative whenever the input and output have opposite turning directions considering the gear system is working as an ordinary train.

The speed of the input and output gears in the planetary train minus the arm velocity will give us the input and output velocity with respect to the arm. Hence, the formula Eq. (9.15) that relates apparent velocity ratio μ_a with the velocity of the input and output gears as well as the arm velocity in the planetary train becomes:

$$\mu_a = \frac{\omega_{in} - \omega_{arm}}{\omega_{out} - \omega_{arm}} \tag{9.15}$$

Arm 2 of the train shown in Fig. 9.7 rotates at angular speed ω_2 while wheel 1 is fixed. To find the speed ratio between the arm and wheel 3 (Eq. 9.18), we compute Eqs. (9.16–9.17):

$$\mu_a = \frac{\omega_{12}}{\omega_{32}} = \frac{-Z_3}{Z_1} = \frac{\omega_1 - \omega_2}{\omega_3 - \omega_2} \tag{9.16}$$

Therefore:

$$\frac{0 - \omega_2}{\omega_3 - \omega_2} = \frac{-Z_3}{Z_1} \tag{9.17}$$

$$\frac{\omega_2}{\omega_3} = \frac{1}{1 + \dfrac{Z_3}{Z_1}} \tag{9.18}$$

9.3.3   Bevel Gear Differential

One of the applications of planetary trains is the differential, a device used mainly in vehicles with four or more wheels. Its main task is to adjust right and left wheel velocities when the vehicle is turning.

Figure 9.8 shows a differential formed by a crown gear (2) that is driven by the driver gear (1) or pinion. The crown also works as a carrier arm for the satellite gears (3) of the planetary train. Gears 4 and 5 are connected to the vehicle wheels by the driver or axle shafts.

Fig. 9.8 Bevel gear
differential

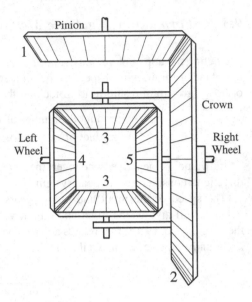

The differential will work differently depending on the vehicle trajectory:

- If the vehicle moves along a straight line, wheels 4 and 5 turn with the same speed as crown 2 and everything moves as a whole block.
- When the vehicle makes a turn, wheels 4 and 5 no longer turn at the same speed. The velocity of the inner wheel decreases while the one of the outside wheel increases. This is possible thanks to satellite gear 3 which starts rotating about its center while moving around toothed wheels 4 and 5. In this case the system behaves as a planetary gear train.

Hence, the differential allows wheels on different sides of a vehicle to have different speeds when making a turn. This is an essential device in gear transmission since if both wheels turned at the same speed, one of them would slip over the pavement.

We will study what happens when one of the wheels is on dry pavement and the other gear is on ice. In that case, the wheel on dry pavement will have much more resistance and only the wheel on ice will spin.

To can calculate the speed of the wheel on ice we will suppose that it is connected to gear 5 (right wheel). The whole turning motion will be transmitted to it while the left wheel stays locked. The speed of the wheel on ice can be obtained by using the tabular method (Table 9.2).

In Table 9.2, gear 4, which is locked, is the sun gear and the crown is the satellite carrier arm. Pinion 1 is not part of the epicyclic gear train. Instead, together with the crown, they are part of an ordinary gear train that we have to study in order to determine the motion of the arm. Knowing that gears 4 and 5 have the same

Table 9.2 Table to determine how many times wheel 5 turns for each revolution of the arm in the bevel gear differential shown in Fig. 9.8

	Arm 2 (Crown)	Gear 3	Gear 4	Gear 5
A	1	1	1	1
B	0	$\dfrac{Z_4}{Z_3}$	-1	$\dfrac{Z_4}{Z_3}\dfrac{Z_3}{Z_5}$
C	1	$1 + \dfrac{Z_4}{Z_3}$	0	$1 + \dfrac{Z_4}{Z_5}$

number of teeth, we can compute how many times wheel 5 turns for each revolution of the arm (row C for gear 5 in Table 9.2) (Eq. 9.19):

$$1 + \frac{Z_4}{Z_5} = 1 + 1 = 2 \tag{9.19}$$

That is, for every turn of the crown, wheel 5 spins twice in the same direction. So, when a stopped vehicle tries to initiate its motion while having one of its wheels on ice and the other on dry asphalt, the former will receive all traction and will spin at twice the speed of the crown while the wheel staying on dry asphalt remains static.

In order to avoid this problem, some cars install locking differentials. This type of differential has a small clutch between one of the wheel axles and the crown. When the vehicle is driving straight everything turns as a whole block and there is no slip between the clutch discs. As the vehicle starts turning there is a small slip between the discs. If this slip increases because one of the wheels loses traction, friction between the discs also increases and speed difference between the two gears and the crown is limited. Hence, the wheel still receives part of the torque.

In order to verify the correct operation of a differential, the left gear of a vehicle is turned at the same speed but opposite direction from the right one. If the differential works correctly the crown should not move. If it moves, the error in the differential can be measured from the rotation of the crown.

9.4 Examples

In this section we will carry out the kinematic analysis of different trains by applying the methods developed in this chapter up to now.

Example 1 Find the velocity ratio, μ, of the train shown in Fig. 9.9. Also find the direction of rotation of wheel M with respect to wheel A.

Data: $Z_1 = Z_4 = Z_{10} = Z_{13} = 15$, $Z_2 = 20$, $Z_3 = Z_7 = 30$, $Z_5 = 14$, $Z_6 = 45$, $Z_8 = Z_9 = Z_{11} = 25$ and $Z_{12} = 35$.

$$\mu = \frac{\omega_1}{\omega_{13}} = \frac{Z_2 Z_4 Z_6 Z_8 Z_9 Z_{11} Z_{13}}{Z_1 Z_3 Z_5 Z_7 Z_8 Z_{10} Z_{12}} = \frac{20 \cdot 15 \cdot 45 \cdot 25 \cdot 25 \cdot 25 \cdot 15}{15 \cdot 30 \cdot 14 \cdot 30 \cdot 25 \cdot 15 \cdot 35} = 1.27$$

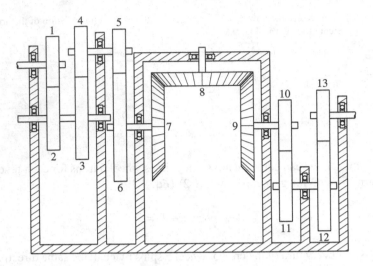

Fig. 9.9 Compound ordinary gear train

Gear 13 turns in the same direction as gear 1.

Example 2 Figure 9.10 shows a gear train with bevel, spur and worm gears. The pinion is installed on a shaft moved by a pulley and a v-belt. Pulley 1 spins at 1000 rpm in the direction specified (counterclockwise looking from left to right). Find the speed and turning direction of wheel 8.

Data: $D_1 = 100\,\text{mm}, D_2 = 200\,\text{mm}, Z_3 = 20, Z_4 = 40, Z_5 = 20, Z_6 = 50, Z_7 = 1$ and $Z_8 = 40$.

First we find the train transmission value (Eq. 9.20):

$$i = \frac{D_1 Z_3 Z_5 Z_7}{D_2 Z_4 Z_6 Z_8} = \frac{100 \cdot 20 \cdot 20 \cdot 1}{200 \cdot 40 \cdot 50 \cdot 40} = 0.0025 \qquad (9.20)$$

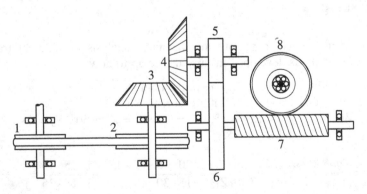

Fig. 9.10 Two pulleys connected by a belt transmit power to a compound ordinary gear train

Hence, the angular speed of wheel 8 (Eq. 9.21) is:

$$\omega_8 = i\omega_1 = 0.0025 \cdot 1000\,\text{rpm} = 2.5\,\text{rpm} \tag{9.21}$$

In order to better understand how the worm gear-crown system works we can compare it with a screw and a nut. When the head of the right-handed thread screw shown in Fig. 9.11a is turned clockwise, the nut moves upward closer to the head. Figure 9.11b shows a left-handed thread screw. When this screw is turned the same way, the nut moves downward, further from the head.

The gear-crown system in Fig. 9.10 works like the right-handed screw. Therefore, the teeth on crown 8 in contact with worm 7 will move upwards when the worm rotates in the same direction as gear 6. Therefore, gear 8 will rotate in a clockwise direction.

Example 3 Calculate the velocity ratio between arm 2 and gear 3 of the planetary train in Fig. 9.12. Gear 1 is stationary. Use the tabular and the algebraic method.
  Data: $Z_1 = 99$, $Z_3 = Z_5 = 100$ and $Z_4 = 101$.

Fig. 9.11 **a** Right-handed screw **b** left-handed screw

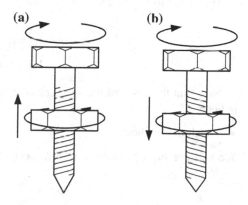

Fig. 9.12 Planetary gear train

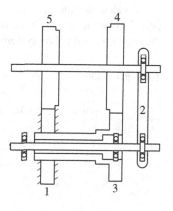

- Tabular method:

Since we have 4 gears plus a carrier arm, we need a table with 5 rows. We will use the sign criterion explained in Sect. 9.3.1.

- Row A: we consider that the whole system moves jointly with the arm while it makes one full revolution in positive direction.
- Row B: we consider that the arm is locked and study the train as an ordinary train knowing that wheel 1 turns one full revolution in the negative direction. This way, when adding up relative movements A and B of wheel 1, we get zero (stationary).
- Row C: we add up the relative motions in A and B to obtain the real velocity of each of the gears and the arm in the planetary train (Table 9.3).

Thus, the speed ratio (Eq. 9.22) we are looking for is:

$$\frac{\omega_2}{\omega_3} = \frac{1}{1 - \frac{Z_1 Z_4}{Z_5 Z_3}} \tag{9.22}$$

This leads to Eq. (9.23):

$$\frac{\omega_2}{\omega_3} = \frac{1}{1 - \frac{99 \cdot 101}{100 \cdot 100}} = 10,000 \tag{9.23}$$

Note that the value of the speed ratio in this planetary train with only four gears is huge.

- Algebraic method:

We will use Eq. (9.15) developed in Sect. 9.3.2:

$$\mu_a = \frac{\omega_{in} - \omega_{arm}}{\omega_{out} - \omega_{arm}} \tag{9.24}$$

To calculate apparent velocity ratio μ_a, we consider gear 1 to be the input and gear 3 to be the output in the ordinary gear train that results from locking the arm in the planetary train. The system is a compound gear train where gears 1 and 4 are the

Table 9.3 Table to determine how many times wheel 3 turns for each revolution of arm 2 in the planetary train shown in Fig. 9.12

	Gear 1	Arm 2	Gear 3	Gear 4	Gear 5
A	1	1	1	1	1
B	−1	0	$-\frac{Z_1 Z_4}{Z_5 Z_3}$	$\frac{Z_1}{Z_5}$	$\frac{Z_1}{Z_5}$
C	0	1	$1 - \frac{Z_1 Z_4}{Z_5 Z_3}$	$1 + \frac{Z_1}{Z_5}$	$1 + \frac{Z_1}{Z_5}$

driver wheels and gears 3 and 5 are the driven ones. This way, Eq. (9.24) remains as follows:

$$\frac{\omega_{12}}{\omega_{32}} = \frac{Z_5 Z_3}{Z_1 Z_4} = \frac{\omega_1 - \omega_2}{\omega_3 - \omega_2} \tag{9.25}$$

By substituting $\omega_1 = 0$ and $\omega_2 = 1$, we can calculate the speed of gear 3 (Eq. 9.26):

$$\omega_3 = 1 - \frac{Z_1 Z_4}{Z_5 Z_3} = 1 - \frac{99 \cdot 101}{100 \cdot 100} = 0.0001 \tag{9.26}$$

Therefore, the velocity ratio between arm 2 and the gear 3 (Eq. 9.27) is:

$$\frac{\omega_2}{\omega_3} = 10,000 \tag{9.27}$$

Example 4 In Fig. 9.13, find the speed of outer wheel 4 when gear 1 spins at 100 rpm clockwise and arm 2 rotates at 200 rpm counterclockwise. Use the tabular and the algebraic method.

Data: $Z_1 = 40, Z_3 = 20$ and $Z_4 = 80$.

• Tabular method:

In this case we have to consider that the arm rotates 200 times instead of just 1. So, in row A of Table 9.4, all gears are locked and rotate counterclockwise jointly with the arm 200 times.

In row B, the arm is locked and the motion of the gears can be studied as an ordinary train. In the planetary train, gear 1 is not steady but rotates 100 times in the opposite direction to the arm rotation. Hence, if the system rotates 200 times

Fig. 9.13 Planetary gear train

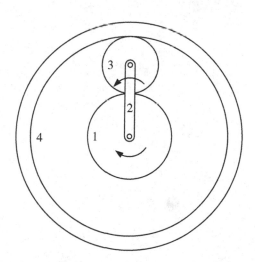

Table 9.4 Table to solve the planetary train shown in Fig. 9.13

	Gear 1	Arm 2	Gear 3	Gear 4
A	−200	−200	−200	−200
B	300	0	$-600 = 300\dfrac{Z_1}{Z_3}$	$-150 = -600\dfrac{Z_3}{Z_4}$
C	100	−200	−800	−350

counterclockwise, gear 1 has to rotate 200 times clockwise plus 100 times due to its own real motion.

Finally in row C the previous relative motions are added up to obtain the real motion in the planetary train. The results are shown in Table 9.4.

So, the speed of outer wheel 4 is 350 rpm counterclockwise when arm 2 rotates at 200 rpm counterclockwise and gear 1 rotates at 100 rpm clockwise.

- Algebraic method:

We can calculate the velocity ratio of gears 1 and 4 with respect to the arm (Eq. 9.28) as:

$$\frac{\omega_{12}}{\omega_{42}} = -\frac{Z_3 Z_4}{Z_1 Z_3} = \frac{\omega_1 - \omega_2}{\omega_4 - \omega_2} \tag{9.28}$$

By substituting $Z_1 = 40, Z_4 = 80, \omega_1 = 100$ rpm and $\omega_2 = -200$ rpm, we can calculate the speed of gear 4 (Eq. 9.29):

$$\frac{100\,\text{rpm} + 200\,\text{rpm}}{\omega_4 + 200\,\text{rpm}} = -\frac{80}{40} \Rightarrow \omega_4 = -350\,\text{rpm} \tag{9.29}$$

So, wheel 4 rotates at 350 rpm counterclockwise.

Chapter 10
Synthesis of Planar Mechanisms

Abstract Design in Mechanical Engineering must give answers to different requests that are based on two cornerstones, analysis and synthesis of mechanisms. Analysis allows determining whether a given system will comply with certain requirements or not. Alternatively, synthesis is the design of a mechanism so that it complies with previously specified requirements. For instance, mechanism synthesis allows finding the dimensions of a four-bar mechanism in which the output link generates a desired function with a series of precision points or a mechanism in which a point follows a given trajectory. Therefore, synthesis makes it possible to find the mechanism with a response previously defined. Currently, there is a set of methods and rules that makes it possible to find the solution to many mechanism design problems. However, since this is quite a newly-developed discipline, there are still many problems that need to be solved. The concept of synthesis was defined in Chap. 1 as follows: synthesis refers to the creative process through which a model or pattern can be generated, so that it satisfies a certain need while complying with certain kinematic and dynamic constraints that define the problem (Fig. 10.1). Other definitions can be added but all of them will somehow express the idea of creating mechanisms that can carry out a certain type of motion or, in a more general way, mechanisms that comply with a set of given requirements. There are several classifications for different types of synthesis but, basically, most authors agree on grouping the synthesis of mechanisms in two main branches: structural synthesis and dimensional synthesis.

10.1 Types of Synthesis

There are several classifications for different types of synthesis but, basically, most authors agree on grouping the synthesis of mechanisms in two main branches: structural synthesis and dimensional synthesis (Fig. 10.1).

© Springer International Publishing Switzerland 2016
A. Simón Mata et al., *Fundamentals of Machine Theory and Mechanisms*,
Mechanisms and Machine Science 40, DOI 10.1007/978-3-319-31970-4_10

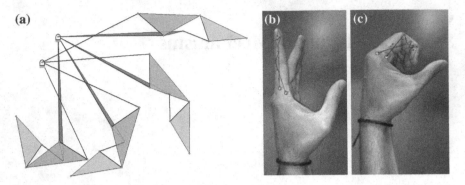

Fig. 10.1 **a** Six-bar linkage with one degree of freedom that can be used to build a finger prosthesis. Overlapping image of the mechanism on a real finger in two different positions (**b** and **c**)

10.1.1 *Structural Synthesis*

This synthesis deals with the topological and structural study of mechanisms. It only considers the interconnectivity pattern of the links so that the results are unaffected by the changes in the geometric properties of the mechanisms.

Structural synthesis includes the following:

- Synthesis of type or Reuleaux synthesis:
  Once the requirements have been defined the following questions are answered: What type of mechanism is more suitable? What type of elements will it be made of? Can it be formed by linkages, gears, flexible elements or cams? (Fig. 10.2) Different configurations are developed according to the pre-established requirements. The criteria to value the different characteristics of the mechanism are set.

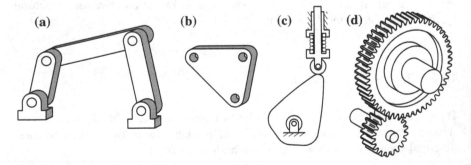

Fig. 10.2 Examples of elements that can be considered in synthesis of type. From *left* to *right*, **a** four bar linkage, **b** tertiary link, **c** cam and **d** pair of gears

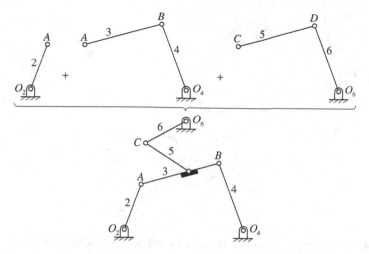

Fig. 10.3 Mechanism generation by addition of two RRR-type dyads

- Synthesis of number or Grübler synthesis:
  In the case of a linkage, it determines the number of links and their configuration. It works with concepts that were defined in the first chapters of this book such as link and link types, kinematic pair, kinematic chain, mechanism, inversion, degree of freedom and mobility criteria among others.

There are several methods to obtain new mechanisms. We will explain one of the most commonly used: the method of dyad addition. We call a group of elements formed by two links and three joints, a dyad. The two links and one of the joints form a kinematic pair and the other two joints connect the pair to two another links.

We can name the different types of dyads according to their joints. For example, RRR-type dyads shown in Fig. 10.3 have three rotational joints. In both cases one of the links is joined to the frame while the other one is joined to a point on a link that is moving. Figure 10.3 shows how new mechanism $\{O_2, A, B, O_4, C, D, O_6\}$ is generated by addition of these two dyads. The first RRR dyad, defined by $\{A, B, O_4\}$, is joined to point A on crank $\{O_2, A\}$ and to point O_4 on the frame. The second RRR dyad, defined by $\{C, D, O_6\}$, is joined to point C on dyad $\{A, B, O_4\}$ and to point O_6 on the frame.

Figure 10.4 shows an RRP-type and an RPR-type dyad. These dyads have two rotational joints and one prismatic joint but in the first case the dyad is formed by a rotational pair and in the second case by a prismatic one. In Fig. 10.4, it can be seen how new mechanism $\{O_2, A, B, C, D, O_6\}$ is generated by addition of dyads. Dyads RRP, defined by $\{A, B, frame\}$, and RPR, defined by $\{C, D, O_6\}$, are added to crank $\{O_2, A\}$.

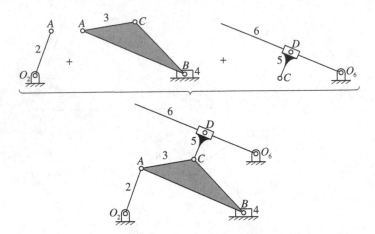

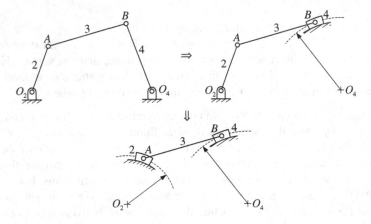

Fig. 10.4 Mechanism generation by addition of an RRP-type and an RPR-type dyads

Fig. 10.5 Mechanism generation by kinematic equivalence

We can also obtain new mechanisms by equivalence kinematics. Figure 10.5 shows how new mechanisms are generated starting from a four-bar linkage and substituting rotation pairs for prismatic ones.

Another commonly used method is the one called degeneration kinematics in which, when we make the length of links 2 and 4 infinite in a four bar mechanism, original rotational pairs 12 and 14 turn into prismatic pairs. Figure 10.6a shows how four-bar mechanism $\{O_2, A, B, O_4\}$ is converted into a crank-shaft mechanism (Fig. 10.6b) or into a double-slide mechanism (Fig. 10.6c).

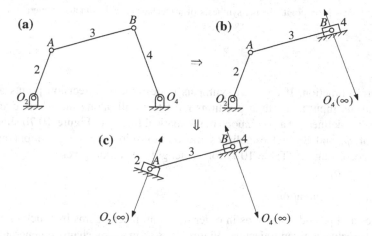

Fig. 10.6 Mechanism generation by kinematic degradation from four-bar (**a**) to crank-shaft (**b**) or into double-slide mechanism (**c**)

10.1.2 Dimensional Synthesis

It focuses on the problem of obtaining the dimensions of a predefined mechanism that has to comply with certain given requirements. It will be necessary to define the dimension of the links and the position of the supports, among others.

Dimensional synthesis can be divided into:

- Function generation.

Pre-established conditions refer to the relation between the input and output motions. These are defined by variables φ nd ψ that identify their positions (Fig. 10.7a). These parameters, normally used in synthesis of mechanisms, are equivalent to angles θ_2 and θ_4 used so far in this book. The relationship between φ and ψ can be defined by means of (Table 10.1) in which n pairs of these values are specified. These pairs can be set manually or according to a mathematical function. In this case, a series of precision points is used to generate the mechanism with exact correspondence between points or with a maximum error measured by means

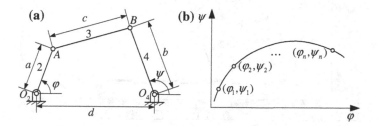

Fig. 10.7 a Four-bar mechanism. **b** Relationship between the input and the output

Table 10.1 Precision points for the synthesis of a mechanism with function generation

φ	φ_1	φ_2	...	φ_n
ψ	ψ_1	ψ_2	...	ψ_n

of an error function. If an exact synthesis is desired, the precision points are not used and the input and output positions have to fulfill, along the whole cycle, the relationship defined by a continuous mathematical function. Figure 10.7b shows the values of φ_i and ψ_i that establishes the relationship between the input and the output according to Table 10.1 for a length of links given $(r_1 = d, r_2 = a, r_3 = c, r_4 = b)$.

- Trajectory generation:

It studies and provides methods in order to obtain mechanisms in which one of the points describes a given trajectory. Figure 10.8 shows a mechanism generated with the WinMecC program. The desired trajectory has been defined by means of eight points. The trajectory followed by a point on the coupler of the obtained mechanism is drawn. The arrows show the error vectors between the points on the obtained trajectory and the desired one.

In general, these problems can be solved with three different types of methods:

- Graphical methods. These methods are very didactic and help us to understand the problem in an easy way. However, they offer a limited range of possibilities.
- Analytical methods. They solve the problem by means of mathematical equations based on the requirements.

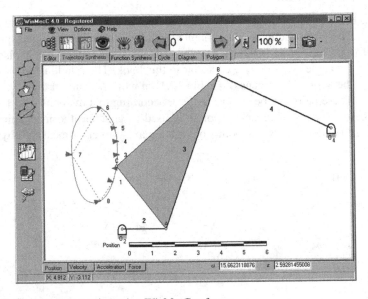

Fig. 10.8 Trajectory generation using WinMecC software

– Optimization-technique-based methods. They can find the optimal solution to the problem by means of the minimization of an objective function and the establishment of a series of restrictions. Different optimization techniques can be used.

10.2   Function Generation Synthesis

It can be defined as the part of synthesis that studies how the position of the input and output links in a mechanism relate to each other. In particular, when using dimensional synthesis, the dimensions of a mechanism are worked out from the imposition of such relationships.

The easiest way to set these relationships is by defining a univocal relationship between a series of n precision points, P_j, of the input link and the correspondent ones of the output link. Another option is to define a continuous function that relates the position of both links, $\psi = f_d(\varphi)$. In both cases, an error function $e(\varphi)$ can be defined between desired, $\psi = f_d(\varphi)$, and generated, $\psi = f_g(\varphi)$, functions to keep the error value within a certain limit (Fig. 10.9).

Function generation can be used to design mechanisms that carry out mathematical operations: addition, differentiation, integration or a combination of them. The first computers were mechanical devices based on this type of mechanisms.

In the mechanism in (Fig. 10.10), the relationship between the positions of links 4 and 6 is defined by mathematical function $\psi = \varphi^2$.

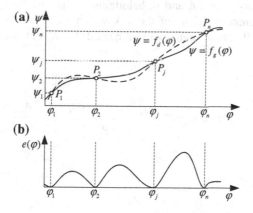

Fig. 10.9   a Precision points $P_1, P_2 \ldots P_n$ on the desired $\psi = f_d(\varphi)$ and generated $\psi = f_g(\varphi)$ functions. **b** Error function

Fig. 10.10 Mechanical computing mechanism to calculate the square of a number

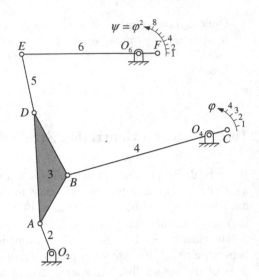

10.2.1 Graphical Methods

As in other areas of the Theory of Mechanisms, different graphical methods have been developed to solve the problem of function generation synthesis. The main benefits of these methods have already been mentioned: they are simple and highly educational. The designer can visually see the procedure and check the result.

In this section, we will explain a method based on the properties of the pole. With this method, in order to obtain the planar motion of a link we have to follow the next steps (Fig. 10.11):

1. Identify any two points, A and B, belonging to the link.
2. Take two different positions of the link which will give us two positions for A and B as well. These points will be named A_1, B_1 and A_2, B_2.

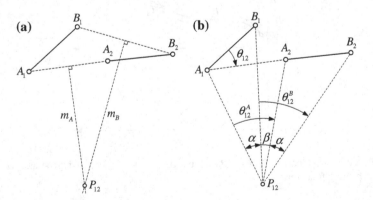

Fig. 10.11 **a** Pole P_{12} at perpendicular bisector intersections. **b** *Triangles $A_1P_{12}B_1$ and $A_2P_{12}B_2$ rotate with angle $\theta_{12} = \theta_{12}^A = \theta_{12}^B$*

Fig. 10.12 Dimensional
synthesis of a four-bar
mechanism in which rotated
angle of output ψ_{12}
corresponds to input rotated
angle φ_{12}

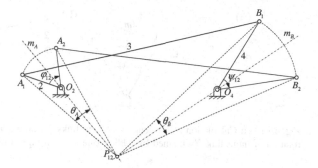

3. Take segments $\overline{A_1A_2}$ and $\overline{B_1B_2}$ and draw their perpendicular bisectors m_A and m_B.
4. Pole P_{12} is at their intersection point (Fig. 10.11a).
5. The link can be moved from position 1 to position 2 by carrying out a turn with angle $\theta_{12} = \theta_{12}^A = \theta_{12}^B$. Since triangles $A_1P_{12}B_1$ and $A_2P_{12}B_2$ in Fig. 10.11b are congruent, we can write (Eq. 10.1):

$$\left.\begin{array}{l}\theta_{12}^A = \alpha + \beta \\ \theta_{12}^B = \beta + \alpha\end{array}\right\} \Rightarrow \theta_{12}^A = \theta_{12}^B = \theta_{12} \tag{10.1}$$

Although the pole has other properties, the latter is enough to develop a graphical method of function generation synthesis. As an example, we will generate a four-bar mechanism in which a rotated angle of the input link between positions 1 and 2, φ_{12}, corresponds to a rotated angle of output link ψ_{12}.
We will follow the next steps (Fig. 10.12):

1. We take an arbitrary point and name it pole P_{12}. We draw line m_A and any point O_2 in it.
2. Next we choose a value for the length of the crank and, taking point O_2 as the origin, we draw points A_1 and A_2 in symmetric positions with respect to line m_A. The angle formed by the points and the origin, has to be equal to specified angle φ_{12}.
3. Points A_1, A_2 and P_{12} are connected so that they form angle $\theta_A = \widehat{A_1P_{12}A_2}$.
4. Taking P_{12} as the origin, we draw a new arbitrary line, m_B, and take point O_4 in it.
5. Taking point P_{12} as the origin, we draw two lines that comply with two conditions, the angle they form is equal to $\theta_B = \widehat{B_1P_{12}B_2} = \theta_A$ and m_B is their bisector.
6. Taking point O_4 as the origin we draw two lines with the condition that the angle they form is equal to ψ_{12} and that m_B is their bisector.
7. The intersection points of these four lines define points B_1 and B_2.

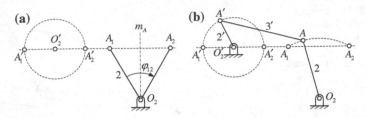

Fig. 10.13 a Calculation of the lengths of new links 2' and 3'. **b** Four-bar mechanism with full rotation of input link 2' obtained by adding one dyad to input link 2 of the original mechanism

8. The solution obtained is mechanism $\{O_2, A_1, B_1, O_4\}$ as well as its second position, given by $\{O_2, A_2, B_2, O_4\}$.

There are other interesting graphical methods like the Overlay one. This method, widely used, allows carrying out the dimensional synthesis of a four-bar mechanism by means of a trial and error procedure.

The results of these synthesis methods, as well as the results obtained with the trajectory generation synthesis methods that will be introduced further on in this chapter, are frequently four-bar mechanisms that do not comply with Grashoff's condition. That is, none of the links in the mechanism can carry out a full revolution. This condition is necessary if the mechanism has to be powered by a motor with continuous rotation.

To solve this problem, there is a method that adds an RRR dyad to the generated mechanism, so that a new input link, which can turn a full revolution, is defined. The steps to follow are:

1. Consider link 2 of the mechanism in Fig. 10.13a and its rotation interval φ_{12}. This is the input link for a mechanism that has already been synthesized. Draw segment $\overline{A_1A_2}$ as well as its perpendicular bisector, m_A.
2. Extend segment $\overline{A_1A_2}$ and draw point O_2' on it.
3. Draw a circle with its center in O_2' and its diameter equal to the length of segment $\overline{A_1A_2}$. This way we obtain points A_1' and A_2' (Fig. 10.13a).
4. Figure 10.13b shows the result, a four-bar mechanism in which link 2' can rotate fully and link 2 is a rocker.

10.2.2 Freudenstein's Method

This analytical method starts with the vector loop equation already used in Chap. 3 to solve kinematic problems by means of Raven's method (Fig. 10.14b). Although this method is valid for any type of mechanism we will develop its explanation by applying it to the four-bar mechanism shown in Fig. 10.14a.

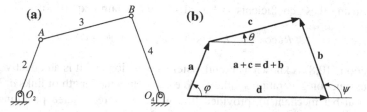

Fig. 10.14 a Four-bar mechanism. **b** Vector loop equation of the four-bar mechanism

Once we have defined the vectors in Fig. 10.14b, we can write its vector loop equation (Eq. 10.2) as

$$\mathbf{a}+\mathbf{b}=\mathbf{d}+\mathbf{b} \tag{10.2}$$

The projections on the ordinate and abscissa axis (Eq. 10.3) result are

$$\left.\begin{array}{l} a\cos\phi+c\cos\theta=d+b\cos\psi \\ a\sin\phi+c\sin\theta=b\sin\psi \end{array}\right\} \tag{10.3}$$

In order to find the relation $\psi=f(\varphi)$, we need to clear variable θ (Eq. 10.4) out of the system:

$$\left.\begin{array}{l} c\cos\theta=d+b\cos\psi-a\cos\phi \\ c\sin\theta=b\sin\psi-a\sin\phi \end{array}\right\} \tag{10.4}$$

Rising both equations to the second power, adding and remembering that (Eq. 10.5):

$$\cos(\psi-\phi)=\cos\psi\cos\phi+\sin\psi\sin\phi \tag{10.5}$$

We obtain (Eq. 10.6):

$$c^2=d^2+b^2+a^2+2bd\cos\psi-2ab\cos(\psi-\phi)-2da\cos\phi \tag{10.6}$$

Next, we define coefficients R_i (Eq. 10.7) as:

$$\left.\begin{array}{l} R_1=\dfrac{d}{a} \\[2mm] R_2=\dfrac{d}{b} \\[2mm] R_3=\dfrac{d^2+b^2+a^2-c^2}{2ab} \end{array}\right\} \tag{10.7}$$

Finally, using these coefficients in Eq. (10.6), we obtain (Eq. 10.8):

$$R_1 \cos \psi - R_2 \cos \phi + R_3 = \cos(\psi - \phi) \qquad (10.8)$$

Equation (10.8) is known as Freudenstein's equation and it is an effective tool to carry out function generation synthesis. We can obtain the length of links a, b, c and d in a four-bar mechanism, provided that we know three related positions of the input and output links. These positions are defined by pairs (φ_1, ψ_1), (φ_2, ψ_2) and (φ_3, ψ_3) which are known as precision points.

When these values are substituted in Freudenstein's equation, we obtain a system with three equations (Eq. 10.9) and three unknowns: R_1, R_2 and R_3:

$$\left. \begin{array}{l} R_1 \cos \psi_1 - R_2 \cos \phi_1 + R_3 = \cos(\psi_1 - \phi_1) \\ R_1 \cos \psi_2 - R_2 \cos \phi_2 + R_3 = \cos(\psi_2 - \phi_2) \\ R_1 \cos \psi_3 - R_2 \cos \phi_3 + R_3 = \cos(\psi_3 - \phi_3) \end{array} \right\} \qquad (10.9)$$

This system is linear and independent. It can easily be solved to obtain unknowns R_1, R_2 and R_3. Using the calculated values in the mathematical definitions of these parameters, we can find the length of links a, b, c and d by assigning an arbitrary value to one of them, for example, $d = 1$. In this case, the size of the mechanism obtained will depend on the value given to d, but it can be escalated to any size.

As said before, this method can be used with other type of mechanisms by following the same steps we have followed in the four-bar mechanism.

If we want to increase the number of precision points, we will also need to increase the number of unknowns in the equation. In the case of four precision points, we can use Freudenstein's equation before defining constants R_i. Hence the equation used has four unknowns: a, b, c and d. Substituting the values of the four precision points in Eq. (10.8), we obtain a system with four equations and four unknowns that can be easily solved.

10.3   Trajectory Generation Synthesis

In this part of the synthesis we will study the relationship between the trajectory described by a point in a link and the motion of another link, usually the input one.

Specifically in the case of dimensional synthesis we look for the dimensions of a given mechanism that complies with the condition that one of its points describes a certain trajectory. Based on these definitions, the pursued objective can be different.

The problem can be addressed from a general approach carrying out a dimensional synthesis of a given type of mechanism in which a certain trajectory is desired for a given point.

The problem can also be approached in a more specific way by carrying out the dimensional synthesis of a mechanism that complies with a certain condition such as a point following a trajectory with a straight segment, a circular arc or any other mathematical curve.

The most interesting problem and, hence, the most commonly addressed one is the synthesis of a mechanism in which the trajectory of a point is constrained to go over a number of precision points. In this case, the difference between the desired trajectory and the one obtained (Fig. 10.9a) is measured by defining an error function.

Several graphical, analytical and optimization-based methods have been proposed in order to solve this problem. We will develop some of them in the following sections.

It is very common to develop the different methods by applying them to a four-bar mechanism and paying special attention to the trajectory described by a point P of the coupler link. The trajectory curves that can be described by point P are called coupler curves and have been studied for a long time. They have been classified into different types depending on their geometry.

10.3.1 Graphical Methods

We will introduce two different methods and apply them to a four-bar mechanism. Due to their simplicity, these methods can be useful in those cases in which the conditions are restrained to two or three precision points.

10.3.1.1 Synthesis for Two Positions of the Coupler Link

The first method that can be used is based on the properties of the pole, which were already explained in this chapter. We can find the desired mechanism by following the next steps (Fig. 10.15):

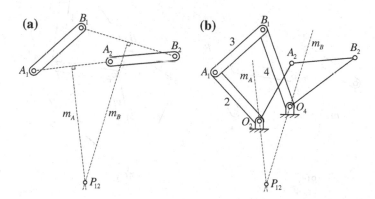

Fig. 10.15 a Defining pole position. **b** Synthesis for two positions of the coupler link

1. Two A_1B_1 and A_2B_2 positions of link AB are considered known.
2. We draw a segment between points A_1 and A_2. Then, we draw its perpendicular bisector, m_A.
3. We join points B_1 and B_2 with a new segment and we draw its perpendicular bisector, m_B.
4. The intersection point of these two bisectors defines the position of pole P_{12} (Fig. 10.15a).
5. We select an arbitrary point on m_A to define point O_2 and do the same to define point O_4 on m_B. The displacement of point A from position A_1 to A_2 can be considered a rotation about point O_2 and the displacement of point B from position B_1 to B_2 can be considered a rotation about point O_4
6. Segments O_2A and O_4B define the lengths of links 2 and 4 respectively (Fig. 10.15b). This way we obtain the mechanism we are looking for, $\{O_2, A, B, O_4\}$.

10.3.1.2   Synthesis for Three Positions of the Coupler Link

This method allows finding a four-bar mechanism in which the coupler link passes through the three specified positions (Fig. 10.16). The steps to follow are the next ones:

1. The three positions of link AB are considered known and they are identified as A_1B_1, A_2B_2 and A_3B_3.
2. We draw a segment between points A_1 and A_2 and then its perpendicular bisector $m_{A_{12}}$. The same way we draw a segment between points A_2 and A_3 and then its perpendicular bisector $m_{A_{23}}$.
3. The intersection point of both bisectors is point O_2 (Fig. 10.16a).

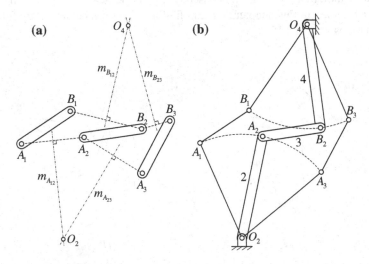

Fig. 10.16 a Defining fixed points O_2 and O_4. **b** Synthesis for three positions of the coupler link

4. We operate the same way drawing segments B_1B_2 and B_2B_3 with their perpendicular bisectors $m_{B_{12}}$ and $m_{B_{23}}$.
5. Their intersection point defines the position of point O_4.
6. Hence we obtain the mechanism we were looking for, $\{O_2, A, B, O_4\}$ (Fig. 10.16b).

10.3.2 Analytical Methods

The trajectory described by a point in a mechanism can be obtained as an analytical expression. In this section we will develop a generic example with a four-bar mechanism and then we will apply the three-precision point method to the same mechanism.

10.3.2.1 Trajectory Generation

In the four-bar mechanism shown in Fig. 10.17, a sextic equation of the following form (Eq. 10.10) can be formulated for the trajectory of point P:

$$F(x, y, a, b, d, g, f, \gamma) = 0 \qquad (10.10)$$

On one side, we define the x and y coordinates of point P by means of segments a and g (Fig. 10.17). On the other side, we define the same coordinates using segments d, b and f. This way we obtain the system of equations (Eq. 10.11):

$$\left.\begin{array}{l} x = a \cos \phi + g \cos(\alpha + \theta) \\ y = a \sin \phi + g \sin(\alpha + \theta) \\ x = d + b \cos \psi + f \cos(\alpha + \theta + \gamma) \\ y = b \sin\psi + f \sin(\alpha + \theta + \gamma) \end{array}\right\} \qquad (10.11)$$

Fig. 10.17 Parameters used to solve a trajectory synthesis problem of a four-bar mechanism

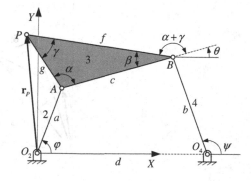

In the system of equations (Eq. 10.11) we want to remove variables φ and ψ. To do so, the first two equations are cleared in terms of $\cos \varphi$ and $\sin \varphi$. After rising the two equations to the second power and adding them, φ disappears.

Similarly, the third and fourth equations can also be cleared in terms of $\cos \psi$ and $\sin \psi$. Rising the two equations to the second power and adding them, variable ψ disappears.

Thus, we obtain (Eq. 10.12):

$$\left.\begin{array}{l} a^2 = x^2 + y^2 - 2gx\,\cos(\alpha+\theta) - 2gy\,\sin(\alpha+\theta) \\ b^2 = (x-d)^2 + y^2 + f^2 - 2f(x-d)\cos(\alpha+\theta+\gamma) \\ \quad -2fy\,\sin(\alpha+\theta+\gamma) \end{array}\right\} \qquad (10.12)$$

In order to obtain Eq. (10.10), we have to remove those terms depending on $\alpha + \theta$ from Eq. (10.12). In the final equation we will have six unknowns (a, b, d, g, f, γ) and two known values (x, y). Therefore, we can solve the problem with a maximum of six precision points.

Equation (10.10) can be used not only to represent the trajectory of a point on the coupler but also to compare it with the desired trajectory in order to calculate the error.

10.3.2.2 Trajectory Generation with Three Precision Points

A method based on complex numbers can be proposed to carry out the synthesis of a four-bar mechanism when we know the three different positions, P_1, P_2 and P_3, of the point of interest, P, on the coupler link (Fig. 10.18).

The position of the point of interest, P, can be defined by setting the following vector loops (Eqs. 10.13 and 10.14):

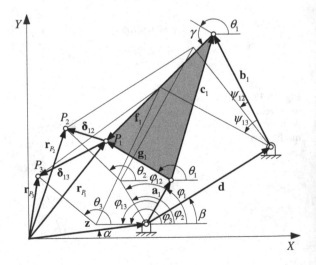

Fig. 10.18 Trajectory generation with three precision points

$$\mathbf{r}_{P_1} = \mathbf{z} + \mathbf{a}_1 + \mathbf{g}_1 \tag{10.13}$$

$$\mathbf{r}_{P_1} = \mathbf{z} + \mathbf{d} + \mathbf{b}_1 + \mathbf{f}_1 \tag{10.14}$$

Since they have no common unknowns, we can analyze each equation separately. We start with Eq. (10.13). Particularizing the expression for positions 1 and 2 yields (Eq. 10.15):

$$\left. \begin{array}{l} \mathbf{r}_{P_1} = z_1 e^{i\alpha} + a e^{i\varphi_1} + g e^{i\theta_1} \\ \mathbf{r}_{P_2} = z_1 e^{i\alpha} + a e^{i\varphi_2} + g e^{i\theta_2} \end{array} \right\} \tag{10.15}$$

In order to obtain displacement δ_{12} (Eq. 10.16), we subtract one equation from the other:

$$\delta_{12} = \mathbf{r}_{P_2} - \mathbf{r}_{P_1} = a(e^{i\varphi_2} - e^{i\varphi_1}) + g(e^{i\theta_2} - e^{i\theta_1}) \tag{10.16}$$

Equation (10.16) can be written as:

$$\delta_{12} = a e^{i\varphi_1}\left(\frac{e^{i\varphi_2}}{e^{i\varphi_1}} - 1\right) + g e^{i\theta_1}\left(\frac{e^{i\theta_2}}{e^{i\theta_1}} - 1\right) = \mathbf{a}_1(e^{i\varphi_{12}} - 1) + \mathbf{g}_1(e^{i\theta_{12}} - 1) \tag{10.17}$$

where angles φ_{12} and θ_{12} (Eq. 10.18) are:

$$\left. \begin{array}{l} \varphi_{12} = \varphi_2 - \varphi_1 \\ \theta_{12} = \theta_2 - \theta_1 \end{array} \right\} \tag{10.18}$$

Displacement δ_{13} (Eqs. 10.19 and 10.20) can be obtained by proceeding the same way:

$$\delta_{13} = \mathbf{r}_{P_3} - \mathbf{r}_{P_1} = a(e^{i\varphi_3} - e^{i\varphi_1}) + g(e^{i\theta_3} - e^{i\theta_1}) \tag{10.19}$$

Thus:

$$\delta_{13} = a e^{i\varphi_1}\left(\frac{e^{i\varphi_3}}{e^{i\varphi_1}} - 1\right) + g e^{i\theta_1}\left(\frac{e^{i\theta_3}}{e^{i\theta_1}} - 1\right) = \mathbf{a}_1(e^{i\varphi_{13}} - 1) + \mathbf{g}_1(e^{i\theta_{13}} - 1) \tag{10.20}$$

where angles φ_{13} and θ_{13} (Eq. 10.21) are:

$$\left. \begin{array}{l} \varphi_{13} = \varphi_3 - \varphi_1 \\ \theta_{13} = \theta_3 - \theta_1 \end{array} \right\} \tag{10.21}$$

Hence, we obtain the system of equations (Eqs. 10.22 and 10.23):

$$\delta_{12} = \mathbf{a}_1(e^{i\varphi_{12}} - 1) + \mathbf{g}_1(e^{i\theta_{12}} - 1) \tag{10.22}$$

$$\delta_{13} = \mathbf{a}_1(e^{i\varphi_{13}} - 1) + \mathbf{g}_1(e^{i\theta_{13}} - 1) \tag{10.23}$$

From which we can clear $\mathbf{a}_1$ (Eq. 10.24) and $\mathbf{g}_1$ (Eq. 10.25):

$$\mathbf{a}_1 = \frac{\begin{vmatrix} \delta_{12} & e^{i\theta_{12}} - 1 \\ \delta_{13} & e^{i\theta_{13}} - 1 \end{vmatrix}}{\begin{vmatrix} e^{i\varphi_{12}} - 1 & e^{i\theta_{12}} - 1 \\ e^{i\varphi_{13}} - 1 & e^{i\theta_{13}} - 1 \end{vmatrix}} \tag{10.24}$$

$$\mathbf{g}_1 = \frac{\begin{vmatrix} e^{i\varphi_{12}} - 1 & \delta_{12} \\ e^{i\varphi_{13}} - 1 & \delta_{13} \end{vmatrix}}{\begin{vmatrix} e^{i\varphi_{12}} - 1 & e^{i\theta_{12}} - 1 \\ e^{i\varphi_{13}} - 1 & e^{i\theta_{13}} - 1 \end{vmatrix}} \tag{10.25}$$

In order to obtain $\mathbf{b}_1$ and $\mathbf{f}_1$, we write Eq. (10.14) for positions 1 and 2 (Eq. 10.26):

$$\left. \begin{array}{l} \mathbf{r}_{P_1} = z e^{i\alpha} + d e^{i\beta} + b e^{i\psi_1} + f e^{i(\theta_1 + \gamma)} \\ \mathbf{r}_{P_2} = z e^{i\alpha} + d e^{i\beta} + b e^{i\psi_2} + f e^{i(\theta_2 + \gamma)} \end{array} \right\} \tag{10.26}$$

Displacement δ_{12} (Eqs. 10.27 and 10.28) is found by subtracting one expression from the other:

$$\delta_{12} = \mathbf{r}_{P_2} - \mathbf{r}_{P_1} = b(e^{i\psi_2} - e^{i\psi_1}) + f(e^{i(\theta_2 + \gamma)} - e^{i(\theta_1 + \gamma)}) \tag{10.27}$$

Thus:

$$\delta_{12} = b e^{i\psi_1}\left(\frac{e^{i\psi_2}}{e^{i\psi_1}} - 1\right) + f e^{i(\theta_1 + \gamma)}\left(\frac{e^{i\theta_2}}{e^{i\theta_1}} - 1\right) = \mathbf{b}_1(e^{i\psi_{12}} - 1) + \mathbf{f}_1(e^{i\gamma_{12}} - 1) \tag{10.28}$$

where angles ψ_{12} and θ_{12} (Eq. 10.29) are:

$$\left. \begin{array}{l} \psi_{12} = \psi_2 - \psi_1 \\ \theta_{12} = \theta_2 - \theta_1 \end{array} \right\} \tag{10.29}$$

Angle θ of vectors $\mathbf{g}_1$ and $\mathbf{f}_1$ is the same since they belong to the same link. Displacement δ_{13} (Eqs. 10.30 and 10.31) is obtained by following the same process:

$$\delta_{13} = \mathbf{r}_{P_3} - \mathbf{r}_{P_1} = b(e^{i\psi_3} - e^{i\psi_1}) + f(e^{i(\theta_3 + \gamma)} - e^{i(\theta_1 + \gamma)}) \qquad (10.30)$$

$$\delta_{13} = be^{i\psi_1}\left(\frac{e^{i\psi_3}}{e^{i\psi_1}} - 1\right) + fe^{i(\theta_1 + \gamma)}\left(\frac{e^{i\theta_3}}{e^{i\theta_1}} - 1\right) = \mathbf{b}_1(e^{i\psi_{13}} - 1) + \mathbf{f}_1(e^{i\theta_{13}} - 1)$$
$$(10.31)$$

where angles ψ_{13} and θ_{13} (Eq. 10.32) are:

$$\left.\begin{array}{l} \psi_{13} = \psi_3 - \psi_1 \\ \theta_{13} = \theta_3 - \theta_1 \end{array}\right\} \qquad (10.32)$$

Hence, we obtain the system of equations (Eq. 10.33):

$$\left.\begin{array}{l} \delta_{12} = \mathbf{b}_1(e^{i\psi_{12}} - 1) + \mathbf{f}_1(e^{i\theta_{12}} - 1) \\ \delta_{13} = \mathbf{b}_1(e^{i\psi_{13}} - 1) + \mathbf{f}_1(e^{i\theta_{13}} - 1) \end{array}\right\} \qquad (10.33)$$

From which $\mathbf{b}_1$ (Eq. 10.34) and $\mathbf{f}_1$ (Eq. 10.35) can be cleared:

$$\mathbf{b}_1 = \frac{\begin{vmatrix} \delta_{12} & e^{i\theta_{12}} - 1 \\ \delta_{13} & e^{i\theta_{13}} - 1 \end{vmatrix}}{\begin{vmatrix} e^{i\psi_{12}} - 1 & e^{i\theta_{12}} - 1 \\ e^{i\psi_{13}} - 1 & e^{i\theta_{13}} - 1 \end{vmatrix}} \qquad (10.34)$$

$$\mathbf{g}_1 = \frac{\begin{vmatrix} e^{i\psi_{12}} - 1 & \delta_{12} \\ e^{i\psi_{13}} - 1 & \delta_{13} \end{vmatrix}}{\begin{vmatrix} e^{i\psi_{12}} - 1 & e^{i\theta_{12}} - 1 \\ e^{i\psi_{13}} - 1 & e^{i\theta_{13}} - 1 \end{vmatrix}} \qquad (10.35)$$

In Eq. (10.36), vectors $\mathbf{z}$ and $\mathbf{d}$ can be deduced from Eqs. (10.13) and (10.14) and $\mathbf{c}_1$ can be obtained from the mechanism vector loop equation:

$$\left.\begin{array}{l} \mathbf{z} = \mathbf{r}_{P_1} - \mathbf{a}_1 - \mathbf{g}_1 \\ \mathbf{d} = \mathbf{r}_{P_1} - \mathbf{z} - \mathbf{b}_1 - \mathbf{f}_1 \\ \mathbf{c}_1 = \mathbf{d} + \mathbf{b}_1 - \mathbf{a}_1 \end{array}\right\} \qquad (10.36)$$

Therefore, we can obtain the dimensions of the links of a four-bar mechanism, when we know the three positions of a point, P, of the coupler link as well as the angles rotated by the links between the positions.

10.4 Optimal Synthesis of Mechanisms

Traditionally, engineers try to solve their design problems, particularly the problems of synthesis of mechanisms that concern us, by considering different alternatives with the intention of reaching the best solution. Thanks to the development of mathematical programming techniques and the rapid advances in computers and software, it is now possible to formulate the design problem as an optimization problem with the objective of minimizing an objective function provided that the design conditions are met. We can, thus, obtain the optimal solution.

In general, the solution of an optimization problem determines the value of the variables $(x_1, x_2, \ldots, x_n)$ that minimize objective function $f(x)$ subject to a set of constraints (Eq. 10.37). This can be written as:

$$\min f(x_1, x_2, \ldots, x_n)$$

Subject to:

$$h_j(x_1, x_2, \ldots, x_n) \leq 0 \quad j = 1, 2, \ldots, m$$

$$g_k(x_1, x_2, \ldots, x_n) = 0 \quad k = 1, 2, \ldots, p$$

$$(10.37)$$

Function $f(x)$ is called objective function and functions $h_j(x)$ and $g_k(x)$ are called constraints of the problem. We can have both inequality and equality constraints. In the context of engineering design, the above mentioned concepts are defined as:

- Objective function: A function that expresses a fundamental aspect of the problem. An extreme value (minimum or maximum) is sought along the process of optimization. This function is often called merit function. Multifunctional functions, in which several features are optimized, can also be formulated. In this case, each one of them is weighted depending on their importance.
- Independent design variables: Such variables represent the geometry of the model. They are usually the dimensions of the mechanism such as the length or width of the links.
- Dependent variables: These are parameters that have to be included in the formulation of the objective function or the constraints but that depend on the design variables.
- Constraints: They are mathematical functions that define the relationships between the design variables that have to be met by every set of values that define a possible design. These relationships can be of three types.

  - Inequality restrictions: They are usually limitations to the behavior of the mechanism or security restrictions to prevent failure under certain conditions.
  - Variable limits: They are a specific case of the previous ones.
  - Equality restrictions: They are conditions that have to be met strictly in order for the design to be acceptable.

Depending on the aspect being considered, design problems can be classified in different ways. The two most commonly used classifications are:

- Classification based on the existence of restrictions.

    - Unconstrained optimization problems.
    - Constrained optimization problems.

- Classification based on the nature of the equations (objective function and constraints).

    - Linear optimization problem, when both the objective function and the constraints are linear.
    - Nonlinear optimization problem, when the objective function and/or the constraints are nonlinear.

Based on the types of constraints that shape the synthesis problem, we face a nonlinear constrained problem that suggests a mathematical programming problem with nonlinear objective functions and nonlinear constraints.

10.5 Analysis of the Objective Function

Previously in this chapter, we defined the function generation synthesis as the establishment of a relationship between the input and output links in a mechanism. Hence, objective function $f(x)$ has to be developed for the particular mechanism to be studied.

10.5.1 Function Generation Synthesis

In this case we will study the function synthesis of a four-bar mechanism (Fig. 10.19) in which we will define a relationship between parameter ψ that defines the position of the output link and parameter φ of the input link.

Fig. 10.19 Four-bar mechanism with the parameters used in function generation synthesis

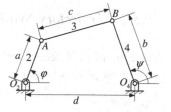

We will use Freudenstein's equation (Eq. 10.8) and the method developed in Appendix B of this book to obtain the relationship between the output angle, ψ, and input one, φ (Eq. 10.38):

$$\psi = 2 \arctan \frac{-B \pm \sqrt{B^2 - 4AC}}{2A} = \psi(a, b, c, d, \varphi) \qquad (10.38)$$

where coefficients A, B and C (Eq. 10.39) are:

$$\left. \begin{array}{l} A = \cos \varphi - R_1 - R_2 \cos \varphi + R_3 \\ B = -2 \sin \varphi \\ C = R_1 - (R_2 + 1) \cos \varphi + R_3 \end{array} \right\} \qquad (10.39)$$

where R_1, R_2 and R_3 are known functions of a, b, c, and d.

The objective function will be the minimization of the quadratic difference between the obtained output angle, ψ, and the desired one, ψ^d. If we calculate the values of the output angle by means of the input angle using (Eq. 10.38), the optimization problem (Eq. 10.40) can be formulated as:

$$\min \sum_{i=1}^{N} \left(\psi_i(X, \varphi_i^d) - \psi_i^d \right)^2$$

$$\qquad (10.40)$$

Subject to:

$$x_i \in [li_i, ls_i] \quad \forall \ x_i \in X = [a, b, c, d]$$

When input angles φ are unknown, these values become design variables in the optimization problem. This problem (Eq. 10.41) is more complex and can be formulated as:

$$\min \sum_{i=1}^{N} \left(\psi_i(X) - \psi_i^d \right)^2$$

$$\qquad (10.41)$$

Subject to:

$$x_i \in [li_i, ls_i] \quad \forall \ x_i \in X = [a, b, c, d, \varphi_1, \varphi_2, \ldots, \varphi_N]$$

10.5.2   Trajectory Synthesis

There are many different methods to measure the error between two curves. In this section we present two of these methods.

10.5.2.1 Mean Square Error

To explain this method, we will address the synthesis of a four-bar mechanism in which a point of the coupler has to pass through a series of positions previously established. The difference between these desired positions and the exact position of the point of the synthesized mechanism will give us the error.

To do so, we have to study the kinematics of the mechanism shown in Fig. 10.20. We need to know the trajectory of the point by means of its coordinates. We use a coordinate system with its origin in O_2 and the x-axis defined by line O_2O_4:

$$r^r_{C_x} = a \cos \varphi + r^n_{CA} \cos \theta - r^t_{CA} \sin \theta \qquad (10.42)$$

$$r^r_{C_y} = a \sin \varphi + r^n_{CA} \sin \theta + r^t_{CA} \cos \theta \qquad (10.43)$$

In Eqs. (10.42) and (10.43), we do not know the values for variable θ. Therefore, we have to express it in terms of known angle φ. This can be done by using the same approach that helped us to obtain Freudenstein's equation (Eq. 10.8) in Sect. 10.2.2 but removing variable ψ instead of θ:

The loop vector (Eq. 10.44) is:

$$\mathbf{a} + \mathbf{c} = \mathbf{d} + \mathbf{b} \qquad (10.44)$$

Projecting the vectors on the X_r and Y_r axis yields (Eq. 10.45):

$$\left. \begin{array}{l} a \cos \varphi + c \cos \theta = d + b \cos \psi \\ a \sin \varphi + c \sin \theta = b \sin \psi \end{array} \right\} \qquad (10.45)$$

In order to find the sought relationship, $\theta = \theta(\varphi)$, we need to remove variable ψ from Eq. (10.45).

Then, we can write the coordinates of point C in terms of the absolute coordinate system $\{O, X, Y\}$ (Eq. 10.46) by multiplying the relative coordinates of point C by the rotation matrix and adding the absolute coordinates of O_2:

Fig. 10.20 Trajectory synthesis of a four-bar mechanism

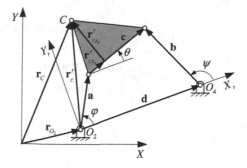

$$\begin{pmatrix} r_{C_x} \\ r_{C_y} \end{pmatrix} = \begin{pmatrix} r_{O_{2x}} \\ r_{O_{2y}} \end{pmatrix} + \begin{pmatrix} \cos\beta & -\sin\beta \\ \sin\beta & \cos\beta \end{pmatrix} \begin{pmatrix} r^r_{C_x} \\ r^r_{C_y} \end{pmatrix} \tag{10.46}$$

Thus, we can express the coordinates of the coupler point in terms of the absolute coordinate system and the design variables are $a, b, c, d, r^r_{C_x}, r^r_{C_y}, \beta, r_{O_{2x}}$ and $r_{O_{2y}}$ for input angle φ.

The goal is to minimize the mean square error between the positions of the point of the coupler and the desired positions (Eq. 10.47). This is expressed as:

$$\min \frac{1}{N} \sum_{i=1}^{N} \left((r^d_{C_x,i} - r_{C_x,i})^2 + (r^d_{C_y,i} - r_{C_y,i})^2 \right) \tag{10.47}$$

where N is the number of desired points.

Once the objective function has been set, we have to add certain constraints that have to be met. In this example, they are the following:

- The values of the design variables have to be within a range.
- It is desirable that four-bar mechanisms have at least one link that can fully rotate, for which Grashoff's condition has to be met.
- Input angles, φ_i, have to be in sequential order. This means that the trajectory points have to be reached as a positive or negative sequence of the input angles.

We can include the last two restrictions in the objective function as $h_1(X)$ and $h_2(X)$ multiplied by constants M_1 and M_2 (Eq. 10.48). The value of these constants is high so that the objective function is penalized if they are not met.

Hence, the function remains:

$$\min \frac{1}{N} \sum_{i=1}^{N} \left((r^d_{C_x,i} - r_{C_x,i}(X))^2 + (r^d_{C_y,i} - r_{C_y,i}(X))^2 \right) + M_1 h_1(X) + M_2 h_2(X)$$

Subject to:

$$x_i \in [li_i, ls_i] \quad \forall \; x_i \in X = [a, b, c, d, r^r_{C_x}, r^r_{C_y}, \beta, r_{O_{2x}}, r_{O_{2y}}, \varphi_1, \varphi_2, \ldots, \varphi_N]$$
$$\tag{10.48}$$

where:

$$h_1(X) = \begin{cases} 0 & \text{if } X \text{ meets Grashoff's condition} \\ 1 & \text{otherwise} \end{cases}$$
$$h_2(X) = \begin{cases} 0 & \text{if } X \text{ meets } \varphi_i \text{ in sequential order condition} \\ 1 & \text{otherwise} \end{cases}$$

10.5.2.2 Turning Function

An intrinsic equation of a curve defines the curve by using properties that do not depend on its location and orientation. Therefore an intrinsic equation defines the curve without specifying its position with respect to an arbitrary coordinate system as we did in the previous section.

Commonly used intrinsic quantities are arc length s, tangential angle θ, curvature κ or radius of curvature ρ. Torsion τ is also used for 3-dimensional curves. Some examples of intrinsic equations are the natural equation that defines a curve by its curvature and torsion, the Whewell equation obtained as a relation between the arc length and the tangential angle and the Cesàro equation obtained as a relation between the arc length and the curvature.

The use of intrinsic properties greatly simplifies the problem. In this section we will use a slightly different definition of the Whewell equation.

First, we define turning function $\Theta_F(s)$ as the relation between the tangential angle and the normalized arc length. This function measures tangential angle θ of shape F at point P, as a function of the normalized arc, $s = l_P/L$, where l_P is the distance along the shape anti-clockwise from reference point O to point P (Fig. 10.21a) and L is the perimeter of the curve. If shape F is convex, its turning function $\Theta_F(s)$ will be monotone increasing. If F is a closed curve, the turning function verifies $\Theta_F(s+n) = \Theta_F(s)+2\pi n$, where n represents the number of cycles or complete turns along the curve.

When the path is polygonal, the turning function is piecewise-constant, with steps corresponding to the vertices of shape F. Figure 10.21a shows polygonal shape F with four vertices and in Fig. 10.21b its turning function is shown.

Some general properties of turning functions, which are also applicable to non-polygonal shapes, are the following:

- If reference point O is changed to position O^*, the turning function shifts horizontally. The normalized arc distance between O and O^*, which can be calculated as $\Delta s_O = \Delta l_O/L$, gives us the shift value where Δl_O is the distance along

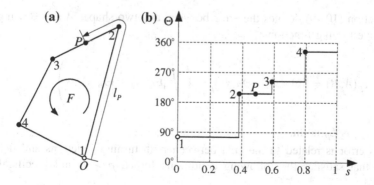

Fig. 10.21 a Polygonal shape F. b Turning function of shape F

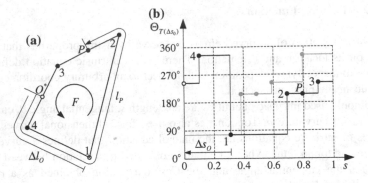

Fig. 10.22 **a** Polygonal shape F with another reference point O^*. **b** Shifted turning function

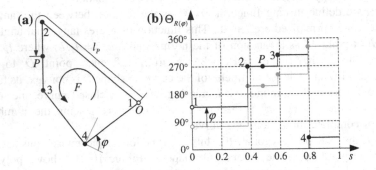

Fig. 10.23 **a** Polygonal shape F rotates angle φ. **b** Rotated turning function

the shape anti-clockwise from reference point O to new reference point O^*. The new turning function, $\Theta_{T(\Delta s_0)}(s) = \Theta(s + \Delta s_0)$, is shown in Fig. 10.22b.

- If curve F is rotated with affine transformation $R(\varphi)$ defined by means of angle φ, the turning function shifts vertically. The new turning function, $\Theta_{R(\varphi)}(s) = \Theta(s) + \varphi$, is shown in Fig. 10.23b.

Equation (10.49) defines the error between any two shapes A and B using their respective turning functions:

$$e_p(A, B) = \|\Theta_A(s) - \Theta_B(s)\|_p = \left(\int_0^1 |\Theta_A(s) - \Theta_B(s)| ds \right)^{\frac{1}{p}} \qquad (10.49)$$

where $\| \, \|_p$ is L_p norm.

This error is related to the area between both turning functions and defines a metric that satisfies the following conditions for all $p > 0$, and specifically for $p = 2$:

- $e_p(A, B) \geq 0$, it is positive everywhere.
- $e_p(A, B) = 0$ if and only if $A = B$, it has the identity property.
- $e_p(A, B) = e_p(B, A)$, it is symmetric.
- $e_p(A, B) + e_p(B, C) \geq e_p(A, C)$, it also obeys the triangle inequality.

Therefore, the smaller the error is, the more similar shapes A and B are. However, relative rotation between both shapes and the reference point used in each one affect the error. To avoid this drawback, the minimum error of function (Eq. 10.49) for any angle φ and normalized arc distance Δs can be found using equation (Eq. 10.50).

$$
\begin{aligned}
e_p(A, B) &= \min_{\Delta s \in R \varphi \in [0, 2\pi]} \left(\int_0^1 |\Theta_A(s + \Delta s) - \Theta_B(s) + \varphi| ds \right)^{\frac{1}{p}} \\
&= \min_{\Delta s \in R \varphi \in [0, 2\pi]} (g(\varphi, \Delta s))^{\frac{1}{p}}
\end{aligned}
\tag{10.50}
$$

When $p = 2$, $e_2(A, B)$ is a quadratic function of φ for any fixed Δs value.

The value for function $g(\varphi, \Delta s)$ (Eq. 10.51) is obtained from equation (Eq. 10.50):

$$
g(\varphi, \Delta s) = \int_0^1 |\Theta_A(s + \Delta s) - \Theta_B(s) + \varphi|^2 ds
\tag{10.51}
$$

The minimum value of $g(\varphi, \Delta s)$ with respect to parameter φ (Eq. 10.52) can be determined as:

$$
\begin{aligned}
\min_{\varphi \in [0, 2\pi]} g(\varphi, \Delta s) &= \int_0^1 |\Theta_A(s + \Delta s) - \Theta_B(s)|^2 ds - \varphi_{opt}^2 \\
\varphi_{opt}(\Delta s) &= \int_0^1 \Theta_B(s) ds - \int_0^1 \Theta_A(s) ds - 2\pi \Delta s
\end{aligned}
\tag{10.52}
$$

where φ_{opt} is the optimum value that shifts turning function $\Theta_A(s + \Delta s)$ vertically. This way, the minimization problem of $e_2(A, B)$ (Eq. 10.53) becomes:

$$
\begin{aligned}
e_2(A, B) &= \min_{\Delta s \in R \varphi \in [0, 2\pi]} \sqrt{\int_0^1 |\Theta_A(s + \Delta s) - \Theta_B(s) + \varphi|^2 ds} \\
&= \min_{\Delta s \in R} \sqrt{\min_{\varphi \in [0, 2\pi]} g(\varphi, \Delta s)}
\end{aligned}
\tag{10.53}
$$

In this example, curves A and B are polygonal shapes defined by their vertex coordinates. Consequently, turning functions $\Theta_A(s)$ and $\Theta_B(s)$ are defined by means of their coordinates $\Theta(s) = \{(s^1, \varphi^1), (s^2, \varphi^2), \ldots, (s^n, \varphi^n)\}$, where n is the number of vertexes of the polygonal shape (Fig. 10.21b). Parameter $\Delta s = s_A^i - s_B^j$ can only take the values of the horizontal shifts that make discontinuities of turning functions $\Theta_A(s + \Delta s)$ and $\Theta_B(s)$ have the same horizontal coordinate. This set of values has an infinite size, depending on the turning function coordinates and their distribution along the interval [0, 1].

With this approach to the optimal trajectory synthesis problem, objective function $f(x)$, is $e_2(A, B)$.

10.6   Optimization Method Based on Evolutionary Algorithms

The method proposed in this section is based on evolutionary algorithms which have proved to be successful to solve optimization problems. These algorithms are based on the mechanisms of natural selection and the laws of natural genetics.

The main benefit of this method stems from the simple implementation of the algorithms, its low computational cost and the absence of needing to know if the search of space is continuous, differentiable or other mathematical constraints that are required in traditional search algorithms.

The evolutionary methods start by generating an initial population which evolves by means of natural mechanisms into new populations that improve the objective function. In this initial population, which can be created randomly, each individual ("chromosome") is a solution to the problem. This individual will consist of parameters ("genes") that represent the design variables that are used to define a problem.

In our case, we represent each individual (Eq. 10.54) as vector $\mathbf{X}$ with n real variables $x_1, x_2, \ldots, x_n$:

$$\mathbf{X} = [x_1, x_2, \ldots, x_n] \quad x_i \in \mathbb{R} \tag{10.54}$$

Objective function $f()$ of the problem to be solved, allows evaluating fitness $f(\mathbf{X})$ of each individual. The optimal solution to the problem will be the individual with minimal fitness value.

The population has to evolve towards populations in which the individuals are better, that is, their fitness values are lower. This is achieved through natural selection, reproduction, mutation and other genetic operators. These operators are implemented as follows:

Selection consists of choosing two individuals of the population which will form a couple for reproduction. This selection can be performed in various ways, the simplest one being purely random with uniform probability distribution. Selection can also be random but applying a weight to each individual depending on the value obtained in the objective function (fitness). In this case, the probability distribution is not uniform but the best individuals are more likely to be chosen.

In our case, we will form couples for reproduction by choosing the best individuals plus another two individuals picked up randomly with uniform distribution of probability in the entire population. These elements will form vector **V** (disturbed vector) (Eq. 10.55) as:

$$\mathbf{X}_i : i \in [1, NP]$$
$$\mathbf{V} = \mathbf{X}_{Best} + F(\mathbf{X}_{r1} - \mathbf{X}_{r2}) \tag{10.55}$$

where $\mathbf{X}_{Best}$ is the best of a population of NP individuals and $\mathbf{X}_{r1}$ and $\mathbf{X}_{r2}$ are the two individuals randomly chosen. Factor F is a real value that regulates the disturbance of the best individual. This scheme is called differential evolution.

Vector **V** can be modified with a genetic operator called mutation that changes the value of its genes. This operator consists of choosing a value randomly among v_j and $v_j \pm \Delta_j$, so that if parameter v_j mutates, range Δ_j will be added or subtracted depending on the direction of the mutation.

Once vector **V** is defined, it is crossed over with individual $\mathbf{X}_i$ of the current population to obtain new individual $\mathbf{X}_i^N$ (offspring) who is a candidate to be part of the following population. This operation is called crossover. Different methods can be used to allocate the genes of the offspring. This is done randomly. In natural reproduction, the genes of the parents are exchanged to form the genes of the child or children. This is better illustrated in the scheme shown in Fig. 10.24.

In order to define the new population, we compare offspring fitness $f(\mathbf{X}_i^N)$ with its antecessor fitness, $f(\mathbf{X}_i)$. The individual with less fitness value is "better" and takes part in the new population. This way the population does not increase or decrease in number.

The scheme of the whole process can be seen in the diagram shown in Fig. 10.25.

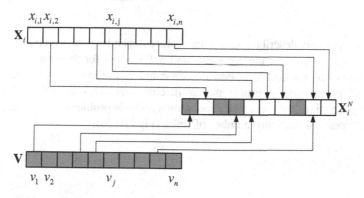

Fig. 10.24 Diagram of the crossover reproduction process

Fig. 10.25 Process scheme of the implementation of the described evolutionary algorithm

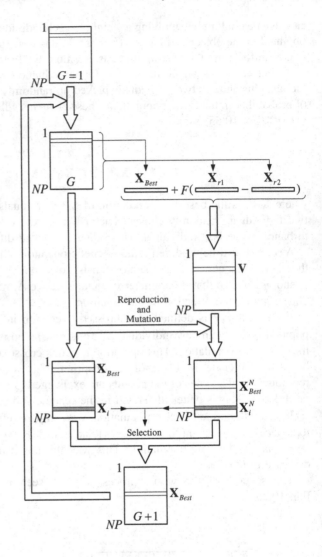

All the operators described above, take place depending on a probability that is defined as $CP \in [0, 1]$ for the crossover and $MP \in [0, 1]$ for the mutation.

The described method has been implemented in the synthesis module of the WinMecC program to optimize planar mechanisms. This module solves both function generation and trajectory generation synthesis problems in mechanisms of different types and with any number of links (Fig. 10.26).

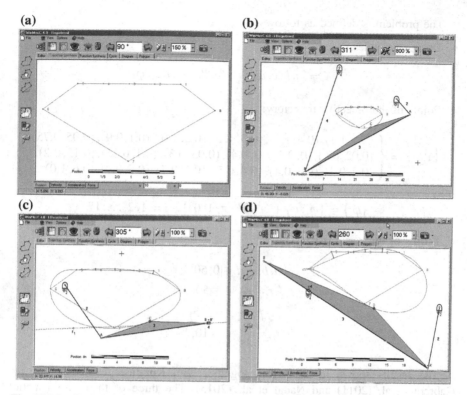

Fig. 10.26 a Definition of the desired path in the WinMecC synthesis module. **b** Synthesis of a four-bar mechanism. **c** Synthesis of a crank-shaft mechanism. **d** Synthesis of a slider-crank mechanism

10.7   Results

In this section we show the results of two problems solved with the proposed evolutionary algorithm. In both cases the turning function has been used to define the objective function of the optimization problems.

10.7.1   Closed Path Generation

In this example, we address the solution of a trajectory generation problem that was originally introduced by Kunjur and Krishnamurty (1997). The authors originally propose that couple point *C* of a four-bar mechanism passes through 18 points that describe a closed curve.

The problem is defined as follows:

- Design variables:

$$X = [a, b, c, d, r^r_{C_x}, r^r_{C_y}, \beta, r_{O_{2x}}, r_{O_{2y}}, \varphi_1]$$

- Points of the objective trajectory:

$$\{\mathbf{r}^d_{C,i}\} = \left\{ \begin{array}{l} (0.5, 1.1), (0.4, 1.1), (0.3, 1.1), (0.2, 1.0), (0.1, 0.9), (0.05, 0.75), \\ (0.02, 0.6), (0, 0.5), (0, 0.4), (0.03, 0.3), (0.1, 0.25), (0.15, 0.2), \\ (0.2, 0.3), (0.3, 0.4), (0.4, 0.5), (0.5, 0.7), (0.6, 0.9), (0.6, 1.0) \end{array} \right\}$$

$$\{\varphi_i\} = \{\varphi_1, \varphi_1 + i \cdot 20° \cdot \pi/180°\} \quad i = 1, 2, \ldots, 17$$

- Limits of the variables:

$$a, b, c, d \in [0, 50]$$
$$r^r_{C_x}, r^r_{C_y} \in [-50, 50]$$
$$\beta, \varphi_1 \in [0, 2\pi]$$
$$r_{O_{2x}}, r_{O_{2y}} \in [-10, 10]$$

The results shown in Table 10.2 together with Kunjur's results, belong to Cabrera et al. (2011) and Nadal et al. (2015). The three of them use genetic algorithms to solve the problem. The first two authors use the mean square error to synthesize the best mechanism while the third one uses turning functions.

Table 10.2 Solutions to the trajectory synthesis problem proposed by Kunjur with different methods

	Kunjur	Cabrera et al. (2011)	Nadal et al. (2015)
a	0.274853	0.297057	0.239834
b	2.138209	0.849372	3.164512
c	1.180253	3.913095	0.941660
d	1.879660	4.453772	2.462696
$r^r_{C_x}$	−0.833592	−2.067338	0.033708
$r^r_{C_y}$	−0.378770	1.6610626	0.483399
β (rad)	4.354224	2.7387359	4.788536
$r_{O_{2x}}$	1.132062	−1.309243	0.569026
$r_{O_{2y}}$	0.663433	2.806964	0.350557
φ_1 (rad)	2.558625	4.853543	2.472660
Mean square error	0.043	0.0196	0.113595
Turning function error	0.250734	0.261502	0.184365

Fig. 10.27 Desired path and paths traced by the coupler of the synthesized mechanisms found with different methods

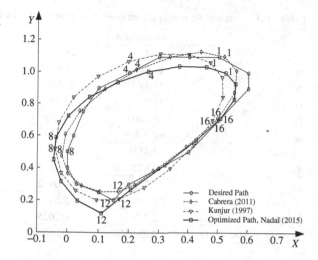

In Fig. 10.27 we can see the curves described by the three different mechanisms compared to the desired curve.

10.7.2 Open Path Generation

This example shows the results of an open path generation synthesis. The aim of the problem is that the coupler point of the synthesized mechanism describes a right angle path during the motion. The original problem was presented by Sedano et al. (2012). The original problem was defined as follows:

- Design variables:

$$\chi = \left[r_1, r_2, r_3, r_4, r_{cx}, r_{cy}, \theta_0, x_0, y_0, \theta_2^1\right]$$

- Target points:

$$\{C_d^i\} = \left\{ \begin{array}{l} (0, 15), (0, 12), (0, 9), (0, 6), (0, 3), (0, 0), \\ (3, 0), (6, 0), (9, 0), (12, 0), (15, 0) \end{array} \right\}$$

As in the previous case, design variables in the original problem included linear and angular positions of the fixed link (Table 10.2 column 1).

Table 10.3 Comparative results for an open path generation example

	Sedano EA	Sedano hybrid	Nadal et al. (2015)
r_1	24.3	28.16	33.06872
r_2	14.81	16.94	21.37403
r_3	24.48	23.05	29.03806
r_4	31.7	32.68	26.49428
r_{cx}	3.126	−0.422	2.51363
r_{cy}	−2.528	−7.1175	−9.72539
θ_0 (rad)	3.4516	3.4216	3.81408
x_0	12.672	19.0533	25.69221
y_0	15.753	12.6922	18.10155
θ_2^1 (rad)	1.68	5.50	5.346
Mean square error	2.439	0.2025	0.1099
Turning function error	0.25202	0.263317	0.02746

Fig. 10.28 The best path traced by the coupler with three different bibliography examples

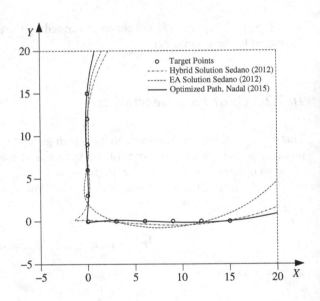

The best mechanism found by the evolutionary algorithm in the last iteration is listed in Table 10.3 along with the best results found by Sedano et al. (2015). In this case, we can observe that the results obtained by the evolutionary algorithm with turning functions, improve those achieved by Sedano et al. The transformed mechanism (Table 10.3 column 4) obtains the best right angle, as it is shown in Fig. 10.28.

References

Cabrera JA, Ortiz A, Nadal F, Castillo JJ (2011) An evolutionary algorithm for path synthesis of mechanisms. Mech Mach Theor 46:127–141

Kunjur A, Krishnamurty S (1997) Genetic algorithms in mechanical synthesis. J Appl Mech Robot 4(2):18–24

Nadal F, Cabrera JA, Bataller A, Castillo JJ, Ortiz A (2015) Turning functions in optimal synthesis of mechanisms. J Mech Des 137(6), Article ID: 062302-10p

Sedano A, Sancibrian R, de Juan A, Viadero F, Egaña F (2012) Hybrid optimization approach for the design of mechanisms using a new error estimator. Math Prob Eng, Article ID: 151590–20p

Appendix A
Position Kinematic Analysis.
Trigonometric Method

Chapter 3 shows the kinematic analysis of several mechanisms by using Raven's method. Writing the equations that solve the problem when using this method is easy. However, finding the solution to such equations can be a complicated task. For that reason, we introduce the trigonometric method in this appendix, which is much simpler to write and solve.

A.1 Position Analysis of a Four-Bar Mechanism

Consider the mechanism shown in Fig. A.1 in which $\overline{O_2O_4}$, $\overline{O_2A}$, $\overline{AB}$ and $\overline{O_4A}$ are the lengths of links 1, 2, 3 and 4 respectively. On the other hand, angles θ_2, θ_3 and θ_4 define the angular position of links 2, 3 and 4 considering the counterclockwise rotations positive.

In order to determine angles θ_3 and θ_4, we need to find the value of distance $\overline{O_4A}$ (Eq. A.1) as well as angles β (Eq. A.3), δ (Eq. A.7) and ϕ (Eq. A.5). The value of distance $\overline{O_4A}$ can be determined in triangle ΔO_2AO_4:

$$\overline{O_4A} = \sqrt{\overline{O_2O_4}^2 + \overline{O_2A}^2 - 2\overline{O_2O_4}\,\overline{O_2A}\cos\theta_2} \qquad (A.1)$$

The same triangle verifies (Eq. A.2):

$$\overline{O_4A}\sin\beta = \overline{O_2A}\sin\theta_2 \qquad (A.2)$$

where:

$$\beta = \arcsin\left(\frac{\overline{O_2A}}{\overline{O_4A}}\sin\theta_2\right) \qquad (A.3)$$

Angles ϕ and δ between bars 3 and 4 and diagonal $\overline{O_4A}$ respectively, can be worked out from triangle ΔABO_4. It verifies (Eq. A.4)

© Springer International Publishing Switzerland 2016
A. Simón Mata et al., *Fundamentals of Machine Theory and Mechanisms*,
Mechanisms and Machine Science 40, DOI 10.1007/978-3-319-31970-4

Fig. A.1 Parameters
involved in the calculation of
the link positions in a four-bar
mechanism by means of the
trigonometric method

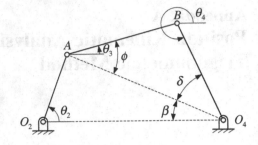

$$\overline{O_4B}^2 = \overline{AB}^2 + \overline{O_4A}^2 - 2\overline{ABO_4A}\cos\phi \qquad (A.4)$$

We can clear ϕ from Eq. (A.4):

$$\phi = \arccos\frac{\overline{AB}^2 + \overline{O_4A}^2 - \overline{O_4B}^2}{2\overline{ABO_4A}} \qquad (A.5)$$

In the same triangle its verified (Eq. A.6):

$$\overline{O_4B}\sin\delta = \overline{AB}\sin\phi \qquad (A.6)$$

Thus:

$$\delta = \arcsin\left(\frac{\overline{AB}}{\overline{O_4B}}\sin\phi\right) \qquad (A.7)$$

Once the values of β, δ and ϕ have been determined, we can obtain θ_3 (Eq. A.8) and θ_4 (Eq. A.9) in the mechanism (Fig. A.1):

$$\theta_3 = \phi - \beta \qquad (A.8)$$

$$\theta_4 = -(\beta + \delta) \qquad (A.9)$$

When angle θ_2 takes values between 180° and 360°, angle β has a negative value and Eqs. (A.8) and (A.9) are also applicable (Fig. A.2).

Fig. A.2 Open four-bar
mechanism with link 2 in a
position between 180° and
360°

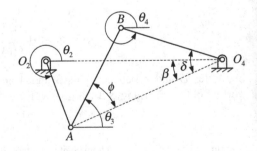

Fig. A.3 Calculation of the position of links 3 and 4 in a crossed four-bar mechanism by means of the trigonometric method

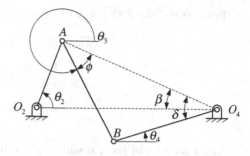

Fig. A.4 Crossed four-bar mechanism with link 2 in a position between 180° and 360°

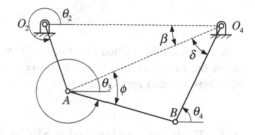

For a crossed four-bar mechanism (Fig. A.3) we will use Eqs. (A.10) and (A.11):

$$\theta_3 = -(\phi + \beta) \tag{A.10}$$

$$\theta_4 = \delta - \beta \tag{A.11}$$

Again, when angle θ_2 takes values between 180° and 360°, angle β has a negative value and Eqs. (A.10) and (A.11) are also applicable (Fig. A.4).

A.2 Position Analysis of a Crank-Shaft Mechanism

Figure A.5 shows a crank-shaft mechanism. x_B and y_B are the Cartesian coordinates of point B with respect a system centered on point O_2 with its X-axis parallel to the piston trajectory. x_B is positive while y_B is negative.

Fig. A.5 Crank-shaft mechanism and the parameters involved in the position analysis with the trigonometric method

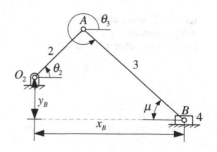

Position of links 3 and 4 can be worked out using Eqs. (A.12)–(A.15):

$$\overline{AB} \sin \mu = \overline{O_2A} \sin \theta_2 - y_B \qquad (A.12)$$

$$\mu = \arcsin \frac{\overline{O_2A} \sin \theta_2 - y_B}{\overline{AB}} \qquad (A.13)$$

$$\theta_3 = -\mu \qquad (A.14)$$

The x position of point B will be given by Eq. (A.15):

$$x_B = \overline{O_2A} \cos \theta_2 + \overline{AB} \cos \theta_3 \qquad (A.15)$$

It can easily be verified that Eq. (A.15) works for any position of input link 2. When the trajectory of point B is above O_2, the sign of y_B is positive and these equations are also applicable.

A.3 Position Analysis of a Slider Mechanism

Consider the slider mechanism in Fig. A.6, where link 3 describes a straight trajectory along link 4 that rotates about O_4 with offset $\overline{O_4B}$.

Similarly as in previous problems, $\overline{O_2O_4}$ and $\overline{O_2A}$ are the lengths of links 1 and 2 respectively while angles θ_2 and θ_4 define the angular position of links 2 and 4. Assuming that we know $\overline{O_2O_4}$, $\overline{O_2A}$ and θ_2, we can obtain unknown values $\overline{AB}$ (Eq. A.17) and θ_4 (Eq. A.20). To do so, we start by obtaining the value of $\overline{O_4A}$ (Eq. A.16):

$$\overline{O_4A} = \sqrt{\overline{O_2O_4}^2 + \overline{O_2A}^2 - 2\overline{O_2O_4}\,\overline{O_2A} \cos \theta_2} \qquad (A.16)$$

We can calculate $\overline{AB}$ as:

$$\overline{AB} = \sqrt{\overline{O_4B}^2 - \overline{O_4A}^2} \qquad (A.17)$$

Fig. A.6 Position analysis of a slider-mechanism by means of the trigonometric method

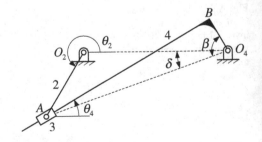

The value of angle δ (Eq. A.18) is:

$$\delta = \arctan \frac{\overline{O_2A} \sin \theta_2}{\overline{O_2O_4} + \overline{O_2A} \cos \theta_2} \tag{A.18}$$

Finally, θ_4 can be determined after first computing the value of β (Eq. A.19):

$$\beta = \arctan \frac{\overline{AB}}{\overline{O_4B}} \tag{A.19}$$

$$\theta_4 = \delta + (90° - \beta) \tag{A.20}$$

If the offset is opposite, point B is below the X-axis and Eq. (A.20) changes to Eq. (A.21):

$$\theta_4 = \delta - (90° - \beta) \tag{A.21}$$

A.4 Two Generic Bars of a Mechanism

Let us consider that we have carried out the kinematic analysis of links 2, 3 and 4 of the mechanism shown in Fig. A.7. We will continue the position analysis of links 5 and 6 considering that the position of point C of link 3 is known.

To find the position of links 5 and 6 we have to define triangle ΔCDO_6 first (Fig. A.8).

The length of side $\overline{O_6C}$ (Eq. A.22) can be calculated by means of the x and y coordinates of points C and O_6:

$$\overline{O_6C} = \sqrt{(x_C - x_{O_6})^2 + (y_C - y_{O_6})^2} \tag{A.22}$$

Fig. A.7 Six-bar mechanism

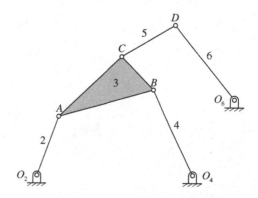

Fig. A.8 Position analysis of
bars 5 and 6 by means of the
trigonometric method

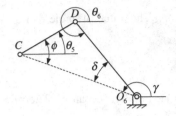

And its angle γ (Eq. A.23) is:

$$\gamma = \arctan \frac{y_C - y_{O_6}}{x_C - x_{O_6}} \tag{A.23}$$

Angle ϕ (Eq. A.25) can be computed by using the law of cosines (Eq. A.24):

$$\overline{DO_6}^2 = \overline{CD}^2 + \overline{O_6D}^2 - 2\overline{CDO_6D}\cos\phi \tag{A.24}$$

$$\phi = \arccos \frac{\overline{CD}^2 + \overline{O_6D}^2 - \overline{DO_6}}{2\overline{CDO_6D}} \tag{A.25}$$

Finally, angle δ (Eq. A.26) is determined by using the law of sines:

$$\delta = \arcsin\left(\frac{\overline{CD}}{\overline{O_6D}}\sin\phi\right) \tag{A.26}$$

Therefore, angles θ_5 and θ_6 are (Eqs. A.27 and A.28):

$$\theta_5 = \phi - (180° - \gamma) \tag{A.27}$$

$$\theta_6 = 180° + \gamma - \delta \tag{A.28}$$

Appendix B
Freudenstein's Method to Solve the Position Equations in a Four-Bar Mechanism

In Chap. 3 we developed the position analysis of a four-bar mechanism by means of Raven's method. In this appendix we explain Freudenstein's method to solve the obtained equations and calculate the value of angles θ_3 and θ_4.

B.1 Position Analysis of a Four-Bar Mechanism by Using Raven's Method

We will apply Raven's method to the four-bar mechanism shown in Fig. B.1.
  The vector loop equation (Eq. B.1) for the position analysis of the mechanism is:

$$r_1 e^{i\theta_1} = r_2 e^{i\theta_2} + r_3 e^{i\theta_3} + r_4 e^{i\theta_4} \qquad (B.1)$$

By converting this equation into its trigonometric form (Eq. B.2):

$$\begin{aligned} r_1(\cos\theta_1 + i\sin\theta_1) = r_2(\cos\theta_2 + i\sin\theta_2) + r_3(\cos\theta_3 + i\sin\theta_3) \\ + r_4(\cos\theta_4 + i\sin\theta_4) \end{aligned} \qquad (B.2)$$

And by separating its real and imaginary parts, we obtain the system (Eq. B.3) with two unknowns (θ_3 and θ_4):

$$\left. \begin{aligned} r_1\cos\theta_1 = r_2\cos\theta_2 + r_3\cos\theta_3 + r_4\cos\theta_4 \\ r_1\sin\theta_1 = r_2\sin\theta_2 + r_3\sin\theta_3 + r_4\sin\theta_4 \end{aligned} \right\} \qquad (B.3)$$

B.2 Freudenstein's Method

We substitute $\theta_1 = 0$ in Eq. (B.3) and isolate θ_3 (Eq. B.4):

© Springer International Publishing Switzerland 2016
A. Simón Mata et al., *Fundamentals of Machine Theory and Mechanisms*,
Mechanisms and Machine Science 40, DOI 10.1007/978-3-319-31970-4

Fig. B.1 Position analysis of
a four-bar mechanism by
means of Raven's method

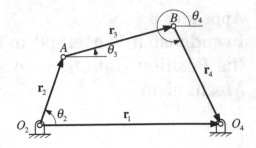

$$\left. \begin{array}{l} r_1 - r_2 \cos\theta_2 - r_4 \cos\theta_4 = r_3 \cos\theta_3 \\ -r_2 \sin\theta_2 - r_4 \sin\theta_4 = r_3 \sin\theta_3 \end{array} \right\} \tag{B.4}$$

We raise each equation to the second power and add them term by term (Eq. B.5):

$$r_1^2 + r_2^2 + r_4^2 - 2r_1r_2\cos\theta_2 - 2r_1r_4\cos\theta_4 + 2r_2r_4(\cos\theta_2\cos\theta_4 + \sin\theta_2\sin\theta_4) = r_3^2 \tag{B.5}$$

By dividing all terms by the coefficient of term $\cos\theta_2\cos\theta_4 + \sin\theta_2\sin\theta_4$, $2r_2r_4$, it yields Eq. (B.6):

$$\frac{r_1^2 + r_2^2 - r_3^2 + r_4^2}{2r_2r_4} - \frac{r_1}{r_4}\cos\theta_2 - \frac{r_1}{r_2}\cos\theta_4 + (\cos\theta_2\cos\theta_4 + \sin\theta_2\sin\theta_4) = 0 \tag{B.6}$$

In order to simplify Eq. (B.6), we use the following coefficients (Eq. B.7):

$$\left. \begin{array}{l} k_1 = \dfrac{r_1}{r_2} \\[2mm] k_2 = \dfrac{r_1}{r_4} \\[2mm] k_3 = \dfrac{r_1^2 + r_2^2 - r_3^2 + r_4^2}{2r_2r_4} \end{array} \right\} \tag{B.7}$$

Thus, Eq. (B.6) remains Eq. (B.8):

$$k_3 - k_2 \cos\theta_2 - k_1 \cos\theta_4 + (\cos\theta_2\cos\theta_4 + \sin\theta_2\sin\theta_4) = 0 \tag{B.8}$$

We substitute $\cos\theta_4$ and $\sin\theta_4$ for their expressions in terms of the half angle tangent (Eq. B.9):

$$k_3 - k_2 \cos\theta_2 - k_1 \frac{1 - \tan^2\frac{\theta_4}{2}}{1 + \tan^2\frac{\theta_4}{2}} + \left(\cos\theta_2 \frac{1 - \tan^2\frac{\theta_4}{2}}{1 + \tan^2\frac{\theta_4}{2}} + \sin\theta_2 \frac{2\tan\frac{\theta_4}{2}}{1 + \tan^2\frac{\theta_4}{2}}\right) = 0$$

$$(B.9)$$

Next, we remove the denominators and group the terms for tan, $\tan^2$ and the independent term (Eq. B.10), all in the same member.

$$(k_3 - k_2 \cos\theta_2 - k_1 - \cos\theta_2) \tan^2\frac{\theta_4}{2} + 2\sin\theta_2 \tan\frac{\theta_4}{2}$$
$$+ (k_3 - k_2 \cos\theta_2 - k_1 + \cos\theta_2) = 0 \qquad (B.10)$$

Again, we rename the different coefficients (Eq. B.11) of the second degree equation:

$$\left.\begin{array}{l} A = k_3 - k_2 \cos\theta_2 - k_1 - \cos\theta_2 \\ B = 2\sin\theta_2 \\ C = k_3 - k_2 \cos\theta_2 - k_1 + \cos\theta_2 \end{array}\right\} \qquad (B.11)$$

Thus, Eq. (B.10) can be written as Eq. (B.12):

$$A \tan^2\frac{\theta_4}{2} + B \tan\frac{\theta_4}{2} + C = 0 \qquad (B.12)$$

Hence, θ_4, which is the unknown that defines the angular position of link 4, is (Eq. B.13):

$$\theta_4 = 2 \arctan \frac{-B \pm \sqrt{B^2 - 4AC}}{2A} \qquad (B.13)$$

where the + and − signs indicate two possible solutions for the open and crossed configurations of the four-bar mechanism respectively.

Similarly, but in this case isolating θ_4 in one of the members, we reach (Eq. B.14) for θ_3, which defines the angular position of link 3. Again, there are two possible solutions depending on the configuration of the four-bar mechanism:

$$\theta_3 = 2 \arctan \frac{-E \pm \sqrt{E^2 - 4DF}}{2D} \qquad (B.14)$$

where the different coefficients (Eq. B.15) of the second degree equation (Eq. B.14) are:

$$\left.\begin{array}{l} D = k_1 - k_4 \cos\theta_2 + k_5 - \cos\theta_2 \\ E = 2\sin\theta_2 \\ F = -k_1 - k_4\cos\theta_2 + k_5 + \cos\theta_2 \end{array}\right\} \quad\text{(B.15)}$$

And k_4 and k_5 (Eq. B.16) are:

$$\left.\begin{array}{l} k_1 = \dfrac{r_1}{r_2} \\[2mm] k_4 = \dfrac{r_1}{r_3} \\[2mm] k_5 = \dfrac{r_1^2 + r_2^2 + r_3^2 - r_4^2}{2r_2 r_3} \end{array}\right\} \quad\text{(B.16)}$$

Appendix C
Kinematic and Dynamic Analysis
of a Mechanism

The conveyor transfer mechanism shown in Fig. C.1 pushes boxes with a mass of 8 kg from one conveyor belt to another. The motor link turns at a constant speed of 40 rpm in counter clockwise direction.

In order to make a complete kinematic and dynamic analysis of the mechanism, we will use all the analysis methods described in this book. We will carry out the analysis at a given position. In general, the most interesting one for dynamic analysis is the position at which the acceleration of the piston is maximum. This way we can determine the forces that act on the links in extreme conditions. The position chosen for this study is $\theta_2 = 350°$.

This analysis includes the following sections:

- Kinematic chain. Study and identification of the kinematic pairs. Number of D.O.F of the mechanism. Kinematic inversion that results from fixing link 4.
- Kinematic graph of slider displacement versus crank rotation.
- Velocity analysis by means of the relative velocity method.
- Velocity analysis by means of the method of Instantaneous Centers of Rotation.
- Acceleration analysis by means of the relative acceleration method.
- Velocity and acceleration analysis by means of Raven's method.
- Calculation of the inertial force and inertial torque of each of the links in the mechanism.
- Dynamic analysis by means of the graphical method.
- Dynamic analysis by means of the matrix method.

C.1 Kinematic Chain

We begin the study of the mechanism by drawing its kinematic diagram as shown in Fig. C.2. This figure also shows the nomenclature that will be used along this study.

Table C.1 shows the different types of kinematic pairs in the mechanism and the degrees of freedom of each pair.

© Springer International Publishing Switzerland 2016
A. Simón Mata et al., *Fundamentals of Machine Theory and Mechanisms*,
Mechanisms and Machine Science 40, DOI 10.1007/978-3-319-31970-4

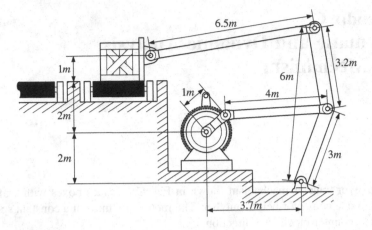

Fig. C.1 Conveyor transfer mechanism

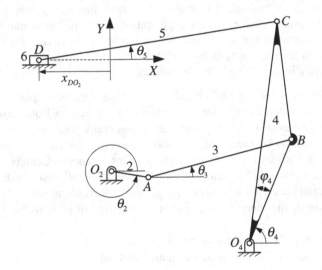

Fig. C.2 Kinematic diagram of the mechanism

Table C.1 Type of kinematic pairs in the mechanism

PAIR	Type	Number of D.O.F.
1–2	Rotation	1
2–3	Rotation	1
3–4	Rotation	1
1–4	Rotation	1
4–5	Rotation	1
5–6	Rotation	1
1–6	Prismatic	1

Fig. C.3 Kinematic inversion of the mechanism when link 4 is fixed

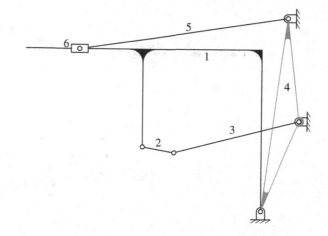

We use Kutzbach's equation to calculate the number of degrees of freedom of the mechanism (Eq. C.1):

- Number of links: $N = 6$
- Kinematic pairs with 1 DOF: $J_1 = 7$
- Kinematic pairs with 2 DOF: $J_2 = 0$

$$DOF = 3(N - 1) - 2J_1 - J_2 = 3(6 - 1) - 2 \cdot 7 - 0 = 1 \qquad (C.1)$$

To better understand the mechanism, we will draw the kinematic diagram of one of its inversions. In this case we will consider link 4 as the frame. This is shown in Fig. C.3.

C.2 Slider Displacement Versus Crank Rotation

We will draw the kinematic graph of point D displacement versus crank rotation by means of the graphical method. To do so, we divide the whole turn of the crank in 12 positions starting from position 0°. This way, we find the 12 positions of point A which correspond to 12 angular positions of the crank in steps of 30°. Knowing the length of the links, we can find the equivalent 12 positions for points B, C and D (Fig. C.4).

We can graph the position of point D versus the crank position. This is shown in Fig. C.5. We can see that the stroke end positions of the piston are close to positions $\theta_2 = 10°$ and $\theta_2 = 195°$. In these positions, the velocity of the piston has to be null. As the velocity is the first time-derivative of displacement, this can be verified by tracing a line tangent to the curve at the end-of-stroke position. If the line is horizontal, the velocity is null.

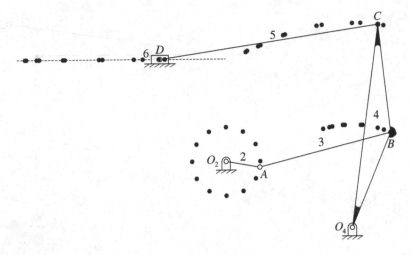

Fig. C.4 Kinematic diagram of the mechanism in a complete turn of the crank in steps of 30°

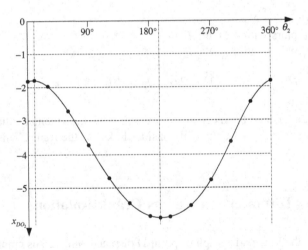

Fig. C.5 Kinematic graph of the slider displacement versus the crank rotation

C.3 Velocity Analysis by Relative Velocity Method

Before starting the velocity analysis, the positions of the links have to be determined. To do so, we will use the trigonometric method explained in Appendix A. Figure C.6 shows the angles and distances used to solve the position problem.

We start with the four-bar mechanism formed by links 1, 2, 3 and 4. Distance $\overline{O_2 O_4}$ (Eq. C.2) and angles θ_1 (Eq. C.3) and γ (Eq. C.4) can be calculated as:

Fig. C.6 Calculation of the position of links 3, 4, 5 an 6 by means of the trigonometric method

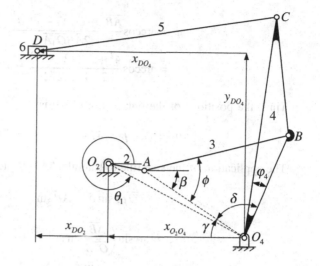

$$\overline{O_2O_4} = \sqrt{(x_{O_4} - x_{O_2})^2 + (y_{O_4} - y_{O_2})^2} = 4.206\ \text{m} \qquad (C.2)$$

$$\theta_1 = 270° + \arctan\frac{3.7}{2} = 331.6° \qquad (C.3)$$

$$\gamma = 180° - 90° - \arctan\frac{3.7}{2} = 28.4° \qquad (C.4)$$

The application of the cosine rule to triangle $\triangle O_2AO_4$ yields (Eq. C.5):

$$\overline{O_4A} = \sqrt{\overline{O_2O_4}^2 + \overline{O_2A}^2 - 2\overline{O_2O_4}\,\overline{O_2A}\cos(\theta_2 - \theta_1)}$$
$$= \sqrt{4.206^2 + 1^2 - 2 \cdot 4.206 \cos 18.4°} = 3.272\ \text{m} \qquad (C.5)$$

And the sine rule on the same triangle yields (Eqs. C.6–B.7):

$$\overline{O_4A}\sin(\beta - \gamma) = \overline{O_2A}\sin(\theta_2 - \theta_1) \qquad (C.6)$$

$$\beta = \gamma + \arcsin\frac{\overline{O_2A}\sin(\theta_2 - \theta_1)}{\overline{O_4A}} = 28.4° + \arcsin\frac{\sin 18.4}{3.272} = 33.9° \qquad (C.7)$$

The application of the cosine rule to triangle $\triangle ABO_4$ yields (Eqs. C.8 and C.9):

$$\overline{O_4B}^2 = \overline{AB}^2 + \overline{O_4A}^2 - 2\overline{AB}\,\overline{O_4A}\cos\phi \qquad (C.8)$$

$$\phi = \arccos \frac{\overline{AB}^2 + \overline{O_4A}^2 - \overline{O_4B}^2}{2\overline{AB}\,\overline{O_4A}}$$

$$= \arccos \frac{4^2 + 3.272^2 - 3^2}{2 \cdot 4 \cdot 3.272} = 47.4°$$

(C.9)

Thus, the positions of the link 3 (Eq. C.10) is:

$$\theta_3 = \phi - \beta = 47.4° - 33.9° = 13.5°$$

(C.10)

The application of the sine rule to triangle $\triangle ABO_4$ yields (Eqs. C.11 and C.12):

$$\overline{O_4B} \sin \delta = \overline{AB} \sin \phi$$

(C.11)

$$\delta = \arcsin\left(\frac{\overline{AB}}{\overline{O_4B}} \sin \phi\right)$$

$$= \arcsin\left(\frac{4 \sin 47.4°}{3}\right) = 79.12°$$

(C.12)

Therefore, the position of link 4 (Eq. C.13) is:

$$\theta_4 = 180° - \beta - \delta = 180° - 33.9° - 79.12° = 67°$$

(C.13)

We continue with the position analysis of the crank-shaft mechanism formed by links 4, 5 and 6 in Fig. C.6.

Although angle φ_4 formed by $\overline{O_4B}$ and $\overline{O_4C}$ has a fixed value and it could be part of the data of the mechanism, in this case we have the length of the sides of triangle $\triangle O_4BC$ instead of angle φ_4 itself. We can easily obtain its value (Eq. C.15) by means of the rule of cosines (Eq. C.14):

$$\overline{BC} = \sqrt{\overline{O_4B}^2 + \overline{O_4C}^2 - 2\overline{O_4B}\,\overline{O_4C} \cos \varphi_4}$$

(C.14)

$$\varphi_4 = \arccos\left(\frac{\overline{O_4B}^2 + \overline{O_4C}^2 - \overline{BC}^2}{2\overline{O_4B}\,\overline{O_4C}}\right)$$

$$= \arccos\left(\frac{3^2 + 6^2 - 3.2^2}{2 \cdot 3 \cdot 6}\right) = 15.1°$$

(C.15)

The projection of triangle $\triangle O_4CD$ over a direction perpendicular to the trajectory of the piston yields the trigonometric equation (Eq. C.16):

$$y_{DO_4} + \overline{CD} \sin \theta_5 = \overline{O_4C} \sin(\theta_4 + \varphi_4)$$

(C.16)

And clearing θ_5 (Eq. C.17), we obtain its value:

$$\theta_5 = \arcsin \frac{\overline{O_4C}\sin(\theta_4 + \varphi_4) - y_{DO_4}}{\overline{CD}}$$
$$= \arcsin \frac{6\sin(67° + 15.1°) - 5}{6.5} = 8.34° \tag{C.17}$$

The projection of the sides of triangle ΔO_4CD over de piston trajectory yields (Eq. C.18):

$$x_{DO_4} = \overline{O_4C}\cos(\theta_4 + \varphi_4) - \overline{CD}\cos\theta_5$$
$$= 6\cos(67° + 15.1°) - 6.5\cos 8.34° = -5.607\,\text{m} \tag{C.18}$$

Hence, the horizontal component of the distance between D and O_2 (Eq. C.19) is:

$$x_{DO_2} = x_{DO_4} - x_{O_2O_4} = -5.607\,\text{m} - (-3.7\,\text{m}) = -1.907\,\text{m} \tag{C.19}$$

Therefore, the positions of the links (Eq. C.20) corresponding to crank position $\theta_2 = 350°$ are:

$$\left.\begin{array}{c} \theta_3 = 13.5° \\ \theta_4 = 67° \\ \theta_5 = 8.34° \\ x_{DO_2} = -1.907\,\text{m} \end{array}\right\} \tag{C.20}$$

The following step is to find the velocity of the links when link 2 rotates at an angular speed of 40 rpm counterclockwise. We have to use the velocity of link 2 in radians per second: 4.19 rad/s.

The velocity of point A (Eq. C.21) can be calculated as:

$$\mathbf{v}_A = \omega_2 \wedge \mathbf{r}_{AO_2} = \begin{vmatrix} \hat{\mathbf{i}} & \hat{\mathbf{j}} & \hat{\mathbf{k}} \\ 0 & 0 & 4.19 \\ 1\cos 350° & 1\sin 350° & 0 \end{vmatrix} = 0.73\hat{\mathbf{i}} + 4.13\hat{\mathbf{j}}$$
$$= 4.19\,\text{cm/s}\angle 80° \tag{C.21}$$

To calculate the angular velocity of links 3 and 4 we have to use the relative velocity vector equation: $\mathbf{v}_B = \mathbf{v}_A + \mathbf{v}_{BA}$.

Vectors $\mathbf{v}_B$ and $\mathbf{v}_{BA}$ can be obtained the following way (Eqs. C.22 and C.23):

$$v_{BA} = \omega_3 \wedge r_{BA} = \begin{vmatrix} \hat{i} & \hat{j} & \hat{k} \\ 0 & 0 & \omega_3 \\ 4\cos 13.5° & 4\sin 13.5° & 0 \end{vmatrix}$$

$$= -4\omega_3(\sin 13.5°\hat{i} - \cos 13.5°\hat{j}) \qquad (C.22)$$

$$v_B = \omega_4 \wedge r_{BO_4} = \begin{vmatrix} \hat{i} & \hat{j} & \hat{k} \\ 0 & 0 & \omega_4 \\ 3\cos 67° & 3\sin 67° & 0 \end{vmatrix} = -3\omega_4(\sin 67°\hat{i} - \cos 67°\hat{j})$$

$$(C.23)$$

By introducing the three velocity vectors, v_A, v_B and v_{BA}, in the relative velocity equation and projecting them on the X and Y Cartesian axles, we reach the system of equations (Eq. C.24):

$$\left.\begin{array}{l} 0.73 - 4\omega_3 \sin 13.5° = -3\omega_4 \sin 67° \\ 4.13 + 4\omega_3 \cos 13.5° = 3\omega_4 \cos 67° \end{array}\right\} \qquad (C.24)$$

The solution to the system of equations (Eq. C.24) yields the velocities of links 3 and 4 (Eq. C.25):

$$\left.\begin{array}{l} \omega_3 = -1.27\,\text{rad/s} \\ \omega_4 = -0.69\,\text{rad/s} \end{array}\right\} \qquad (C.25)$$

Using these values, we can calculate velocities v_B (Eq. C.26) and v_{BA} (Eq. C.27):

$$v_B = 1.91\hat{i} - 0.81\hat{j} = 2.08\,\text{m/s}\angle 336.96°\,13.5 \qquad (C.26)$$

$$v_{BA} = 1.185\hat{i} - 4.939\hat{j} \qquad (C.27)$$

The velocity of point C (Eq. C.28) can be determined by using the value of ω_4:

$$v_C = \omega_4 \wedge r_{CO_4} = \begin{vmatrix} \hat{i} & \hat{j} & \hat{k} \\ 0 & 0 & \omega_4 \\ 6\cos(67° + 15.1°) & 6\sin(67° + 15.1°) & 0 \end{vmatrix} \qquad (C.28)$$

$$= 4.12\hat{i} - 0.58\hat{j} = 4.16\,\text{m/s}\angle 352.1°$$

We use vector equation (C.29) to calculate the angular velocity of link 5 and the linear velocity of link 6.

$$\mathbf{v}_D = \mathbf{v}_C + \mathbf{v}_{DC} \qquad (C.29)$$

Since points C and D are two points of the same link, their relative velocity is given by Eq. (C.30):

$$\mathbf{v}_{DC} = \boldsymbol{\omega}_5 \wedge \mathbf{r}_{DC} = \begin{vmatrix} \hat{\mathbf{i}} & \hat{\mathbf{j}} & \hat{\mathbf{k}} \\ 0 & 0 & \omega_5 \\ 6.5\cos(\theta_5 + 180°) & 6.5\sin(\theta_5 + 180°) & 0 \end{vmatrix} \qquad (C.30)$$
$$= -6.5\omega_5(\sin(\theta_5 + 180°)\hat{\mathbf{i}} - \cos(\theta_5 + 180°)\hat{\mathbf{j}})$$

The velocity of point D (Eq. C.31) has the same direction as the trajectory. Therefore, its vertical component is null:

$$\mathbf{v}_D = v_D\hat{\mathbf{i}} \qquad (C.31)$$

By substituting the three velocity vectors, $\mathbf{v}_C$, $\mathbf{v}_D$ and $\mathbf{v}_{DC}$, in Eq. (C.29) we obtain the system of equations (Eq. C.32):

$$\left.\begin{array}{r} 4.12 - 6.5\omega_5 \sin 188.3° = v_D \\ -0.58 + 6.5\omega_5 \cos 188.3° = 0 \end{array}\right\} \qquad (C.32)$$

Hence, the values of the velocities of links 5 and 6 (Eq. C.33) are:

$$\left.\begin{array}{l} \omega_5 = -0.09 \,\text{rad/s} \\ v_6 = v_D = 4.04 \,\text{m/s} \end{array}\right\} \qquad (C.33)$$

And the vector velocity of point D (Eq. C.34) is:

$$\mathbf{v}_D = 4.04\hat{\mathbf{i}} = 4.04 \,\text{m/s}\angle 0° \qquad (C.34)$$

Figure C.7 shows the velocity polygon of the mechanism. We can see how absolute velocities start at velocity pole O and relative velocities connect the end points of the absolute velocity vectors. It can also be seen that triangle Δobc in the polygon is similar to ΔO_4BC in the mechanism since their sides are perpendicular.

C.4   Instantaneous Center Method for Velocities

To calculate the ICRs in the mechanism, we start by identifying the ICRs which correspond to real joints. In this case, the known ICRs are: I_{12}, I_{23}, I_{34}, I_{14}, I_{45}, I_{16}, and I_{56} (Fig. C.8).

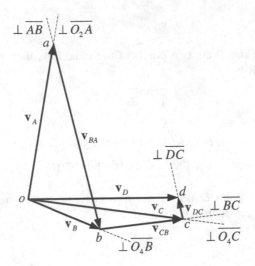

Fig. C.7 Velocity polygon

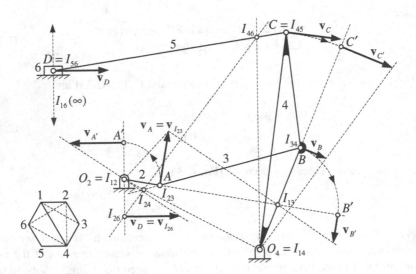

Fig. C.8 Velocity calculation by means of the ICR method

Then we draw a polygon with as many vertexes as links in the mechanism. Each of the sides or diagonals of the polygon represents one ICR. A solid line is used to draw those ICRs that are already known while those which are unknown are drawn as dotted lines.

In this case, to calculate the velocity of points B, C and D (Eqs. C.35–C.39), we have to obtain ICRs I_{16}, I_{24} and I_{46} by using Kennedy's theorem.

$$I_{26}\begin{cases} I_{16}I_{12} \\ I_{24}I_{46} \end{cases} \quad I_{24}\begin{cases} I_{23}I_{34} \\ I_{14}I_{12} \end{cases} \quad I_{46}\begin{cases} I_{45}I_{56} \\ I_{14}I_{16} \end{cases}$$

Figure C.8 shows the graphical development of the method and the vector obtained for each velocity.

$$\left.\begin{array}{l} v_{23} = v_A \\ v_{23} = \overline{I_{13}I_{23}}\omega_3 = 3.33\,\text{m} \cdot \omega_3 \end{array}\right\} \rightarrow \omega_3 = 1.26\,\text{rad/s} \quad (\text{C.35})$$

$$v_B = \overline{BI_{13}}\omega_3 = 2.08\,\text{m/s} \quad (\text{C.36})$$

$$\left.\begin{array}{l} v_{24} = \overline{I_{12}I_{24}}\omega_2 = 0.58\,\text{m} \cdot \omega_2 \\ v_{24} = \overline{I_{14}I_{24}}\omega_4 = 3.52\,\text{m} \cdot \omega_4 \end{array}\right\} \rightarrow \omega_4 = 0.69\,\text{rad/s} \quad (\text{C.37})$$

$$v_C = \overline{I_{14}C}\omega_4 = 4.14\,\text{m/s} \quad (\text{C.38})$$

$$\left.\begin{array}{l} v_{26} = \overline{I_{12}I_{26}}\omega_2 = 0.96\,\text{m} \cdot \omega_2 \\ v_{26} = v_D \end{array}\right\} \rightarrow v_D = 4.04\,\text{m/s} \quad (\text{C.39})$$

C.5 Acceleration Analysis with the Relative Acceleration Method

We know that the motor link turns at a constant rate of 40 rpm. Therefore, its angular acceleration is null ($\alpha_2 = 0$). In order to calculate the acceleration of the links, we start with the acceleration of point A (Eq. C.40). The tangential component will be zero as it depends on the angular acceleration value. Therefore, it will have only one normal component:

$$\mathbf{a}_A = \mathbf{a}_{AO_2}^n = \omega_2 \wedge v_A = \begin{vmatrix} \hat{\mathbf{i}} & \hat{\mathbf{j}} & \hat{\mathbf{k}} \\ 0 & 0 & 4.19 \\ 0.73 & 4.13 & 0 \end{vmatrix} \quad (\text{C.40})$$
$$= -17.3\hat{\mathbf{i}} + 3.06\hat{\mathbf{j}} = 17.55\,\text{m/s}^2\angle 170°$$

To calculate the angular acceleration of links 3 and 4, we use the vectors (Eqs. C.41 and C.42):

$$\mathbf{a}_B = \mathbf{a}_A + \mathbf{a}_{BA} \quad (\text{C.41})$$

$$\mathbf{a}_B^n + \mathbf{a}_B^t = \mathbf{a}_A^n + \mathbf{a}_A^t + \mathbf{a}_{BA}^n + \mathbf{a}_{BA}^t \quad (\text{C.42})$$

$$\mathbf{a}_B^n = \boldsymbol{\omega}_4 \wedge \mathbf{v}_B = \begin{vmatrix} \hat{\mathbf{i}} & \hat{\mathbf{j}} & \hat{\mathbf{k}} \\ 0 & 0 & -0.69 \\ 1.91 & -0.81 & 0 \end{vmatrix} = -0.559\hat{\mathbf{i}} - 1.318\hat{\mathbf{j}} \qquad (C.43)$$

$$\mathbf{a}_B^t = \boldsymbol{\alpha}_4 \wedge \mathbf{r}_{BO_4} = \begin{vmatrix} \hat{\mathbf{i}} & \hat{\mathbf{j}} & \hat{\mathbf{k}} \\ 0 & 0 & \alpha_4 \\ 3\cos 67^\circ & 3\sin 67^\circ & 0 \end{vmatrix} = -3\alpha_4 \sin 67^\circ \hat{\mathbf{i}} + 3\alpha_4 \cos 67^\circ \hat{\mathbf{j}}$$

$$(C.44)$$

$$\mathbf{a}_{BA}^n = \boldsymbol{\omega}_3 \wedge \mathbf{v}_{BA} = \begin{vmatrix} \hat{\mathbf{i}} & \hat{\mathbf{j}} & \hat{\mathbf{k}} \\ 0 & 0 & -1.27 \\ 1.185 & -4.939 & 0 \end{vmatrix} = -6.272\hat{\mathbf{i}} - 1.506\hat{\mathbf{j}} \qquad (C.45)$$

$$\mathbf{a}_{BA}^t = \boldsymbol{\alpha}_3 \wedge \mathbf{r}_{BA} = \begin{vmatrix} \hat{\mathbf{i}} & \hat{\mathbf{j}} & \hat{\mathbf{k}} \\ 0 & 0 & \alpha_3 \\ 4\cos 13.5^\circ & 4\sin 13.5^\circ & 0 \end{vmatrix}$$

$$= -4\alpha_4 \sin 13.5^\circ \hat{\mathbf{i}} + 4\alpha_3 \cos 13.5^\circ \hat{\mathbf{j}} \qquad (C.46)$$

Substituting these vectors (Eqs. C.43–C.46) in Eq. (C.42) and projecting them on the Cartesian axles, we reach the system of equations (Eq. C.47):

$$\left. \begin{array}{l} -0.559 - 3\alpha_4 \sin 67^\circ = -17.3 - 6.272 - 4\alpha_3 \sin 13.5^\circ \\ -1.311 + 3\alpha_4 \cos 67^\circ = +3.06 - 1.506 + 4\alpha_3 \cos 13.5^\circ \end{array} \right\} \qquad (C.47)$$

The solution yields the angular speed of links 3 and 4 (Eq. C.48).

$$\left. \begin{array}{l} \alpha_3 = 1.98 \,\text{rad/s}^2 \\ \alpha_4 = 9 \,\text{rad/s}^2 \end{array} \right\} \qquad (C.48)$$

Once α_4 is known, we can calculate the acceleration of point B (Eq. C.49):

$$\mathbf{a}_B = -25.4\hat{\mathbf{i}} + 9.24\hat{\mathbf{j}} = 27.03 \,\text{m/s} \angle 160^\circ \qquad (C.49)$$

The acceleration of point C can be calculated by means of (Eq. C.50):

$$\mathbf{a}_C = \mathbf{a}_B + \mathbf{a}_{CB} \qquad (C.50)$$

As points B and C belong to the same link, the components of the relative acceleration (Eq. C.51) are:

$$\left. \begin{array}{l} \mathbf{a}_{CB}^n = \boldsymbol{\omega}_4 \wedge \mathbf{v}_{CB} \\ \mathbf{a}_{CB}^t = \boldsymbol{\alpha}_4 \wedge \mathbf{r}_{CB} \end{array} \right\} \qquad (C.51)$$

Substituting the values previously obtained in Eq. (C.51), we calculate the acceleration vector of point C (Eq. C.52):

$$\mathbf{a}_C = -53.89\hat{\mathbf{i}} + 4.59\hat{\mathbf{j}} = 54.06\,\text{m/s} \angle 175° \qquad (C.52)$$

To determine the angular acceleration of link 5 and the linear acceleration of link 6, we use the vector equation (C.53):

$$\mathbf{a}_D^n + \mathbf{a}_D^t = \mathbf{a}_C^n + \mathbf{a}_C^t + \mathbf{a}_{DC}^n + \mathbf{a}_{DC}^t \qquad (C.53)$$

where:

$$\left.\begin{array}{l} \mathbf{a}_D^n = 0 \\ \mathbf{a}_D^t = a_D\hat{\mathbf{i}} \end{array}\right\} \qquad (C.54)$$

$$\left.\begin{array}{l} \mathbf{a}_{DC}^n = \boldsymbol{\omega}_5 \wedge \mathbf{v}_{DC} \\ \mathbf{a}_{DC}^t = \boldsymbol{\alpha}_5 \wedge \mathbf{r}_{DC} \end{array}\right\} \qquad (C.55)$$

Substituting vectors $\mathbf{a}_C$ (Eq. C.52), $\mathbf{a}_D$ (Eq. C.54) and $\mathbf{a}_{DC}$ (Eq. C.55) in Eq. (C.53) and projecting them onto the Cartesian axles, we reach to the system of equations (Eq. C.56):

$$\left.\begin{array}{l} a_D = -53.89 + 0.052 \quad 6.5\alpha_5 \sin 188.3° \\ 0 = 4.59 + 0.0076 + 6.5\alpha_5 \cos 188.3° \end{array}\right\} \qquad (C.56)$$

The solution of the system yields the accelerations of links 5 and 6 (Eq. C.57):

$$\left.\begin{array}{l} \alpha_5 = 0.72\,\text{rad/s}^2 \\ a_D = -53.14\,\text{m/s}^2 \end{array}\right\} \qquad (C.57)$$

Thus, the vector acceleration of point D (Eq. C.58) is:

$$\mathbf{a}_D = -53.14\hat{\mathbf{i}} = 53.14\,\text{m/s}^2 \angle 180° \qquad (C.58)$$

Figure C.9 shows the acceleration polygon of the mechanism. It can be noticed that triangle $\triangle obc$ of the acceleration polygon is similar to triangle $\triangle O_4BC$ of the mechanism. In the acceleration polygon, the sides of triangle $\triangle obc$ are not perpendicular to the sides of triangle $\triangle O_4BC$ like in the velocity polygon. Angle ϕ_4 (Eq. C.59) between the sides of both triangles can be calculated as:

$$\phi_4 = \arctan\frac{a_C^t}{a_C^n} = \arctan\frac{a_B^t}{a_B^n} = \arctan\frac{a_{CB}^t}{a_{CB}^n} = \arctan\frac{\alpha_4}{\omega_4^2} \qquad (C.59)$$

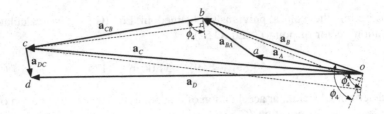

Fig. C.9 Acceleration polygon

Therefore, triangle Δobc in the polygon is similar to triangle ΔO_4BC in the mechanism and rotated angle ϕ_4.

C.6 Raven's Method

The number of needed vector loop equations depends on the number of unknowns. In this case, the position unknowns are θ_3, θ_4, θ_5 and r_6. As each vector equation allows solving two unknowns and we have four, we will need 2 vector equations (Eq. C.60) (Fig. C.10).

$$\left.\begin{array}{c} \mathbf{r}_1 + \mathbf{r}_4 = \mathbf{r}_2 + \mathbf{r}_3 \\ \mathbf{r}_{1'} + \mathbf{r}_5 + \mathbf{r}_6 = \mathbf{r}_{4'} \end{array}\right\} \tag{C.60}$$

Using the complex exponential form for the vectors, vector equation (Eq. C.60) can be written as (Eq. C.61):

Fig. C.10 Kinematic diagram of the mechanism with the two vector loop equations used to solve the problem

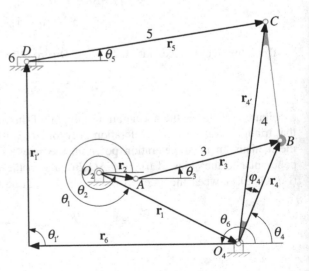

$$\left.\begin{array}{r} r_1 e^{i\theta_1} + r_4 e^{i\theta_4} = r_2 e^{i\theta_2} + r_3 e^{i\theta_3} \\ r_{1'} e^{i\theta_{1'}} + r_5 e^{i\theta_5} + r_6 e^{i\theta_6} = r_{4'} e^{i(\theta_4 + \varphi_4)} \end{array}\right\} \tag{C.61}$$

By separating the real and imaginary parts we obtain a system (Eq. C.62) with four equations and four unknowns: θ_3, θ_4, θ_5 and r_6:

$$\left.\begin{array}{r} r_1 \cos\theta_1 + r_4 \cos\theta_4 = r_2 \cos\theta_2 + r_3 \cos\theta_3 \\ r_1 \sin\theta_1 + r_4 \sin\theta_4 = r_2 \sin\theta_2 + r_3 \sin\theta_3 \\ r_{1'} \cos\theta_{1'} + r_5 \cos\theta_5 + r_6 \cos\theta_6 = r_{4'} \cos(\theta_4 + \varphi_4) \\ r_{1'} \sin\theta_{1'} + r_5 \sin\theta_5 + r_6 \sin\theta_6 = r_{4'} \sin(\theta_4 + \varphi_4) \end{array}\right\} \tag{C.62}$$

Using Freudenstein's equation, explained in Appendix B of this book, the first two equations of the system yields (Eqs. C.63 and C.64):

$$\theta_3 = 2 \arctan \frac{-B \pm \sqrt{B^2 - 4AC}}{2A} \tag{C.63}$$

$$\theta_4 = 2 \arctan \frac{-E + \sqrt{E^2 - 4DF}}{2D} \tag{C.64}$$

where A, B, C, D, E and F coefficients (Eq. C.65) are:

$$\left.\begin{array}{l} A = k_3 \cos\theta_1 - k_2 \cos(\theta_2 - \theta_1) + k_1 - \cos\theta_2 \\ B = 2(\sin\theta_2 - k_3 \sin\theta_1) \\ C = k_3 \cos\theta_1 - k_2 \cos(\theta_2 - \theta_1) + k_1 + \cos\theta_2 \\ D = k_3 \cos\theta_1 - k_5 \cos(\theta_2 - \theta_1) + k_4 + \cos\theta_2 \\ E = 2(-\sin\theta_2 + k_3 \sin\theta_1) \\ F = k_3 \cos\theta_1 - k_5 \cos(\theta_2 - \theta_1) + k_4 - \cos\theta_2 \end{array}\right\} \tag{C.65}$$

And where k_1, k_2, k_3, k_4 and k_5 geometrical data (Eq. C.66) are:

$$\left.\begin{array}{l} k_1 = \frac{r_1^2 + r_2^2 + r_3^2 - r_4^2}{2 r_2 r_3} \\[2mm] k_2 = \frac{r_1}{r_3} \\[2mm] k_3 = \frac{r_1}{r_2} \\[2mm] k_4 = \frac{r_1^2 + r_2^2 + r_4^2 - r_3^2}{2 r_2 r_4} \\[2mm] k_5 = \frac{r_1}{r_4} \end{array}\right\} \tag{C.66}$$

Using Freudenstein's equation, explained in Appendix B of this book, the last two equations of the system (Eq. C.62) yields (Eqs. C.67 and C.68):

$$\theta_5 = \arcsin \frac{r_{4'} \sin(\theta_4 + \varphi_4) - r_{1'}}{r_5} \tag{C.67}$$

$$r_6 = r_5 \cos \theta_5 - r_{4'} \cos(\theta_4 + \varphi_4) \tag{C.68}$$

Using Eqs. (C.63), (C.64)–(C.67), (C.68), we can plot the position of the links relative to the positions of link 2 along one full turn. Figure C.11a shows angles θ_3, θ_4 and θ_5 and Fig. C.11b shows distance r_6.

These figures illustrate the benefits of mathematical methods over graphical ones. The latter would only yield the solution to one of the points in such curves and the problem has to be solved again when there are any changes in the geometric parameters of the mechanism. Conversely, the expressions in Raven's method yield a solution for all the points in the curve and they do not need to be modified whenever geometrical data are modified.

The solution of the obtained equations for $\theta_2 = 350°$ yield (Eq. C.69) the following values for the position unknowns:

Fig. C.11 a Angular position of links 3, 4 and 5 in terms of θ_2, **b** Plot of the linear position of link 6 versus θ_2

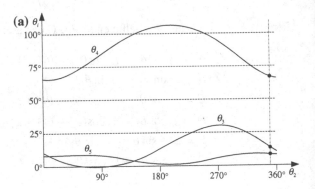

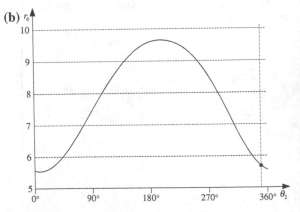

$$\left.\begin{array}{l} \theta_3 = 13.5° \\ \theta_4 = 67.03° \\ \theta_5 = 8.3° \\ r_6 = 5.607\,\text{m} \end{array}\right\} \qquad (C.69)$$

The position along the horizontal path of link 6 with respect to the coordinate system origin (Eq. C.71) will be given by the position of point D (Eq. C.70):

$$r_6 = -x_{DO_4} = -x_{DO_2} - x_{O_2O_4} = -x_{DO_2} - (-3.7\,\text{m}) = 5.607\,\text{m} \qquad (C.70)$$

So for $\theta_2 = 350°$ the X coordinate of link 6 is:

$$x_{DO_2} = -1.907\,\text{m} \qquad (C.71)$$

By differentiating with respect to time (Eq. C.61), we obtain (Eq. C.72):

$$\left.\begin{array}{l} ir_2\omega_2 e^{i\theta_2} + ir_3\omega_3 e^{i\theta_3} = ir_4\omega_4 e^{i\theta_4} \\ v_6 e^{i\pi} + ir_5\omega_5 e^{i\theta_5} = ir_{4'}\omega_4 e^{i(\theta_4 + \varphi_4)} \end{array}\right\} \qquad (C.72)$$

We separate the real and imaginary parts in Eq. (C.72), which yields the equation system (Eq. C.73) with four unknowns: ω_3, ω_4, ω_5 and v_6:

$$\left.\begin{array}{l} -r_2\omega_2 \sin\theta_2 - r_3\omega_3 \sin\theta_3 - \quad r_4\omega_4 \sin\theta_4 \\ r_2\omega_2 \cos\theta_2 + r_3\omega_3 \cos\theta_3 = r_4\omega_4 \cos\theta_4 \\ -v_6 - r_5\omega_5 \sin\theta_5 = -r_{4'}\omega_4 \sin(\theta_4 + \varphi_4) \\ r_5\omega_5 \cos\theta_5 = r_{4'}\omega_4 \cos(\theta_4 + \varphi_4) \end{array}\right\} \qquad (C.73)$$

From the first two algebraic equations in Eq. (C.73) we can obtain the expressions for ω_3 (Eq. C.74) and ω_4 (Eq. C.75):

$$\omega_3 = \frac{r_2 \sin(\theta_4 - \theta_2)}{r_3 \sin(\theta_3 - \theta_4)}\omega_2 \qquad (C.74)$$

$$\omega_4 = \frac{r_2 \sin(\theta_3 - \theta_2)}{r_4 \sin(\theta_3 - \theta_4)}\omega_2 \qquad (C.75)$$

Finally, from the third and fourth algebraic equation we reach expressions for ω_5 (Eq. C.76) and v_6 (Eq. C.77):

$$\omega_5 = \frac{r_{4'}}{r_5} \frac{\cos(\theta_4 + \varphi_4)}{\cos\theta_5}\omega_4 \qquad (C.76)$$

$$v_6 = -r_5\omega_5 \sin\theta_5 + r_{4'}\omega_4 \sin(\theta_4 + \varphi_4) \qquad (C.77)$$

Fig. C.12 a Angular velocities of links 3, 4 and 5 in terms of θ_2, **b** Plot of the linear velocity of link 6 versus θ_2

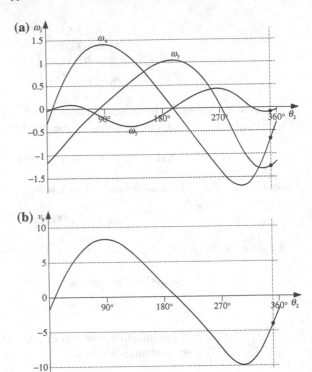

Using Eqs. (C.74) and (C.75), we can plot the kinematic curve of the link velocity versus the position of link 2. These curves are shown in Fig. C.12a, b.

Again, equations (Eqs. C.74–C.77) can be particularized for $\theta_2 = 350°$ yielding the values for the velocity unknowns (Eq. C.78):

$$\left.\begin{array}{l} \omega_3 = -1.27\,\text{rad/s} \\ \omega_4 = -0.69\,\text{rad/s} \\ \omega_5 = -0.09\,\text{rad/s} \\ v_6 = -4.04\,\text{m/s} \end{array}\right\} \tag{C.78}$$

Once more, Eq. (C.72) can be time-differentiated again in order to find accelerations (Eq. C.79):

$$\left.\begin{array}{l} (-r_2\omega_2^2 + ir_2\alpha_2)e^{i\theta_2} + (-r_3\omega_3^2 + ir_3\alpha_3)e^{i\theta_3} = (-r_4\omega_4^2 + ir_4\alpha_4)e^{i\theta_4} \\ (-r_5\omega_5^2 + ir_5\alpha_5)e^{i\theta_5} + a_6e^{i\theta_6} = (-r_{4'}\omega_4^2 + ir_{4'}\alpha_4)e^{i(\theta_4 + \varphi_4)} \end{array}\right\} \tag{C.79}$$

By separating real and imaginary parts we reach, once more, a system (Eq. C.80) with four equations and four unknowns: α_3, α_4, α_5 and a_6.

$$
\left.\begin{array}{l}
-r_2\omega_2^2\cos\theta_2 - r_2\alpha_2\sin\theta_2 - r_3\omega_3^2\cos\theta_3 - r_3\alpha_3\sin\theta_3 = -r_4\omega_4^2\cos\theta_4 - r_4\alpha_4\sin\theta_4 \\
-r_2\omega_2^2\sin\theta_2 + r_2\alpha_2\cos\theta_2 - r_3\omega_3^2\sin\theta_3 + r_3\alpha_3\cos\theta_3 = -r_4\omega_4^2\sin\theta_4 + r_4\alpha_4\cos\theta_4 \\
-r_5\omega_5^2\cos\theta_5 - r_5\alpha_5\sin\theta_5 + a_6\cos\theta_6 = -r_{4'}\omega_4^2\cos\theta_{4'} - r_{4'}\alpha_4\sin(\theta_4+\varphi_4) \\
-r_5\omega_5^2\sin\theta_5 + r_5\alpha_5\cos\theta_5 + a_6\sin\theta_6 = -r_{4'}\omega_4^2\sin\theta_{4'} + r_{4'}\alpha_4\cos(\theta_4+\varphi_4)
\end{array}\right\}
$$

(C.80)

Again, we start by considering the first two algebraic equations in the system (Eq. C.80), which yield the angular accelerations of links 3 and 4 (Eq. C.81).

$$
\left.\begin{array}{l}
\alpha_3 = \dfrac{-r_2\alpha_2\sin\theta_2 + r_4\alpha_4\sin\theta_4 - r_2\omega_2^2\cos\theta_2 - r_3\omega_3^2\cos\theta_3 + r_4\omega_4^2\cos\theta_4}{r_3\sin\theta_3} \\[2ex]
\alpha_4 = \dfrac{-r_2\alpha_2\sin(\theta_3-\theta_2) + r_2\omega_2^2\cos(\theta_3-\theta_2) + r_3\omega_3^2 - r_4\omega_4^2\sin(\theta_4-\theta_3)}{r_4\sin(\theta_4-\theta_3)}
\end{array}\right\}
$$

(C.81)

Finally, a_6 and α_5 (Eq. C.82) are obtained from the last two algebraic equations in the system (Eq. C.80):

$$
\left.\begin{array}{l}
\alpha_5 = \dfrac{r_{4'}\alpha_4\cos(\theta_4+\varphi_4) - r_{4'}\omega_4^2\sin(\theta_4+\varphi_4) + r_5\omega_5^2\sin\theta_5}{r_5\cos\theta_5} \\[2ex]
a_6 = r_{4'}\alpha_4\sin(\theta_4+\varphi_4) + r_{4'}\omega_4^2\cos(\theta_4+\varphi_4) - r_5\alpha_5\sin\theta_5 - r_5\omega_5^2\cos\theta_5
\end{array}\right\}
$$

(C.82)

These expressions (Eqs. C.81 and C.82) can be particularized for $\theta_2 = 350°$ yielding the values for the unknowns (Eq. C.83):

$$
\left.\begin{array}{l}
\alpha_3 = 1.98\,\text{rad/s}^2 \\
\alpha_4 = 9\,\text{rad/s}^2 \\
\alpha_5 = 0.72\,\text{rad/s}^2 \\
a_6 = 53.14\,\text{m/s}^2
\end{array}\right\}
$$

(C.83)

C.7 Mass, Inertia Moments, Inertia Forces and Inertia Pairs

We assume that we know the value of the mass and the moment of inertia of the links. Their values are included in Table C.2.

Figure C.13 shows the center of mass of each link. Their position (Eq. C.84) is given by the following distances:

Table C.2 Mass and moment of inertia of the links

Link	2	3	4	5	6
M_k (kg)	15.31	61.26	154.75	99.54	85
I_k (kg m^2)	1.278	81.68	495.76	358.479	–

Fig. C.13 Position of the centers of mass of the links

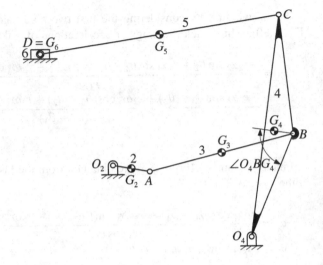

$$\overline{O_2G_2} = 0.5\,\text{m}$$
$$\overline{AG_3} = 2\,\text{m}$$
$$\overline{BG_4} = 0.52\,\text{m} \qquad\qquad\qquad (\text{C.84})$$
$$\angle O_4BG_4 = 75.4°$$
$$\overline{DG_5} = 3.25\,\text{m}$$

The acceleration of the center of mass of each link (Eq. C.85) has been determined by Raven's Method yielding the following results:

$$\left.\begin{aligned}
\mathbf{a}_{G_2} &= -8.64\hat{\mathbf{i}} + 1.52\hat{\mathbf{j}} = 8.78\,\text{m/s}^2\angle170° \\
\mathbf{a}_{G_3} &= -21.35\hat{\mathbf{i}} + 6.15\hat{\mathbf{j}} = 22.22\,\text{m/s}^2\angle163.94° \\
\mathbf{a}_{G_4} &= -25.84\hat{\mathbf{i}} + 4.57\hat{\mathbf{j}} = 26.24\,\text{m/s}^2\angle170° \\
\mathbf{a}_{G_5} &= -53.55\hat{\mathbf{i}} + 2.16\hat{\mathbf{j}} = 53.59\,\text{m/s}^2\angle177.69° \\
\mathbf{a}_{G_6} &= -53.14\hat{\mathbf{i}} = 53.14\,\text{m/s}^2\angle180°
\end{aligned}\right\} \qquad (\text{C.85})$$

Once the masses, moments of inertia and accelerations of each center of mass have been determined, we can calculate the forces (Eq. C.86) and moments (Eq. C.87) due to inertia:

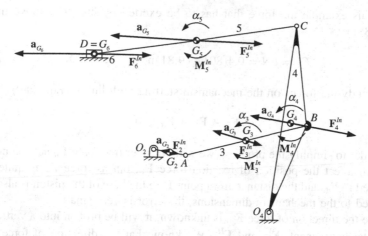

Fig. C.14 Forces and moments due to inertia in the mechanism

$$F_2^{In} = 15.31 \cdot 8.78 = 134.42\,\text{N}$$
$$F_3^{In} = 61.26 \cdot 22.22 = 1361.2\,\text{N}$$
$$F_4^{In} = 154.75 \cdot 26.24 = 4060.64\,\text{N}$$
$$F_5^{In} = 99.54 \cdot 53.59 = 5334.35\,\text{N}$$
$$F_6^{In} = 85 \cdot 53.14 = 4516.9\,\text{N}$$

(C.86)

$$M_2^{In} = 1{,}278 \cdot 0 = 0$$
$$M_3^{In} = 81.68 \cdot 1.98 = 161.73\,\text{Nm}$$
$$M_4^{In} = 495.76 \cdot 9 = 4461.84\,\text{Nm}$$
$$M_5^{In} = 358.48 \cdot 0.72 = 258.1\,\text{Nm}$$
$$M_6^{In} = 0$$

(C.87)

Figure C.14 shows the force and moment that acts on each link due to inertia. We can see that the force is opposite to the linear acceleration of the center of mass and the moment is opposite to the angular acceleration of the link.

C.8 Force Analysis. Graphical Method

In order to calculate the torque that is acting on link 2 to equilibrate the mechanism, we will consider the inertia of the links as well as the force needed to move the 80 kg box. We will consider a friction coefficient of 0.4 and will neglect the inertia force of the box. Obviously, in a real problem the inertia of the box would have to be considered.

So, in this example the force that has to be exerted by link 6 to move the box (Eq. C.88) is:

$$F_R = \mu N = 0.4(80\,\text{kg} \cdot 9.81\,\text{m/s}^2) = 314.2\,\text{N} \qquad (C.88)$$

We study the forces on the mechanism starting with link 6 (Eq. C.89).

$$\mathbf{F}_{56} + \mathbf{F}_{16} + \mathbf{F}_R = 0 \qquad (C.89)$$

In order to simplify the problem, we will consider that force $\mathbf{F}_R$ acts on point D. This will affect the position of reaction force $\mathbf{F}_{16}$ but as force $\mathbf{F}_R$ is quite small compared to $\mathbf{F}_{i6}$ and the distance from point D to the base of the piston is also small compared to the mechanism dimensions, the error is very small.

Since the direction of force $\mathbf{F}_{56}$ is unknown, it will be broken into a vertical and horizontal component, $\mathbf{F}_{56}^V$ and $\mathbf{F}_{56}^H$. We know that the direction of force $\mathbf{F}_{16}$ is perpendicular to the slider trajectory. So, this force will be equilibrated by $\mathbf{F}_{56}^V$. The value of $\mathbf{F}_{56}^H$ (Eq. C.91) can be calculated by means of the force equilibrium of the horizontal components of the forces acting on the link (Eq. C.90) as shown in Fig. C.15b.

$$\left.\begin{array}{l} 0 = \Sigma F_x = F_R + F_6^{In} + F_{56}^H \\ 0 = \Sigma F_y = F_{16} + F_{56}^V \end{array}\right\} \qquad (C.90)$$

$$\mathbf{F}_{56}^H = 4831.14\,\text{N}\angle 180° \qquad (C.91)$$

Figure C.16 shows the free body diagram of link 5. As we already know, force $\mathbf{F}_{56}^H$ is equal to force $\mathbf{F}_{65}^H$ but with opposite direction.

Fig. C.15 **a** Free body diagram of link 6, **b** horizontal components acting on link 6

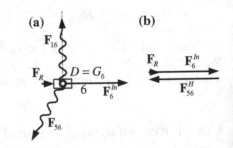

Fig. C.16 Free body diagram of link 5

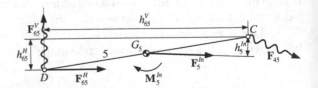

Next, we analyze the moment equilibrium at point C (Eqs. C.92–C.94) assuming the direction of $\mathbf{F}_{65}^V$ to be upwards:

$$\sum_j M_{j_z}^C = 0 \rightarrow h_5^{In} F_5^{In} - h_{65}^V F_{65}^V + h_{65}^H F_{65}^H - M_5^{In} = 0 \tag{C.92}$$

$$0.6\,\text{m} \cdot 5334.35\,\text{N} - 6.43\,\text{m} \cdot F_{65}^V + 0.94\,\text{m} \cdot 4831.14\,\text{N} - 258.1\,\text{Nm} = 0 \tag{C.93}$$

$$\mathbf{F}_{56}^V = 1163.88\,\text{N}\angle 90° \tag{C.94}$$

Then, $\mathbf{F}_{65}$ will be (Eq. C.95):

$$\left.\begin{aligned}
F_{65} &= \sqrt{\left(F_{65}^V\right)^2 + \left(F_{65}^H\right)^2} \\
&= \sqrt{4831.14^2 + 1163.88^2} = 4969.32\,\text{N} \\
\theta_{F_{65}} &= \arctan \frac{F_{65}^V}{F_{65}^H} \\
&= \arctan \frac{1163.88}{4831.14} = 13.5°
\end{aligned}\right\} \tag{C.95}$$

where distances h_5^{In}, h_{65}^V and h_{65}^H (Eq. C.96) are:

$$\left.\begin{aligned}
h_5^{In} &= \overline{G_5 C} \sin(360° - \theta_{F_5^{In}} + \theta_5) = 3.25 \sin 10.65° \\
h_{65}^V &= \overline{DC} \cos \theta_5 = 6.5 \cos 8.34° \\
h_{65}^H &= \overline{DC} \sin \theta_5 = 6.5 \sin 8.34°
\end{aligned}\right\} \tag{C.96}$$

Back to link 6, we know the vertical forces acting on it, as $\mathbf{F}_{65}^V = -\mathbf{F}_{56}^V = \mathbf{F}_{16}$. The equilibrium of forces acting on link 5 (Eq. C.97) yields the value of force $\mathbf{F}_{45}$ (Eq. C.98) as shown in Fig. C.17.

$$\mathbf{F}_{56}^H + \mathbf{F}_{56}^V + \mathbf{F}_5^{In} + \mathbf{F}_{45} = 0 \tag{C.97}$$

$$\mathbf{F}_{45} = 10{,}207.9\,\text{N}\angle 185.36° \tag{C.98}$$

Similarly, the equilibrium equations of links 3 and 4 yield the value of the forces transmitted by the links.

Fig. C.17 Polygon of forces acting on link 5

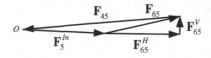

Fig. C.18 Free body diagram of link 3

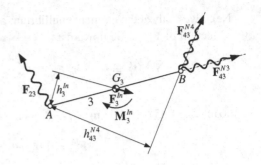

In link 3, we break force $\mathbf{F}_{43}$ into components $\mathbf{F}_{43}^{N3}$ and $\mathbf{F}_{43}^{N4}$ (Eq. C.101), with directions $\overline{AB}$ and $\overline{O_4B}$ respectively (Fig. C.18). We consider the equilibrium of moments about point A (Eqs. C.99 and C.100) assuming that $\mathbf{F}_{43}^{N4}$ (Eq. C.101) goes upwards.

$$\sum_j M_{j_z}^A = 0 \rightarrow -h_3^{In} F_3^{In} + h_{43}^{N4} F_{43}^{N4} - M_3^{In} = 0 \qquad (C.99)$$

$$-0.986\,\text{m} \cdot 1361.2\,\text{N} + 3.215 \cdot F_{43}^{N4} - 161.73\,\text{Nm} = 0 \qquad (C.100)$$

$$F_{43}^{N4} = 467.77\,\text{N} \qquad (C.101)$$

Where distances h_3^{In} and h_{43}^{N4} can be measured on the drawing of the mechanism or be determined as (Eq. C.102):

$$\left.\begin{array}{l} h_3^{In} = \overline{G_3A}\cos(90° + \theta_{F_3^{In}} - \theta_3) = 2\cos 60.44° \\ h_{43}^{N4} = \overline{BA}\sin(\theta_4 - \theta_3) = 4\sin 53.5° \end{array}\right\} \qquad (C.102)$$

Fig. C.19 Free body diagram of link 4

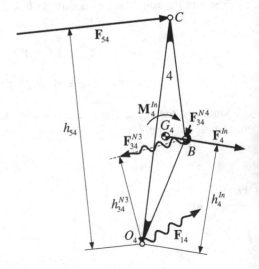

Figure C.19 shows the free body diagram of link 4. We can calculate F_{43}^{N3} (Eq. C.105) by means of the equilibrium equation of moments with respect to point O_4 (Eqs. C.103 and C.104). We assume the direction of F_{34}^{N3} to be oriented to the left.

$$\sum_j M_{j_z}^{O_4} = 0 \rightarrow -h_4^{In}F_4^{In} + h_{34}^{N3}F_{34}^{N3} - h_{54}F_{54} - M_4^{In} = 0 \qquad (\text{C.103})$$

$$0 = -2.923\,\text{m} \cdot 4060.64\,\text{N} + 2.412\,\text{m} \cdot F_{34}^{N3}$$
$$- 5.84\,\text{m} \cdot 10{,}207.9\,\text{N} - 4461.84\,\text{Nm} \qquad (\text{C.104})$$

$$F_{34}^{N3} = 31{,}486\,\text{N} \qquad (\text{C.105})$$

where distances h_4^{In}, h_{34}^{N3} and h_{54} (Eq. C.106) can be determined as:

$$\left.\begin{array}{l} h_4^{In} = \overline{O_4B}\cos(90° + \theta_{F_4^{In}} - \theta_4) = 3\cos 13° \\ h_{54} = \overline{O_4C}\cos(90° + \theta_{F_{54}} - \theta_{4'}) = 6\cos 13.26° \\ h_{34}^{N3} = \overline{O_4B}\sin(\theta_4 - \theta_3) \end{array}\right\} \qquad (\text{C.106})$$

The force equilibrium analysis of the forces acting on link 4 (Eq. C.107) yields F_{14} (Eq. C.108). Figure C.20 shows the force polygon.

$$\mathbf{F}_{54} + \mathbf{F}_4^{In} + \mathbf{F}_{34}^{N4} + \mathbf{F}_{34}^{N3} + \mathbf{F}_{14} = 0 \qquad (\text{C.107})$$

$$\mathbf{F}_{14} = 18{,}454\,\text{N}\angle 24° \qquad (\text{C.108})$$

Also, the analysis of the force equilibrium of link 3 (Eq. C.109) yields force $\mathbf{F}_{23}$ (Eq. C.110). The force polygon of forces acting on link 3 is shown in Fig. C.21.

$$\mathbf{F}_3^{In} + \mathbf{F}_{43} + \mathbf{F}_{23} = 0 \qquad (\text{C.109})$$

$$\mathbf{F}_{23} = 31{,}520\,\text{N}\angle 193.5° \qquad (\text{C.110})$$

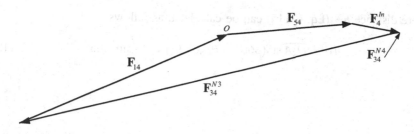

Fig. C.20 Polygon of forces acting on link 4

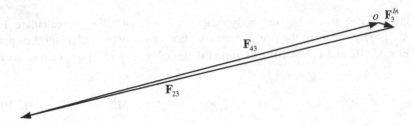

Fig. C.21 Polygon of forces acting on link 3

Fig. C.22 Free body diagram of link 2

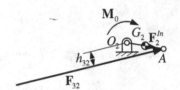

Finally, Fig. C.22 shows the equilibrium analysis of link 2, from which we find the value of the equilibrating torque (Eq. C.114) as well as force F_{12} (Eq. C.112). Since the value of F_2^{In} is very small compared to F_{32}, we will neglect it in Eq. (C.111). Therefore:

$$F_{32} + F_{12} = 0 \tag{C.111}$$

$$F_{12} = -F_{32} = F_{23} \tag{C.112}$$

Figure C.22 shows the free body diagram of link 2.

Torque M_0, which acts on link 2 to equilibrate the mechanism, can be obtained with (Eq. C.113):

$$\sum_j M_{j_z}^{O_2} = 0 \rightarrow h_{32}F_{32} + M_0 = 0 \tag{C.113}$$

$$M_0 = -0.398\,\text{m} \cdot 31{,}520\,\text{N} = -12{,}568.6\,\text{Nm} \tag{C.114}$$

where distance h_{32} (Eq. C.115) can be calculated as follows:

$$h_{32} = \overline{O_2A}\sin(360° - \theta_{F_{32}} - \theta_2) = 1 \cdot \sin 23.5° \tag{C.115}$$

C.9 Dynamic Analysis. Matrix Method

We start the dynamic analysis of the mechanism by writing the equations of the force and moment equilibrium of each link (Eqs. C.116–C.20). Figure C.23 shows radius vectors $\mathbf{p}_i$, $\mathbf{q}_i$ and $\mathbf{r}_i$ used in the moment equations:

- Link 2:

$$\left.\begin{array}{r}\mathbf{F}_{32} - \mathbf{F}_{21} = m_2\mathbf{a}_{G_2}\\ \mathbf{p}_2 \wedge \mathbf{F}_{32} - \mathbf{q}_2 \wedge \mathbf{F}_{21} + \mathbf{M}_0 = I_{G_2}\alpha_2\end{array}\right\} \tag{C.116}$$

- Link 3:

$$\left.\begin{array}{r}\mathbf{F}_{43} - \mathbf{F}_{32} = m_3\mathbf{a}_{G_3}\\ \mathbf{p}_3 \wedge \mathbf{F}_{43} - \mathbf{q}_3 \wedge \mathbf{F}_{32} = I_{G_3}\alpha_3\end{array}\right\} \tag{C.117}$$

- Link 4:

$$\left.\begin{array}{r}\mathbf{F}_{54} - \mathbf{F}_{43} + \mathbf{F}_{14} = m_4\mathbf{a}_{G_4}\\ \mathbf{p}_4 \wedge \mathbf{F}_{54} - \mathbf{q}_4 \wedge \mathbf{F}_{43} + \mathbf{r}_4 \wedge \mathbf{F}_{14} = I_{G_4}\alpha_4\end{array}\right\} \tag{C.118}$$

Fig. C.23 Radius vectors used in the moment equilibrium equations

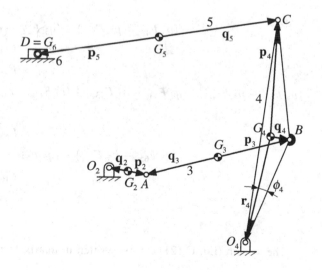

- Link 5:

$$\left.\begin{array}{c} \mathbf{F}_{65} - \mathbf{F}_{54} = m_5 \mathbf{a}_{G_5} \\ \mathbf{p}_5 \wedge \mathbf{F}_{65} - \mathbf{q}_5 \wedge \mathbf{F}_{54} = I_{G_5} \alpha_5 \end{array}\right\} \qquad (C.119)$$

- Link 6:

$$\mathbf{F}_{16} - \mathbf{F}_{65} + \mathbf{F}_R = m_6 \mathbf{a}_{G_6} \qquad (C.120)$$

The moment equilibrium equation of link 6 is not necessary as we suppose the forces are concurrent at point D and $\alpha_6 = 0$.

Equations C.116–C.20 yield a system of 14 algebraic equations (Eq. C.121) and 14 unknowns where $F_{16_x} = 0$ as the direction of force $\mathbf{F}_{16}$ has to be perpendicular to the sliding trajectory. So, in this example it will only have a vertical component.

Projecting each force equation on the X and Y axles and finding the vector products in the torque equations, we reach the following system of equations:

$$\left.\begin{array}{r} F_{32_x} - F_{21_x} = m_2 a_{G_{2x}} \\ F_{32_y} - F_{21_y} = m_2 a_{G_{2y}} \\ (p_{2_x} F_{32_y} - p_{2_y} F_{32_x}) - (q_{2_x} F_{21_y} - q_{2_y} F_{21_x}) + M_0 = I_{G_2} \alpha_2 \\ F_{43_x} - F_{32_x} = m_3 a_{G_{3x}} \\ F_{43_y} - F_{32_y} = m_3 a_{G_{3y}} \\ (p_{3_x} F_{43_y} - p_{3_y} F_{43_x}) - (q_{3_x} F_{32_y} - q_{3_y} F_{32_x}) = I_{G_3} \alpha_3 \\ F_{54_x} - F_{43_x} + F_{14_x} = m_4 a_{G_{4x}} \\ F_{54_y} - F_{43_y} + F_{14_y} = m_4 a_{G_{4y}} \\ (p_{4_x} F_{54_y} - p_{4_y} F_{54_x}) - (q_{4_x} F_{43_y} - q_{4_y} F_{43_x}) + (r_{4_x} F_{14_y} - r_{4_y} F_{14_x}) = I_{G_4} \alpha_4 \\ F_{65_x} - F_{54_x} = m_5 a_{G_{5x}} \\ F_{65_y} - F_{54_y} = m_5 a_{G_{5y}} \\ (p_{5_x} F_{65_y} - p_{5_y} F_{65_x}) - (q_{5_x} F_{54_y} - q_{5_y} F_{54_x}) = I_{G_5} \alpha_5 \\ F_{16_x} - F_{65_x} = m_6 a_{G_{6x}} - F_R \\ F_{16_y} - F_{65_y} = m_6 a_{G_{6y}} \end{array}\right\}$$

$$(C.121)$$

The system (Eq. C.121) can be written in matrix form (Eq. C.122):

$$[L]\mathbf{q} = \mathbf{F} \qquad (C.122)$$

where:

$$[L] = \begin{pmatrix}
-1 & 0 & 1 & 0 & 0 & 0 & 0 & 0 & 0 & 0 & 0 & 0 & 0 & 0 \\
0 & -1 & 0 & 1 & 0 & 0 & 0 & 0 & 0 & 0 & 0 & 0 & 0 & 0 \\
q_{2y} & -q_{2x} & -p_{2y} & p_{2x} & 0 & 0 & 0 & 0 & 0 & 0 & 0 & 0 & 0 & 1 \\
0 & 0 & -1 & 0 & 1 & 0 & 0 & 0 & 0 & 0 & 0 & 0 & 0 & 0 \\
0 & 0 & 0 & -1 & 0 & 1 & 0 & 0 & 0 & 0 & 0 & 0 & 0 & 0 \\
0 & 0 & q_{3y} & -q_{3x} & -p_{3y} & p_{3x} & 0 & 0 & 0 & 0 & 0 & 0 & 0 & 0 \\
0 & 0 & 0 & 0 & -1 & 0 & 1 & 0 & 1 & 0 & 0 & 0 & 0 & 0 \\
0 & 0 & 0 & 0 & 0 & -1 & 0 & 1 & 0 & 1 & 0 & 0 & 0 & 0 \\
0 & 0 & 0 & 0 & q_{4y} & -q_{4x} & -p_{4y} & p_{4x} & -r_{4y} & r_{4x} & 0 & 0 & 0 & 0 \\
0 & 0 & 0 & 0 & 0 & 0 & -1 & 0 & 0 & 0 & 1 & 0 & 0 & 0 \\
0 & 0 & 0 & 0 & 0 & 0 & 0 & -1 & 0 & 0 & 0 & 1 & 0 & 0 \\
0 & 0 & 0 & 0 & 0 & 0 & q_{5y} & -q_{5x} & 0 & 0 & -p_{5y} & p_{5x} & 0 & 0 \\
0 & 0 & 0 & 0 & 0 & 0 & 0 & 0 & 0 & 0 & -1 & 0 & 0 & 0 \\
0 & 0 & 0 & 0 & 0 & 0 & 0 & 0 & 0 & 0 & 0 & -1 & 1 & 0
\end{pmatrix}$$

$$\text{(C.123)}$$

$$\mathbf{q} = \begin{pmatrix}
F_{21_x} \\
F_{21_y} \\
F_{32_x} \\
F_{32_y} \\
F_{43_x} \\
F_{43_y} \\
F_{54_x} \\
F_{54_y} \\
F_{14_x} \\
F_{14_y} \\
F_{65_x} \\
F_{65_y} \\
F_{16_y} \\
M_0
\end{pmatrix} \qquad \text{(C.124)}$$

$$\mathbf{F} = \begin{pmatrix}
m_2 a_{G_{2x}} \\
m_2 a_{G_{2y}} \\
I_{G_2} \alpha_2 \\
m_3 a_{G_{3x}} \\
m_3 a_{G_{3y}} \\
I_{G_3} \alpha_3 \\
m_4 a_{G_{4x}} \\
m_4 a_{G_{4y}} \\
I_{G_4} \alpha_4 \\
m_5 a_{G_{5x}} \\
m_5 a_{G_{5y}} \\
I_{G_5} \alpha_5 \\
m_6 a_{G_{6x}} - F_R \\
m_6 a_{G_{6y}}
\end{pmatrix} \qquad \text{(C.125)}$$

The analytical expressions of the radius vectors in matrix $[L]$ are defined in Eqs. (C.126)–(C.129). Angles θ_2, θ_3, θ_4, φ_4 and θ_5 are the ones defined in Sect. 3.6 for the analysis with Raven's Method. Angle ϕ_4 is defined in Fig. C.23.

- Link 2:

$$\left.\begin{array}{l} \mathbf{p}_2 = \overline{AG_2}(\cos\theta_2\hat{\mathbf{i}} + \sin\theta_2\hat{\mathbf{j}}) \\ \mathbf{q}_2 = -\overline{O_2G_2}(\cos\theta_2\hat{\mathbf{i}} + \sin\theta_2\hat{\mathbf{j}}) \end{array}\right\} \qquad (C.126)$$

- Link 3:

$$\left.\begin{array}{l} \mathbf{p}_3 = \overline{BG_3}(\cos\theta_3\hat{\mathbf{i}} + \sin\theta_3\hat{\mathbf{j}}) \\ \mathbf{q}_3 = -\overline{AG_3}(\cos\theta_3\hat{\mathbf{i}} + \sin\theta_3\hat{\mathbf{j}}) \end{array}\right\} \qquad (C.127)$$

- Link 4:

$$\left.\begin{array}{l} \mathbf{p}_4 = \overline{CO_4}(\cos(\theta_4+\varphi_4)\hat{\mathbf{i}} + \sin(\theta_4+\varphi_4)\hat{\mathbf{j}}) - \overline{O_4G_4}(\cos(\theta_4+\phi_4)\hat{\mathbf{i}} + \sin(\theta_4+\phi_4)\hat{\mathbf{j}}) \\ \mathbf{q}_4 = \overline{BO_4}(\cos\theta_4\hat{\mathbf{i}} + \sin\theta_4\hat{\mathbf{j}}) - \overline{O_4G_4}(\cos(\theta_4+\phi_4)\hat{\mathbf{i}} + \sin(\theta_4+\phi_4)\hat{\mathbf{j}}) \\ \mathbf{r}_4 = -\overline{O_4G_4}(\cos(\theta_4+\phi_4)\hat{\mathbf{i}} + \sin(\theta_4+\phi_4)\hat{\mathbf{j}}) \end{array}\right\}$$

$$(C.128)$$

- Link 5:

$$\left.\begin{array}{l} \mathbf{p}_5 = -\overline{DG_5}(\cos\theta_5\hat{\mathbf{i}} + \sin\theta_5\hat{\mathbf{j}}) \\ \mathbf{q}_5 = \overline{CG_5}(\cos\theta_5\hat{\mathbf{i}} + \sin\theta_5\hat{\mathbf{j}}) \end{array}\right\} \qquad (C.129)$$

If we find the values of these vectors (Eqs. C.126–C.129) for $\theta_2 = 350°$ we obtain (Eq. C.130):

$$\left.\begin{array}{l} \mathbf{p}_2 = 0.4924\hat{\mathbf{i}} - 0.0868\hat{\mathbf{j}}\,\mathrm{m} \\ \mathbf{q}_2 = -\mathbf{p}_2 \\ \mathbf{p}_3 = 1.9447\hat{\mathbf{i}} + 0.4672\hat{\mathbf{j}}\,\mathrm{m} \\ \mathbf{q}_3 = -\mathbf{p}_3 \\ \mathbf{p}_4 = 0.1692\hat{\mathbf{i}} + 3.1052\hat{\mathbf{j}}\,\mathrm{m} \\ \mathbf{q}_4 = 0.5144\hat{\mathbf{i}} - 0.0766\hat{\mathbf{j}}\,\mathrm{m} \\ \mathbf{r}_4 = -0.6598\hat{\mathbf{i}} - 2.8373\hat{\mathbf{j}}\,\mathrm{m} \\ \mathbf{p}_5 = -3.2157\hat{\mathbf{i}} - 0.4712\hat{\mathbf{j}}\,\mathrm{m} \\ \mathbf{q}_5 = -\mathbf{p}_5 \end{array}\right\} \qquad (C.130)$$

The analytical expressions of the acceleration vector of the center of mass of each link (Eqs. C.131–C.34) are:

$$\mathbf{a}_{G_2} = \overline{O_2 G_2}\alpha_2(-\sin\theta_2\hat{\mathbf{i}} + \cos\theta_2\hat{\mathbf{j}}) - \overline{O_2 G_2}\omega_2^2(\cos\theta_2\hat{\mathbf{i}} + \sin\theta_2\hat{\mathbf{j}}) \qquad (C.131)$$

$$\begin{aligned}\mathbf{a}_{G_3} &= \overline{O_2 A}\alpha_2(-\sin\theta_2\hat{\mathbf{i}} + \cos\theta_2\hat{\mathbf{j}}) - \overline{O_2 A}\omega_2^2(\cos\theta_2\hat{\mathbf{i}} + \sin\theta_2\hat{\mathbf{j}}) \\ &\quad + \overline{G_3 A}\alpha_3(-\sin\theta_3\hat{\mathbf{i}} + \cos\theta_3\hat{\mathbf{j}}) - \overline{G_3 A}\omega_3^2(\cos\theta_3\hat{\mathbf{i}} + \sin\theta_3\hat{\mathbf{j}})\end{aligned} \qquad (C.132)$$

$$\begin{aligned}\mathbf{a}_{G_4} &= \overline{O_4 G_4}\alpha_4(-\sin(\theta_4 + \phi_4)\hat{\mathbf{i}} + \cos(\theta_4 + \phi_4)\hat{\mathbf{j}}) \\ &\quad - \overline{O_4 G_4}\omega_4^2(\cos(\theta_4 + \phi_4)\hat{\mathbf{i}} + \sin(\theta_4 + \phi_4)\hat{\mathbf{j}})\end{aligned} \qquad (C.133)$$

$$\begin{aligned}\mathbf{a}_{G_5} &= \overline{CO_4}\alpha_4(-\sin(\theta_4 + \varphi_4)\hat{\mathbf{i}} + \cos(\theta_4 + \varphi_4)\hat{\mathbf{j}}) \\ &\quad - \overline{CO_4}\omega_4^2(\cos(\theta_4 + \varphi_4)\hat{\mathbf{i}} + \sin(\theta_4 + \varphi_4)\hat{\mathbf{j}}) \\ &\quad + \overline{G_5 C}\alpha_5(-\sin(\theta_5 + 180°)\hat{\mathbf{i}} + \cos(\theta_5 + 180°)\hat{\mathbf{j}}) \\ &\quad - \overline{G_5 C}\omega_5^2(\cos(\theta_5 + 180°)\hat{\mathbf{i}} + \sin(\theta_5 + 180°)\hat{\mathbf{j}})\end{aligned} \qquad (C.134)$$

$$\begin{aligned}\mathbf{a}_{G_6} &= \overline{CO_4}\alpha_4(-\sin(\theta_4 + \varphi_4)\hat{\mathbf{i}} + \cos(\theta_4 + \varphi_4)\hat{\mathbf{j}}) \\ &\quad - \overline{CO_4}\omega_4^2(\cos(\theta_4 + \varphi_4)\hat{\mathbf{i}} + \sin(\theta_4 + \varphi_4)\hat{\mathbf{j}}) \\ &\quad + \overline{DC}\alpha_5(-\sin(\theta_5 + 180°)\hat{\mathbf{i}} + \cos(\theta_5 + 180°)\hat{\mathbf{j}}) \\ &\quad - \overline{DC}\omega_5^2(\cos(\theta_5 + 180°)\hat{\mathbf{i}} + \sin(\theta_5 + 180°)\hat{\mathbf{j}})\end{aligned} \qquad (C.135)$$

If we find the values of all the elements in the vector $\mathbf{F}$ for $\theta_2 = 350°$ and solve the system (Eq. C.122), we obtain the values (Eq. C.136) for the unknowns (Eq. C.124):

$$\left.\begin{aligned} \mathbf{F}_{21} &= 32{,}239\hat{\mathbf{i}} + 7386\hat{\mathbf{j}}\,\mathrm{N} \\ \mathbf{F}_{32} &= 32{,}106\hat{\mathbf{i}} + 7409\hat{\mathbf{j}}\,\mathrm{N} \\ \mathbf{F}_{14} &= 16{,}636\hat{\mathbf{i}} + 7550\hat{\mathbf{j}}\,\mathrm{N} \\ \mathbf{F}_{54} &= 10{,}161\hat{\mathbf{i}} + 944\hat{\mathbf{j}}\,\mathrm{N} \\ \mathbf{F}_{65} &= 4833\hat{\mathbf{i}} + 1173\hat{\mathbf{j}}\,\mathrm{N} \\ \mathbf{F}_{16} &= 1173\hat{\mathbf{j}}\,\mathrm{N} \\ M_0 &= -12{,}872\,\mathrm{Nm} \end{aligned}\right\} \qquad (C.136)$$

Fig. C.24 Instantaneous
motor torque M_0 versus crank
angle θ_2

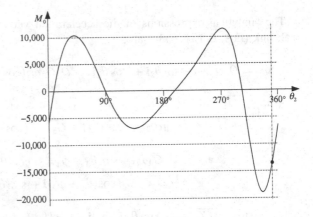

Fig. C.25 Shaking force
curve versus crank angle θ_2

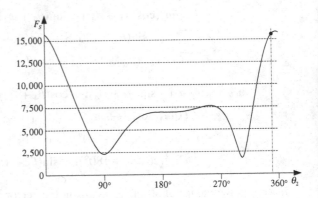

One of the main advantages of the Matrix Method for the dynamic analysis when
compared to the graphical method is the ability to calculate the value of the
unknowns along a complete cycle.

With the latter we can only find one solution for one position of the crank in the
curve and the expressions cannot be used again if there are any changes in the
geometrical data of the mechanism.

Figure C.24 shows a curve with the value of the instantaneous motor torque for
the different positions of the crank, θ_2. The value of the motor torque is given by M_0
and it is the torque that is necessary to apply to motor link 2 in order to obtain the
desired speed and acceleration. In this case, $\omega_2 = 4.19 \, \text{rad/s}$ and $\alpha_2 = 0$.

Moreover, we can obtain the curve of the magnitude of the shaking force versus the positions of the crank (Fig. C.25). The shaking force is given by Eq. (C.137):

$$\mathbf{F}_S = \mathbf{F}_{21} + \mathbf{F}_{41} + \mathbf{F}_{61} \tag{C.137}$$

In Fig. C.25 we can see that the maximum value for the shaking force is at a position close to $\theta_2 = 350°$. In this case, this position coincides with the maximum acceleration of link 6.

Printed in the United States
By Bookmasters